ECOLOGY OF MIGRATORY ANIMALS

RECENT ADVANCES IN ENVIRONMENTAL ECOLOGY SERIES—8

Ecology of Migratory Animals

A. P. DIWAN
D. K. ARORA

ANMOL PUBLICATIONS PVT LTD
NEW DELHI-110 002

ANMOL PUBLICATIONS PVT LTD
4374/4B, Ansari Road
Darya Ganj, New Delhi-110 002

First Edition, 1995

ISBN 81-7488-117-4

Printed in India

Published by J. L. Kumar for Anmol Publications Pvt. Ltd., New Delhi
and Printed at Amit Enterprises, 518/14, Anup Mkt., Maujpur, Delhi.

Preface

One of the characteristics of animals which differentiate them from inanimate objects is their ability to move their organs, limbs, or even the whole body either by conscious action or involuntarily. Most animals are capable of some kind of observable movement. It may be the simple waving of the tentacles of a sedentary coral polyp, the swimming of fish, or the fast, coordinated dash of a cheetah for pursuit of prey.

Locomotion in animals has many different functions, including finding food, escaping from predators, finding partners for sexual reproduction and also for changing habitate periodically. Single-celled animals (protozoans) move by the action of hair-like cilia or flagella, or by amoeboid crawling. Multicelled animals, however, move from place to place by a wide variety of methods like crawling swimming, fiying, etc. To achieve these various kinds of locomotion, muscles and skeletons work in harmony to provide the power.

This book deals with the different types of migration a different animals both in marine and terrestrial environment as well as their locomotory organs, which help them to change their habitat periodically.

In the preparation of such a title the authors have obviously had a great deal of help from a great many people. We would like to take this opportunity to thank the many friends and colleagues with whom we have thrashed out our ideas. We thanks all of them most warmly for helping to create the present title.

Any worthwhile criticism and suggestion for improvement would be thankfully acknowledged.

Authors

Contents

1

Introduction

One of the characteristics of animals which, for most non-scientists, differentiate them from inanimate objects is their ability to move organs, limbs, or even the whole body either by conscious action or involuntarily. Most animals are capable of some kind of observable movement whether it be the simple waving of the tentacles of a sedentary coral polyp, the swimming of fish, or the fast, coordinated dash of a cheetah pursuit of prey.

Most animals possess the power of locomotion, that is, the ability to move from place to place. Locomotion in animals has many different functions, including finding food, escaping from predators, finding partners for sexual reproduction and so on. Single-celled animals (protozoans) move by the action of hairlike cilia or flagella, by amoeboid crawling and by several other techniques. Multicelled animals, however, move from place to place by a wide variety of methods. Earthworms or snails crawl, mammals may walk, run, swim or fly, while birds are considered to be the masters of the air and fish are wonderfully suited to propelling themselves through water. And, to achieve the various kinds of locomotion, muscles and skeletons work in harmony to provide the power. Their remains much to be learnt about the techniques of locomotion.

The forces required for most movement of multicelled animals are provided by the contraction of muscles. There are two main types of muscle: smooth and striated. Smooth muscles are concerned with the movements of certain internal organs, such as the rhythmic contractions of the gut. These are involuntary movements and the individual neither has conscious control over them nor is normally aware of the movements. Striated muscles are so named because of the banded appearance of their fibers when viewed under the light microscope. They are responsible for most externally visible movements, such as the movements of limbs and other appendages. Contractions of most striated muscles are under conscious control and the individual is aware of the resulting movement or other action.

Fish, reptiles, amphibians, birds and mammals (vertebrates) have three main types of striated muscle: red and white skeletal muscle, and heart muscle. Red muscles have a good blood supply and obtain their energy by oxidizing foodstuffs to corbon dioxide and water. They have plenty of mitochondria, the sub-cellular sites of this process of oxidation. While muscles have a sparser blood supply, their cells have fewer mitochondria and they obtain energy by converting the carbohydrate glycogen to lactic acid. This process does not require an immediate oxygen supply although an oxygen debt is built up which has to be repaid later (see Gas Exchange and Circulation). The different uses of the types of striated muscle have been demonstrated in experiments with dogfish. These fish can swim slowly for hours on end, using only a narrow strip of red muscle sited along the sides of the body. They can swim very much faster for short periods, however, by using the larger quantity of white muscle in the tail, building up an oxygen debt as they do so.

Skeletons

Muscles often act on skeletons which may be essential parts of the apparatus of movement. A vertebrate body without a skeleton would be like a blob of jelly, incapable of precise movement. Typical vertebrate skeletons are constructed of bone, which consists of calcium phosphate crystals (70%) embedded in the protein collagen (30%). The shells of crabs and other crustaceans consist of 60-90 per cent calcium carbonate with protein and a fibrous carbohydrate (chitin) Mollusk shells consist of calcium carbonate with (usually) less than 5 per cent protein. Insect cuticle contains no inorganic crystals, but is a composite of protein and chitin. In most cases the composite structure (crystals plus protein) is important because the crystalline material alone would be too brittle and the protein alone would be flexible and rather weak.

Bones and shells can be likened to fiberglass, which is a strong, tough material although it consists of glass, which is brittle, and a plastic resin which is not particularly strong. Bone and insect cuticle are about as strong and stiff as hardwood. Mollusk shell is weaker but stiffer.

Antagonistic Action of Muscles. The only active movement that a muscle can make is to shorten. It can be stretched, but it cannot lengthen itself and it cannot of itself twist or bend. How, then, can muscles cause the complex movements of living animals? One of the most important principles involved is antagonism. Two muscles are antagonistic if each reverses the effect of the other. Contraction of muscle A causes a movement and stretches muscle B. Contraction of B then reverses the movement and stretches A. Almost every muscle has an antagonist.

In many animals (those with so-called hydrostatic skeletons) antogonism depends on the incompressibitity of water. Body tissues and fluids cannot be compressed like air in a bicycle pump, but have essentially constant volume. Imagine a sausage-shaped animal with two sets of muscles. One set runs hoop-like around the body (circular muscles) and the other set lengthwise along the body (longitudinal muscles). Contraction of the circular muscles makes the sausage thinner, but also makes it longer because its volume cannot be changed. Contraction of the longitudinal muscles makes it shorter and fatter.

Thus, the two sets of muscles are antagonistic. Now suppose that the longitudinal musles of the left side contract while the circular muscles keep constant length. The animal bends to the left and the longitudinal museles on the right are stretched. Thus, the muscles of the two sides can act as antagonists. Our simple animal can lengthen, shorten and bend from side to side although it has no skeleton. Earthworms are like this, but have their bodies divided into segments. One group of segments can lengthen or bend to the left while another group shortens or bend to the right. Internal partitions cross the liquid-filled body cavity between segments.

In some animals, such as nematodes (roundworms), the circular muscles are replaced by an inextensible cuticle which prevents swelling. These animals can bend from side to side. or in a vertical plane, by contracting appropriate longitudinal muscles, but they cannot lengthen or shorten.

Snakes are similar in shape to earthworms and nematodes but, like other vertebrates, they havs a backbone of bony, jointed vertebrae. Contraction of longitudinal muscles on one side bends the snake and stretches the muscles on the other side because the vertebrae cannot be compressed.

A few muscles do not have other muscles as antagonists. For example, the two halves of a claim shell are joined by an elastic hinge, which makes them spring open when the muscles holding them together relax. Part of the muscles that closes the clam is specialized to maintain tension for long periods with little expenditure of energy.

These principles of muscle contraction, antagonism and their actions on various kinds of skeleton are used by whole animals to move them through their environment.

Crawling

Crawling is usually taken to mean movement overland without legs. Most major groups of animals have members which practice it. The feature common to all is that forward movement is caused by rhythmic waves of activity traveling along the body. In a crawling earthworm, segments in some parts of its body are short and fat and others long and thin. As it crawls, the thickenings move backwards along its body. When a snake crawls, bends start at the head and travel to the tail.

When an earthworm is burrowing, the thick part of the body is jammed firmly in the burrow. Segments in front of the thickening elongate, pushing the head forward, and segments just behind the thickening shorten pulling the tail forward. These changes propel the body forward and also cause the thickening to travel backwards along it. Similar movements also serve for crawling on the surface of the ground, aided by the ratchet-like effect of tiny bristles (chaetae) that project from the body. They points bacwards, making it easier for the body to slidn forwards than backwards.

A snake cannot shorten or lengthen its body because it has a backbone. Instead, it curves its body into bends around stones and other irregularities on the ground. As the bends travel backwards along the body they push against the stones and drive the snake forwards.

Snails crawl on a long foot that is covered with slimy mucus. Although some of the small species are propelled by cilia, most make use of waves of muscular contraction that travel along the foot. Like eartworms, they depend on a ratchet, but here the rachet is the mucus, which resists gentle pressure as if it were a solid but, if pressed hard enough, gives way and behaves like a liquid. The muscular waves of snails' feet are so proportioned that the stress under the parts that are pushing forward is enough to liquefy the mucus, but the stress under the parts that are pushing back is not.

Walking and Running

Soft-bodied animals, such as earthworms and slugs, can crawl and burrow effectively, but complex jointed skeletons (see box) are generally required for walking and running Humans and other terrestrial mammals walk to travel slowly and run to go faster. A walking person does not lift one foot until the other has been set down, so there are brief stages when both feet are on the ground simultaneously. In contrast, running involves stages when both feet are off the ground. A similar distinction applies to four-footed animals: in walking, but not in running, there are always at least one forefoot and one hindfoot on the ground. Another important difference between human walking and running is that, in walking, each leg is kept fairly straight while the foot is on the ground. In running, the knees bend much more.

Most four-footed mammals walk at low speeds, but use several kinds of run at higher speeds. As horses and dogs increase speed they change successively from a walk to a trot to a center and finally to a gallop, but mady mammals omit one or more of these gaits. The sequence of footfalls in the gallop, in which the two forefeet are set down at one stage of the stride and the two hindfeet at another, makes it possible for back muscles as well as leg muscles to contribute to the work.

Walking and running have evolved to keep · energy costs as low as possible, in very different ways. The principle of human walking is the principle of the pendulum, which requires very little energy to keep it swinging, although the bob is repeatedly rising and falling, speeding up and slowing down. As the bob swings down it loses height (and so potential energy) but gains speed (and so kinetic energy). As it swings up the reverse hapens. Energy is swapped back and forth between the potential and kinetic forms. Similarly, a walking person slows down and rises in the first half of each step but speeds up and falls in the second half. Ideally, the total mechanical energy (potential plus kinetic) should remain constant. This is not quite achieved, however, and the leg muscles have to do work at one stage of the step (increasing the mechanical energy of the body) and act like brakes at another (getting rid of mechanical energy by converting it to heat). Nevertheless, human walking is reasonably economical of energy. This is also true of the somewhat more complicated walking of four-footed mammals.

The principle of running is the principle of the bouncing ball, which will make many bounces without a fresh input of energy. When the ball hits the ground it is brought to a halt, losing its kinetic energy, but the energy is returned in the elastic recoil that throws it back into the air. The principal elastic structures in the legs that give them their bounce are the tendons, especially the Achilles tendon, which connects the calf muscles to the heel. It has been calculated that the Achilles tendon of a running human stretches and recoils by about 18mm (0.7 in) at each footfall.

Swimming

Organisms that swim range from flagellate protozoans a few micrometers log to giant Blue whales more than 30m (100ft) long. Their mechanisms of propulsion use the principles of the oar, of the propeller, of bode undulation and of jet propulsion.

The essential difference between oars and propellers is that oars push in the direction of movement, but propellers exert forces at right angles to the movement of their blades. Thus, the thrust of an oar is derived from drag as it pushes backwards on the water, but the thrust of a propeller is derived from lift created as the blades (hydrofoils) move through the water. When a hydrofoil moves through water, tilted at a small angle of attack to its direction of motion, the force that the water exerts on it has two components: drag acting backwards along the direction of motion and lift acting at right angels to it. If the hydrofoil is suitably shaped and the angle of attack not too large, lift is much larger than drag.

To see how this applies to animals, compare a water beetle to a penguin. Both swim under water using paddle-like limbs. The water beetle has legs fringed with bristles and uses them as oars. It pushes the legs backwards with the bristles spread and then brings them forward again with the bristles folded so as to retard its motion as little as possible. The penguin swims with its wings; it does not move them backwards and forwards as an oar, but up and down, using them as hydrofoils. Because it is moving forwards, the wing tips take an undulating path through the water. The angles of attack are adjusted so that lift acts forwards and up in the downstroke, forwards and down in the upstroke. The upward and downward components cancel out and the net effect is forward thrust.

Many fish row themselves along using their fins, especially when swimming slowly. Ciliate protozoans and many of the larvae in marine plankton row themselves along with vast numbers of tiny cilia. Whales are propelled by up and down movements of the tail flukes, in much the same way as penguins are propelled by their wings. Some fish, including tunnies, have hydrofoil-shaped tails, but move them from side to side, not up and down.

Many animals swim by undulation, making movements like a crawling snake. Just as the irregularities of the ground make it easier for the snake's body to slide forwards along its own length than to slide sideways, so does water resist sideways movement of a swimmer's body more than lengthwise movement. Therefore undulations travelling backwards along the body push the animal forwards. Water snakes, eels, leeches, flagellate protozoans and spermatozoans (male gametes) swim in this way.

The swimming techniques of fish grade imperceptibly from the undulation of eels to the hydrofoil action of tunnies. Many fish swim slowly by undulating their fins and for some, such as the sea horse, it is their only method of swimming. Cuttlefish (which are mollusks, not fish) swim slowly by undulating their fins and quickly by jet propulsion. They are related to the squids, which also use jet propulsion. They draw water into their mantle cavities and squirt it out forcefully forwards (to swim backwards) or backwards (to swim forwards).

Buoyancy. Muscle, bone, shell and most of the other issues that make up animals are denser than water, making them tend to sink. Flatfish are denser than water and swim upwards to counteract sinking. Tunnies and many sharks swim with fins extended like airplane wings to give lift at the front end of the body, while their tails give lift at the rear. Many other swimming animals have their densities adjusted to match the water by some buoyant material. For example, some deep-sea sharks have huge quantities of squalene (a hydrocarbon) in their livers, the coelacanth and some of the oceanic lantern fish have wax esters widely distributed in their bodies and the cranchid squids have swollen body cavities filled with an (impure) ammonium chloride solution which is less dense than sea water although it has the same osmotic concentration.

Large quantities of all these materials are needed because their densities are not particularly low. Gas-filled floats can be much more compact. Many bony fish have gas-filled sacs (swimbladders) in their body cavities. These have flexible walls and are compressed by their increasing pressure as the fish swims deeper, so that more gas has to be added if the density of the fish is to be kept constant. Oxygen and other gases can be secreted from the blood into the swimbladder as repuired, or withdrawn again, but these processes are very slow. Cuttle-fish and some other cephalopods have rigid shells divided into chambers, from which they extract the water to leave a partial vacuum. Because the shell is rigid there is no need to adjust for depth changes but, if the mollusk were to swim too deep, the pressure would make its shell collapse.

Swimming of Dolphins

Dolphins are among the fastest of swimming animals, propelling themselves by up-and-down movements of their tail flukes, but their apparent speed is sometimes deceptive. A dolphin keeping pace with a ship may be getting a free ride by resting in the wave that the ship is pushing along with its bows. The fastest record of unaided swimming is 40 km/h (25 mph), achieved by a trained dolphin chasing a lure which was towed by a winch across a lagoon in Hawaii. It has been calculated that this required at least twice as much power as can be produced by a human athlete of similar size. It has been suggested that the shape of dolphins and the properties of their skin have evolved to minimize drag, but it also seems that dolphin muscles must be more powerful than human ones.

Gliding

Insects, birds and bats are the only animals that can make long flights, but some other animals glide for short distances. Flying phalangers from Australia have webs of skin between their fore and hind legs that enable them to glide from tree to tree. The same is true of the flying squirrels of Asia and North America and the flying "lemur" of Southeast Asia.

Flying fish, such as Exocoetus spp. have large fins that can be spread as wings. They emerge from the sea and travel for some distance with fins spread but with their tail still beating the water to increase their speed. Then they take off and glide for several seconds. The South American hatchet fish, such as Carnegiella spp. are small river-dwelling fish which are capable of leaping from the water and gliding for distances of about Iom (33 ft). As well as large pectoral fins that function as wings, they have huge fin muscles and may beat their fins like insect wings. There seem to be no films that confirm this, but a buzzing noise has been heard as they fly past.

The weight of a glider (animal or machine) is supported by lift on its wings, but drag tends to slow it down. Nevertheless, it can maintain constant speed in still air by gliding on a downward gradient. Lift depends on speed and on the angle of attack of the wings, so the glider can travel quickly with a small angle of attack or slowly with a large one and get the lift needed to support its weight in either case. There is a minimum speed below which it cannot get enough lift and another higher speed at which drag is at a minimum and it can glide farthest.

As a general rule, the larger the aircraft the greater these speeds are. For instance, the minimum drag speed is about 90 km/h (55 mph) for a typical glider and 30 km/h (18 mph) for a model glider. The equivalent speeds have been shown to be 44 km/h (27 mph) for the small vulture Coragyps (which is about the size of the model glider), and 10 km/h (6 mph) for a monarch butterly. The full-sized glider can glide with a mlnimum gradient of 1 in 40, the model 1 in 10, the vulture 1 in 11 and the butterfly 1 in 4. Notice the similarity between the vulture and the model. In contrast, flying phalangers glide badly, with minimum gradients of 1 in 3.

Soaring. In air that is rising quickly enough, a glider can rise instead of falling. Rising air can be found over hillsides or cliffs, where the wind is deflected upwards by the slope. It is also found in thermals, which consist of hot air rising from ground heated by the sun. In slope soaring, pilots glide back and forth along hillsides to take advantage of the rising air. In thermal soaring, they circle in thermals to gain height. In favourable conditions it is possible to soar for long distances, gaining height over hillsides or in thermals and losing height while gliding.

Vultures soar to keep airborne as they scan the ground for carcasses, but they can also travel long distances by soaring. A Ruppell's griffon vulture that had been feeding on the Serengeti plains. Tanzania, was followed back to its nest by a glider pilot. It travelled 75 km (47 mi) in 96 minutes without flapping its wings, only stopping to circle in five thermals. Storks migrate between Europe and East Africa by thermal soaring, going by way of Suez to avoid crossing the Mediterranean where there are no thermals. Monarch butterflies also use thermals on their migrations between Canada and Mexico. Gulls slope soar along coastal cliffs.

Albatrosses seldom flap their wings, but soar for hours over the oceans by a special technique that depends on the wind being slower close to the sea than it is higher up. They glide downwind, losing height but gaining speed, then turn into the wind and rise. As they rise they slow down relative to the sea, but, because they are rising into faster wind, the speed of the wind over their wings remains high enough to give the lift indeed to keep them rising. After rising about 12m (40 ft) they turn and repeat the process.

Gliding birds tilt their bodies to suit different gliding angles by swinging their wings forward or sweeping them back, altering their position relative to the center of mass of the body. They bank to make a turn by giving one wing a higher angle of attack than the other, so that it gets more lift. They normally fly with their feet retracted among the feathers, but they lower them when a braking effect is required.

The extinct pterosaurs probably soared like gulls and albatrosses. These winged reptiles, which lived at the time of the dinosaurs, in the Jurassic and Cretaceous periods (190-65 million years ago) included the largest known flying animals. Among them, Pteranodon had a wing span of 8m (26 ft), and a few fragements of a considerably larger pterosaur have been found. The largest wing span among birds is about 3.4m (11 ft), in the Wandering albatross. Pterosaurs seem to have had relatively small wing muscles but Pteranodon was so lightly built that its total mass was probably only about 17 kg (38 lb) and it could have taken off by spreading its wings and facing into a light breeze.

Powered Flight

When soaring is not feasible, power is needed for flight. Aircraft use propellers or jet engines, but animals flap their wings. There are two main kinds of flapping flight. In hovering flight, the animal's body is more or less stationary and its wings move very quickly. Hummingbirds and many moths hover in front of flowers while feeding on nectar, and dragonflies hover while watching for insect prey. In fast forward flight of birds, bats and large insects, the body travels forward at a speed which is comparable to the maximum speed of the wing tips. Small insects such as flies (Diptera) fly in a manner which is intermediate between hovering and fast forward flight.

In fast forward flight the wings beat up and down. taking a wavy path as the animal moves forward through the air. On the downstroke, lift acts forwards and upwards, propelling the animal forward and supporting its weight. For the upstroke, the angle of attack is decreased so that little or no lift is produced except possibly near the base of the wing (lift acting upwards, at right angles to the path of the wing, would have a backward component and so would tend to slow the animal down). In contrast to fast forward flight where lift is produced only on the downward stroke, in a penguin swimming with its wings, lift is produced on the upward as well as the downward stroke.

Animals use several different hovering techniques. In hummingbirds, moths, bees and many other insects, the wings are beaten horizontally, backward and forwards. They turn upside-down for the backward stroke and the angle of attack is adjusted so that the lift always acts upwards.

Most hummingbirds best their wings at frequencies of 20 to 50 cycles per second, a littte above the lower limit of the range of frequencies of audible sound. For this reason, they emit a humming sound as they hover. Small insects best their wings at higher frequencies, as indicated by the buzz of bees and the high-pitched whine of mosquitoes.

Each beat, up or down, of a mosquito's wing lasts less than one thousandth of a second. No ordinary muscle can contract so fast, but mosquito wings are worked by a special kind called fibrillar muscle. Other flies (Diptera), wasps etc. (Hymenoptera), beetles (Coleoptera) and bugs (Hemiptera) also have fibrillar flight muscls. The special property of this muscle is that, so long as it kept active by occasional nerve impulses (action potentials), it vibrates at the resonant frequency of the system.

A fibrillar muscle attached to a turning fork would keep the fork vibrating. Tuning forks and many other mechanical systems have resonant properties because they have mass and elastic stiffness. Similarly, the mass of an insect's wings and the elastic properties of the thorax combine to make a resonant system which fibrillar muscle can drive. A simple experiment illustrates this; if the wings of a fly are shortened they beat at a higher frequency because their mass has been reduced. In many insects, the elasticity of the resonant system seems to be mainly the elasticity of the muscles themselves. In others, the elastic properties of the skeleton are also important. Many have parts of the cuticle of the thorax made of resilin, a protein with properties like rubber. These serve to adjust the resonant frequency to the required wing-beat frequency.

The resonant qualities of the thorax are important even in insects with non-fibrillar wing muscles because a resonant system needs relatively little power to keep it vibrating. The kinetic energy of the wings is not lost at the end of each stroke, but is made available for the next stroke by the elastic rebound.

Unsteady Aerodynamics

It has been said that science has proved bumblebees cannot fly. Science has, of course, proved no such thing. What scientists have shown is that conventional aerodynamics is incapable of explaining the flight of many insects. The lift on their wings, calculated for each stage of the wing beat cycle from the standard aerodynamic equations, is far too small to support their weight. The explanation is that the equations refer only to wings or propeller blades moving steadily at constant speeds. Flapping wings stop and accelerate again between one stroke and the next, making it possible to exploit "unsteady" effects unknown in engineering.

There is still a lot of uncertainty about a aerodynamics of insect flight. There is no satisfactory, quantitative explanation of the hovering flight of dragonflies and hoverfiies, which use rather small, roughly vertical wing movements instead of the usual large horizontal ones.

One of the unsteady effects used by insects (and some birds) was discovered in 1973 by Professor Torkel Weig-Fogh of Cambridge University, who called it the "clap and fling." The wings are clapped together, back to back, at the end of the unstroke. The air flowing into the growing space between them, as they separate again, sets up a circulation of air which gives greatly enhanced lift. For a short time, the lift is far greater than would act on the wings if they were moving steadily at the same speed. The clap and fling is used by many insects and also by some birds when hovering or flying slowly. The wings of pigeons can be heard clapping together, when they take off.

2

Locomotion in Different Vertebrates

IN CYCLOSTOMATA

Swimming and Buoyancy

Lampreys and hagfishes swim like most other fishes, but I have deferred discussion of swimming to this chapter because it has been studied more thoroughly in selachians.

It moves its tail from side to side as it swims forward so the tail takes a wavy path through the water. (I have been careful, in drawing the diagram, to made the amplitude and wavelength of the path realistic fractions of body length.) The tail is a hydrofoil, shaped like an aeroplane wing, and is held at an angle of attack to its path. As it moves to the right, the left on it acts forward and to the left. As it moves to the left in the return stroke, the lift acts forward and to the right. The components to left and right cancel out and the fish is driven forward through the water. The lift on the tail peoples to fish but the drag hinders it, so the larger the ratio of lift to drag the more efficient swimming will be. To get a high lift-to-drag ratio you need a hydrofoil with a high aspect ratio: that is, with a high ratio of span (the distance from tip to tip) to chord (from front edge to back edge). The tails of sharks such as Lamna have reasonably high aspect retios. These sharks also have well-streamlined bodies, seeming well adapted for efficient swimming.

The tail propels the fish by driving water backwards. It is a reasonable approximation to say that the water it pushes on forms a cylinder, of diameter equal to the span of the tail. The more water it pushes on, the more efficient will swimming be. This leads to the same conclusion as the previous paragraph: a tail with large span will propel a fish more efficiently than a smaller-span tail of the same area. The energy cost of shark swimming has not been measured but teleost fishes swim much more economially than squids, apparently because they push on more water.

There is little published information about shark swimming speeds, but 1 m carcharhinus in a large aquarium have been filmed swimming at $0.5-4$ m s^{-1}.

Seems a fairly realistic representation of swimming by sharks such as Lamna and Carcharhinus, whose tails have relatively high aspect ratios. It is much less realistic for slender selachians such as dogfish, still less for lampreys and eels. These fish do not have hydrofoil-like tails, and their swimming action does not even approximate to simple side-to-side movement of the tail. They throw themselves into waves which travel backwards along the body. The movements are like the swimming movements of nemrtode worms (but the resulting water movements are very different because the Reynolds number is to much higher.

Rays swim by undulating their huge pectoral fins. Waves are passed backward along the fin, driving the fish forwards.

The principal muscles of rays are the pectoral fin muscles, but more typical fishes have small fin muscles, and huge axial muscles on either side of the vertebral column. These muscles (the fillets that we eat) make up more than 50 per cent of the mass of the body. They are mostly white, but most fish have a narrow lateral strip of red muscle. (The colours remian after cooking, so you can see this if you eat fish.) In dogfish (Scyliorhinus) 18 per cent of the swimming muscle is red.

The white muscle fibres resemble the white muscle of squid mantle they have few mitochondria and a relatively poor blood supply. The red fibres have many mito-chondria and a good blood supply, like the yellow muscle of squid mantle. Thus the white muscle seems adapted for anaerobic metabolism and the red for aerobic. The red colour is due to myoglobin, a haemoglobin-like compound that seems to be involved in transport of oxygen within the cells.

Long before the distinction between aerobic and anaerobic muscle was known for squid, Dr. Quentin Bone demonstrated the distinct functions of red and white muscle in dogfish. He destroyed the fish's brain but kept the body alive by pumping aerated water over the gills. Dogfish prepared in this way still make swimming movements, if fixed by the head with the tail free in a tank of water. The movements continue at a slow, more or less steady rate, but if the tail is pinched the fish swims faster or makes a strong, sustained bend. Bone stuck electrodes into the red and white parts of the swimming muscle and recorded potentials from them. During slow swimming movements potentials were detected in the red muscle, but not in the white. (Dogfish red muscle does not seem to propagate action potentials, and it is likely that these potentials, were action potentials in the nerves rather than the muscles.) When the fish was pinched and so stimulated to make more virorous movements, muscle action potentials were detected in the white muscle. It is not certain whether the red muscle was active as well because the potentials in the white muscle are much larger than those detected (during slow swimming) in the red. They are large enough to affect the electrode in the thin layer of red muscle and mask any activity in the red muscle itself. These experiments are interpreted as evidence that the red muscle alone is used for slow and sustained swimming but that the much greater bulk of white muscle comes into use for violent movements and bursts of speed. This has been confirmed in more recent experiments in which recordings have been made from the red and white muscles of intact teleost fishes, swimming in water tunnels like the one.

Bone also investigated the effects of exercise on the stores of food materials in the red and white muscles. Some fish were kept unexercised for one or two days before they were killed for analysis. Others were killed in a state of exhaustion, after they had been made to swim vigorously for up to ten minutes. Yet others had their brains destroyed and were set swimming in the manner that has been described. They were kept swimming slowly for up to 50 h, and then killed. Exhaustion by a few minutes' exercise halved the glycogen content of the white muscle but had no apparent effect on the composition of the red. Prolonged exercise halved the fat content of the red muscle but had no significant effect on the white. It seems that glycogen fuels the anaerobic metabolism of the white muscle and that fat fuels the aerobic metabolism of the red.

There seems to be a general rule that vertebrate striated muscles are most efficient, and can exert most power, when shortening at about 30 per cent of their maximum rates. The rates at which carp (Cyprinus, a teleost) red muscle shortens during swimming have been measured. Films were made of carp swimming at various speeds in a water tunnel to find out how much and how fast they bent. The fish were killed, their bodies were bent into positions like those seen in swimming and they were left until rigor mortis had set in, tightening the muscles. Muscle fibres were dissected out and their sarcomere lengths measured. This made it possible to calculate the rates at which sarcomere lengths changed during swimming. In other experiments, muscle fibres from freshly killed carp were tested to discover their rates of shortening under different loads. The rate of shortending during swimming at the highest speed that the red muscle could sustain was about 30 per cent of the maximum rate. Experiments with carp muscle have also shown that the white fibres (used in fast swimming) have much higher maximum rates of shortening than red fibres from the same fish. Also, the red fibres run parallel to the vertebral column, but the white fibres of teleosts are twisted into helices like the strands of a rope. The consequence of this is that the body bends more, for a given percentage shortening of the white muscles, than for the same shortening of the red ones: it is as though the muscles worked through a gearbox, with the white muscles using a higher gear. Selachians have a different arrangement of white muscle fibres, which gives a much less pronouuced gearbox effect.

The swimming muscles are divided into segmental blocks (myomeres) separated by connective tissue partitions. Cooking breaks down the partitions, so cooked fish separates easily into flakes, which are individual myomeres.

Nearly all selachians live in the sea, in water of density about 1026 kg/m^3. Many of their tissues are denser, including muscle (about 1060 kg/m^3), cartilage (1100 kg/m^3) and scales (probably about 2000 kg/m^3). Consequently, most selachians are denser than the water they live in. For example, dogfish have a density of about 1075 kg/m, about the same as the squid Loligo.

Such fish must swim perpetually, if they are not to sink. Dogfish spend a lot of time resting on the sea bottom but sharks such as Lamna seem to swim all the time. Dogfish get most of the lift they need to prevent sinking by using their pectoral fins as hydrofoils, but these fins are anterior to the centre of mass. The moments about the centre of mass are balanced by lift on the tail, which is shaped so as to drive water downwards (producing lift) as well as backwards (producing thrust). The vertebral column bends up into the upper lobe of the tail leaving the lower (hypochordal) lobe stiffened only by cartilages called pterygiophores (described below). Such tails are called heterocercal. The forces on the tails of dead selachians have been measured in experiments in which they have been moved sideways through water, at appropriate speeds to imitate swimming. The vertical forces depended on the speed, but had values around those required for equilibrium by the swimming fish.

If the pectoral fins were held always at the same angle of attack, they would produce more lift at higher speeds. The lift that the fish needs is the same at all speeds, because its weight in water is constant. The fish needes muscles to adjust the angle of attack of the fins (and also of the tail) as it changes speed. Fin movements are also used to steer the fish up or down, and rays use them for swimming.

The fins are segmented structures, with muscles which develop in the embryo by separating off the myotomes. The increase in length of the fin as the embryo grows generally fails to keep pace with the growth of the body, so that a dorsal fin derived from fourteen segments (for instance), may only extend over six myotomes in the adult fish. Each segment of the fin contains two or more rods of cartilage (pterygiophores) connected end-to-end. The distal pterygiophores lie between sheets to packed ceratotrichia, which are very thin rods of a form of collagen known as elastoidin. The ceratotrichia extend to the edge of the fin, and stiffen it. There are a great many of them in each segment, but there is only one muscle on each side of the fin in each segment. It is attached by a broad tendon to the sheet of ceratotrichia, and when it contracts it bends the lin over to its side.

In early selachians, and also in rays, the pterygiophores extend near the margins of the fins. In other modern selachians, the pterygiophores are confined to the base of the fin and most of the area of the fin is stiffened only by ceratotrichia. The long-based pectoral fin of Cladoselache must have been less mobile than the narrow-based ones of modern sharks.

Not all selachians are substantially denser than seawater. Some, such as the Portuguese shark (Centroscymnus, a deep-sea shark) and the Basking shark (Cetorhinus), are very close indeed to the density of seawater. Their pectoral lins do not have to produce appreciable vertical forces except for manoeuvring, and are appropriately small. Their tails are heterocercal but presumably produce more or less horizontal forces. They owe their low density to oil that consists of liqids (density around 920 kg m^{-3}) and the hydrocarbon squalene (density 860 kg m^{-3}). The quantities required are large and the liver, which contains most of the buoyant material, is enormous. In section 6.5 we calculated the volume of gas-filled shell needed to float the mollusc Nautilus. A similar calculation shows that to reduce the density of a dogfish 1075 kg m^{-3}) to that of sea-water 1026 kg m^{-3} you would have to add enough squalene to increase its volume by 30 per cent.

I argued in section that buoyancy organs are advantageous to the slow-moving cephalopods Nautilus and Helicocranchia, but that it is better for fast-swimming squid to prevent themselves from sinking by using their fins as hydrofoils. The same argument shows how squalene may be advantageous only to slow-swimming selachians.

Basking sharks swim to filter plankton at about one metre per second, which seems slow for 10 metre fish, The speed of Centroscymnus is not known.

IN PISCES

Buoyancy

Most teleosts have a swimbladder, a sac of gas in the body cavity which reduces their density. Teleosts without swimbladders, or with the swimbladder deflated, generally have densities between 1060 and 1090 kg m^{-3}, about the same as typical selachians. (The bone of teleost skeletons is much denser than selachian cartilage, but their high urea content must make selachian body fluids denser than teleost ones.) Many teleosts with swimbladders can float almost motionless in mid-water, which shows that their densities are very close indeed to the density of the water.

A 1 kg teleost, of density 1080 kg m^{-3} without a swimbladder, would have a volume of $1/1080 = 0.000926$ m^3 (926 cm^3). A swimbladder containing 74 cm^3 gas would increase its volume to 0.00100 m^3 without appreciably increasing its mass, giving it a density of 1000 kg m^{-3}, the same as fresh water. Similarly, a 49 cm^3 swimbladder would match its density to that of sea water, 1026 kg m^{-3}. These volumes are much smaller than the volumes of squalene that would be needed to have the same effect on its density. A fish with a swimbladder is much less bulky than it would have to be if it depended on squalene for buoyancy, and can be expected to need less energy for swimming at the same speed. However, swimbladders have disadvantages that will soon become apparent.

The swimbladder develops in the embryo as an outgrowth of the gut. Adults of the more primitive teleosts retain a duct (the pneumatic duct) connecting it to the oesophagus. Many of the more advanced teleosts including the Acanthopterygii lose the duct as they develop, so that the swimbladder becomes a closed bag of gas.

Gases have been taken from the swimbladders of fish caught at various depths, and analysed. Swimbladders of fish from shallow water generally contain around 80 per cent nitrogen and 20 per cent oxygen; they contain a mixture of gases very similar to atmospheric air. Swimbladders of fish from greater depth generally contain higher proportions of oxygen. However, trout and their relatives (Salmonoidei) commonly have very high proportions of nitrogen in their swimbladders, no matter how deep they are living. Ciscoes (Leucichthys spp.) from depths over 150 m in Lake Huron generally have over 95 per cent nitrogen in their swimbladders.

Most swimbladders have extensible, flexible walls, so the pressure of gas inside matches that of the water around the fish. The pressure in the water (and so in the swimbladder) is 1 atm at the surface and increases by 1 atm for every 10 m of depth, so at D m it is (1+0.1 D) atm.

If the water were in equilibrium with the atmosphere, the partial pressures of dissolved oxygen and nitrogen in it would be about 0 2 atm and 0.3 atm, at all depths. About 0.8 atm nitrogen is indeed found at all depths, but the partial pressure of oxygen is always a good deal less than 0.2 atm at depths more than about 200 m. Thus the total of the partial pressures of dissolved gases in the water is 1 atm or less, at all depths. The partial pressures of gases in arterial blood are generally about the same as in the water, because of the exchange of gases that occurs at the gills, so their total is also 1 atm or less, at all depths. However, the pressure in the swimbladder (the total of the partial pressures of the individual gases) is (1+0.1 D) atm. Thus at least some of the gases must have higher partial pressures in the swimbladder than in the blood, and will tend to diffuse out. The deeper the fish is swimming, the larger the partial pressure differences and the faster gas will be lost from the swimbladder. Teleosts that have pneumatic ducts and live near the surface may make good the loss by taking an occasional gulp of air at the surface, but others have to secrete new gas, as will be explained.

Secretion of gas from a low partial pressure in the blood to a high partial pressure in the swimbladder is essentially a process of compression, and requires energy. If diffusion losses from the swimbladder can be kept small, energy will be saved. Most swimbladders have sparse blood supplies, except in the specialized regions of secretion and resorption. Many have walls made highly impermeable to diffusion of gases by a layer of extraordinary crystals. These crystals are made of guanine, a purine. They have the form of broad sheets, so thin as to be very flexible. Electron micrographs show that they are only 20 nm thick, and often show them folded into pleats. The crystals have such low diffusion constants that diffusion of gases is effectively limited to the narrow spaces around and between them.

Rates of diffusion of oxygen through swimbladder walls have been measured by means of an oxygen electrode (Fig. 2.1). The partial pressure of oxygen in an oxygen electrode is effectively zero. Oxygen electrodes are normally used to determine unknown partial pressures of oxygen, by measuring the rate at which oxygen diffuses in through a plastic membrane of known properties. They can, however, be used to determine the diffusion constants of membranes of unknown properties, from the rates at which oxygen diffuses in from a fluid in which it has a known partial pressure. A piece of swimbladder wall was removed from a fish and fastened over the plastic membrane of an oxygen electrode, which was then exposed in turn to moist air and moist oxygen. (Moist gases were used so that the swimbladder wall would not dry out.) The electrode current was measured in each case and the rates of diffusion of oxygen were calculated. The oxygen had to diffuse through the plastic membrane as well as through the swimbladder wall, but account was taken of this in the calculations.

The experiments were performed with the swimbladders of shallow-water marine fish and also of the conger eel (Conger) which lives at depths down to 300 m. As might have been expected, the conger swimbladder was the least permeable. Its diffusion constant was about 6×10^{-7} mm^2 atm s^{-1}, which only about one fortieth of the values usually found for other tissues. Some other teleosts live at much greater depths, and many of them have swimbladders. Two species caught at 1400 m and 2800 m had ten times as much guanine, per unit area of swimbladder wall, as the conger.

Since pressure increases with depth, swimbladders are compressed as fish swim deeper and expand when they swim nearer the surface. Fish may need to secrete or resorb gas to compensate for depth changes, as well as secreting to replace diffusion losses.

The process of resorption is straightforward, since diffusion from the swimbladder is in any case inevitable. Many teleosts have means of speeding it up when required. The eel Anguilla has a pneumatic duct which could be used to pass gas to the mouth and out of the fish, but which serves as a gas-resorbing organ (Fig. 2.1 a). Normally this duct is slender, and through it has plenty of blood vessels little blood flows through them. When gas is to be resorbed it swells and fills with has from the main cavity of the swimbladder. Its blood vessels dilate, and it has been estimated from measurements of oxygen concentration in them and other vessels that as much as 20 per cent of the blood pumped by the heart may travel through them. Advanced teleosts with no swimbladder duct commonly have a pocket in the swimbladder wall, known as the oval, where resorption occurs. It has plenty of blood vessels but these are generally constricted so that little blood flows through them and the oval is closed off from the main cavity of the swimbladder by a sphincter. During resorption the blood vessels dilate and the sphincter opens.

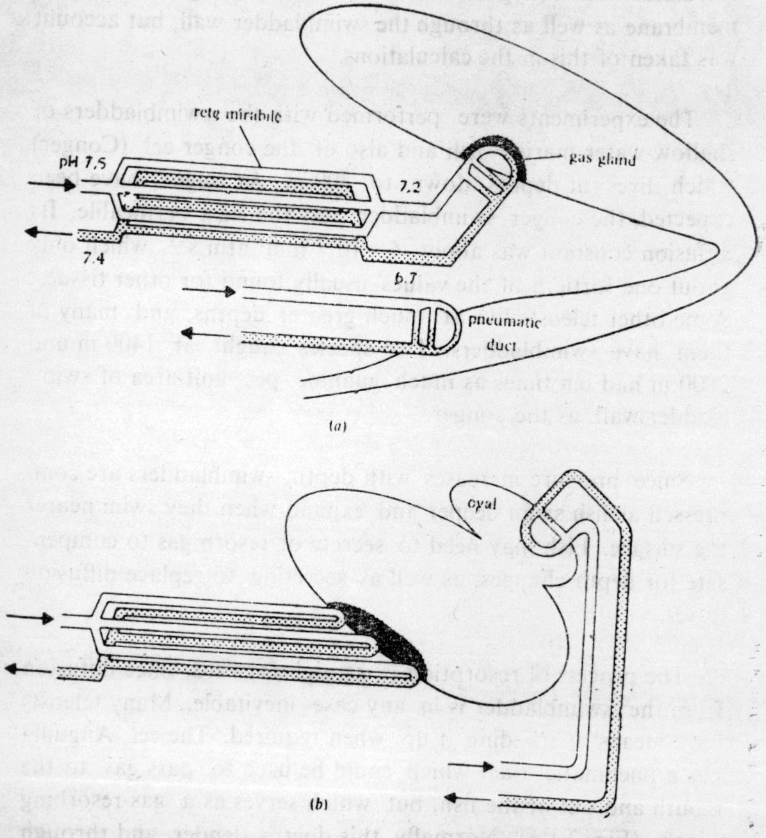

FIG. 2.1. Diagrams of the swimbladder and its blood supply in (a) the eel Anguilla, and (b) a typical advanced teleost with no swimbladder duct. The actual numbers of capillaries are, of course, very much larger than the numbers shown.

Some teleosts have no obvious locality for gas secretion, which probably occurs over large areas of the swimbladder wall. Others have conspicuous red patches known as gas glands. Their function has been demonstrated by injecting cod (Gadus morhua) with yohimbine (a drug which stimulates gas secretion). The swimbladders of injected cod were cut open and transparent plastic was laid over the gas gland. Bubbles of gas could be seen collecting under the plastic.

It will be shown that, at least in eels, the process of gas secretion into the swimbladder involves secretion of lactic acid into the blood passing through the gas gland. This lactic acid tends to release gases. Because it is an electrolyte it must reduce the solubility of all gases in blood, just as the salts in seawater make gases less soluble in it than in fresh water. This effect is small, but it is the only known basis for secretion of nitrogen and other inert gases. Because it is an acid, lactie acid tends to release carbon dioxide from bicarbonates in the blood. Also because it is an acid, lactic acid tends to release oxygen from haemoglobin: Fig. 2.2 shows the effect of carbon dioxide (which

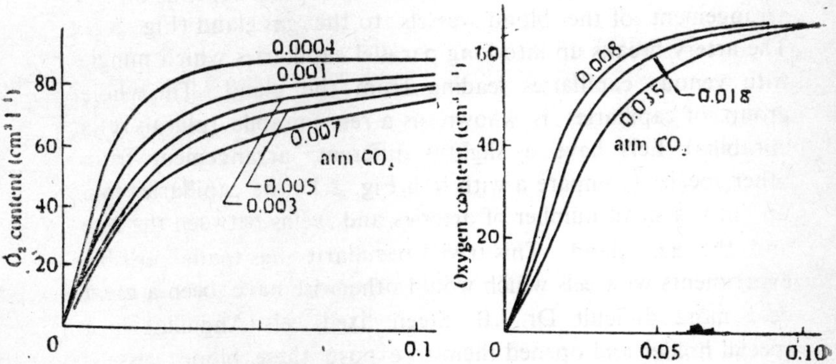

FIG. 2.2. Graphs of oxygen content against partial pressure of oxygen for the blood of (a) rainbow trout (Salmo gairdneri) at 15°C and (b) the African lungfish (Protopterus aethiopicus) at 25°C. In each case graphs are given for several different partial pressures of carbon dioxide.

reduces the pH) on the amount of oxygen that teleost blood will hold. The gas secreted into most swimbladders is mainly oxygen, and the effect on haemoglobin is the most important effect of the lactic acid secreted by the gas gland.

Nevertheless, this effect could not by itself release oxygen from the blood at substantial depths. Telecot blood typically contains enough haemoglobin to combine with about 10 cm^3 oxygen (measured at atmospheric pressure) per 100 cm^3 blood. Additional oxygen can be taken up in physical solution in the blood; if the partial pressure of oxygen is X atm. the amount is 4X cm^3 per 100 cm^3 blood. Oxygen is taken up at the gills from water in which its partial pressure is 0.2 atm (or less), so at a depth of 20 m, where the pressure is 3 atm, the blood is capable of dissolving an additional $4(3.0-0.2)=11$ cm^3 oxygen. Even if all the 10 cm^3 oxygen per 100 cm^3 blood could be released from the haemoglobin it would not pass into an oxygen-filled swimbladder, but would be taken up in physical solution in the blood.

Secretion of oxygen at substantial depths depends on the arrangement of the blood vessels to the gas gland (Fig. 2.1). The artery breaks up into long parallel capillaries which mingle with venous capillaries leading from the gland. The whole group of capillaries is known as a rete mirabile (plural: retia mirabilia). Eels have a slightly different arrangement from other species (compare a with b in Fig. 2.1): the capillaries join up into a small number of arteries and veins between the rete and the gas gland. This trivial peculiarity has made possible experiments with eels which would otherwise have been a great deal more difficult Dr. J.B. Steen fixed eels (Anguilla) in a special holder and opened them to expose these blood vessels. He injected the drug yohimbine to stimulate gas secretion and when this was in progress withdrew for analysis small (40 mm^3) samples of blood from the four positions indicated by numbers in Fig. 2.1 (a). These numbers give the mean pH of samples from each position. Oxygen, carbon dioxide and lactic acid content were also determined, and their values were also in accordance with the theory of gas secretion which will be outlined. It is apparent from the pH values shown in Fig. 2.1 (a) that acid is secreted into the blood passing through the gas gland. This is lactic acid: the blood entering the gland contains 8 mmol l^{-1} lactic acid, and the blood leaving it 13 mmol l^{-1}. There is also diffusion of ions in the rete, between the venous and the arterial capillaries: this accounts for the fall in pH along the arterial capillaries and the rise along the venous ones.

As blood passes through the gas gland and lactic acid is added to it, the partial pressure of the oxygen in it is increased, mainly through the effect of the acid on haemoglobin. When this blood passes into the venous capillaries of the rete it contains oxygen at a higher partial pressure than in the parallel arterial capillaries. Oxygen diffuses from it into the arterial capillaries, raising the oxygen concentration and partial pressure of the blood there before it arrives at the gas gland. The process is repeated with this blood, and the concentration and partial pressure of oxygen build up higher and higher at the gas gland end of the rete. Eventually the partial pressure exceeds the partial pressure in the swimbladder, and oxygen is released. It can be shown theoretically that the maximum partial pressure that can be built up (and so the depth at which secretion is possible) should increase sharply as the length of the rete increases. In large Anguilla the retia (and their capillaries) may be 10 mm long, but teleosts living at depths of 2000 m or more tend to have longer retia. Retia 25 mm long have beeen found in several deep-sea fishes. These lengths are immensely greater than those of ordinary capillaries, such as the ones in Anguilla muscle which are generally 0.5 mm long or less.

The fall in the lactic acid concentration and the consequent rise in the pH of the blood as it travels along the venous capillaries of the rete might be expected to have an adverse effect, partly cancelling out the effect of the rise in concentration in the gas gland. In fact this adverse effect is probably negligible because of a rather surprising property of eel blood. When the pH is descreased, for instance by adding lactic acid, the partial pressure of oxygen is affected very rapidly indeed; the effect has gone half way to completion in 0.05 s. Reversal of the effect by an increase in pH is much slower, and about 15 s is needed for it to go half way. It can be calculated that each blood corpuscle takes a second or thereabouts to travel the length of the rete. This is plenty of time for a fall in pH to take effect, but not for recovery when the pH rises again.

Secretion and resorption are necessarily slow processes, as can be understood by considering the quantities of gas involved. Consider a 1 kg fish with a 50 cm^3 swimbladder. Suppose that it moves deeper in the sea, keeping the volume of the Swimbladder constant. Every increase in depth of 10 m increases the pressure by 1 atm and requires secretion of a quantity of gas which would have a volume of 50 cm^3 at atmospheric pressure. Experiments on brook trout (Salvelinus) showed that 1 kg specimens in well-aerated water at 20°C use about 50 cm^3 oxygen h^{-1} when resting and up to 150 cm^3 h^{-1} when active. The latter is presumably the maximum rate at which the gills can take up oxygen and the blood can transport it to the muscles. Even if one third of the blood pumped by the heart could be directed to the gas gland, it seems most unlikely that oxygen could be supplied fast enough to allow secretion of more than 50 cm^3 h^{-1} (measured at atmospheric pressure). This rate would fill an empty swimbladder in an hour, or maintain the volume of the swimbladder in a descent at the very slow rate of 10 metres per hour.

To find out how fast they could secrete and resorb gases, cod (Gadus) were kept in a pressure tank with a viewing window. A pump could raise the pressure in the tank to a maximum of 7.5 atm, corresponding to a depth of 65 m. Fish from shallow water subjected to high pressures were initially denser than the water (because their swimbladders were compressed) but recovered their buoyancy over periods of many hours. The progress of their recovery was monitored by occasional tests taking 3 minutes or less, in which the pressure was adjusted until the fish had the same density as the water and could float with only slight fin movements. The rates of gas secretion were even slower than suggested above, enough only to compensate for depth increases at the rate of one metre per hour. Rates of gas resorption depended on depth: the fish resorbed faster at greater depths, where the differences in partial pressure between swimbladder gases and dissolved gases were greater. At a depth of 5 metres they could compensate for depth reductions only at the rate of 1 metre per hour but at 65 metres they could compensate at 20 metres per hour.

Some teleosts make large depth changes every day. Herring (Clupea) spend the night near the surface and the day at depths which vary with locality, but may exceed 150 m. Since swimbladders reflect underwater sound, these movements can be followed by echo-sounding. Even bigger vertical migrations are made by the small lantern fishes (Myctophidae) which are very plentiful in the oceans. Many of them spend the night within 50 m of the surface and the day at 300 m or more: their movements have been observed from deep submersible vehicles (research submarines). The changes of depth at dawn and dusk seem much too fast for compensation by secretion and resorption of swimbladder gases.

When herring and such species of lantern fishes as have gas-filled swimbladders are caught at night at the surface, they have enough gas in the swimbladders to give them about the same density as seawater. It is not known how much gas they have in the swimbladder at their daytime depths, but it seems uhlikely that enough extra gas is secreted to compensate for the increased depth. More probably the quantity of gas is kept constant, so that the swimbladder is compressed during the day to small fraction of its night-time volume. If so, it would be almost totally ineffective by day. The fish would be denser than seawater and would have to depend on the hydrofoil action of fins to avoid sinking further. In these circumstances the advantage of a swimbladder over materials like squalene might disappear. Some lantern fishes have gas-filled swimbladders only as juveniles. As they grow the swimbladder regresses and wax esters are deposited around it. These are quite different chemically from squalene, but have about the same density. Some lantern fishes with no gas in the swimbladder have enough wax ester to give them about the same density as seawater. Their density is virtually unaffected by changes of pressure.

A great many bottom-living teleosts have also lost the swimbladder. The angler fish Lophius and the flatfishes (Pleuronectiformes) are among them. A fish which has the same density as water may rest in contact with the bottom but it needs no vertical force to support it, so there are no frictional forces to hold it in place. It will move with every water movement, and must swim to keep stationary in a current. If it were denser than the water it would use more power when it swam but would not need to swim all the time and so might make a net saving of energy. This is particularly likely to be the case for fish which find their food near bottoms exposed to currents, in rivers and on tidal shores.

Teleosts that have the same density as the water can hang almost motionless in mid-water, and they can execute a remarkable range of manoeuvres by moving their fins. They are not limited to swimming forwards like sharks, but can move backwards, or vertically up and down (with the body horizontal), and they can turn on the spot. Such manoeuvrability is especially useful in cluttered environments, for example among corals. The fins may make rowing movements, or waves may be passed along them. All the fins, including the tail fin, may be used.

Selachian fins are relatively thick and rigid, though the pectoral fins of rays are mobile enough for waves to be passed along them in swimming. Teleost fins are much more mobile. They consist of widely spaced rays joined only by very thin webs of tissue and (unlike selachian fins) can be folded and unfolded like a fan.

Fig. 2.3 .shows part of a teleost fin. The pterygiophores are bone. The rays are not simple collagenous rods like the ceratotrichia of selachians but are branched and consist of short lengths of bone joined by collagen fibres. Each consists of two halfrays, one on each side of the fin. Except at the ends of the fin, each ray articulates with a separate group of pterygiophores. The most distal pterygiophore consists of two halves, tightly bound together by collagen fibres. The bases of the two halves of the ray lie on either side of it, attached by ligaments so as to form a rather loose hinge joint. This joint bends when the fin is folded like a fan, and extends as it unfolds (Fig 2.3 b). The joint between the distal radial and the next one is also a hinge joint but its axis is longitudinal so that it allows the ray to swing from side to side (Fig. 2.3. d). This movement occurs when the fin is undulated: each ray rocks from side to side a little out of phase with its neighbours so that the fin is thrown into waves which travel forwards or backwards along it.

Fig. 2.3. (a, c) shows the muscles which move the rays. The erectors unfold the fin and the depressors fold it. The left inclinator swings the ray to the left and the right one to the right. An inclinator can be assisted by the erector and depressor of its side contracting together, but they also tend to have the effect shown in Fig. 2.3 (e). The half-ray is pulled basally but cannot simply slide past its mate since the two half-rays are attached quite closely to each other at the edge of the fin. The ray therefore bends to the side. This is particularly apt to happen if the muscles are moving the fin against strong resisting forces.

When they are swimming at all fast teleosts keep their fins (other than the caudal fin) folded flat against the body. Unlike most sharks, they do not rely on spread pectoral fins to provide lift. They spread their fins to brake. Pectoral fins can be spread at right angles to the direction of motion and so make particularly effective brakes, but other fins may help.

If the pectoral fins were simple membranes, they would be bent back at the edges by drag when they were used as brakes. Because of their complex structure, with the muscles attached to half-rays, they tend to bend in the manner shown in

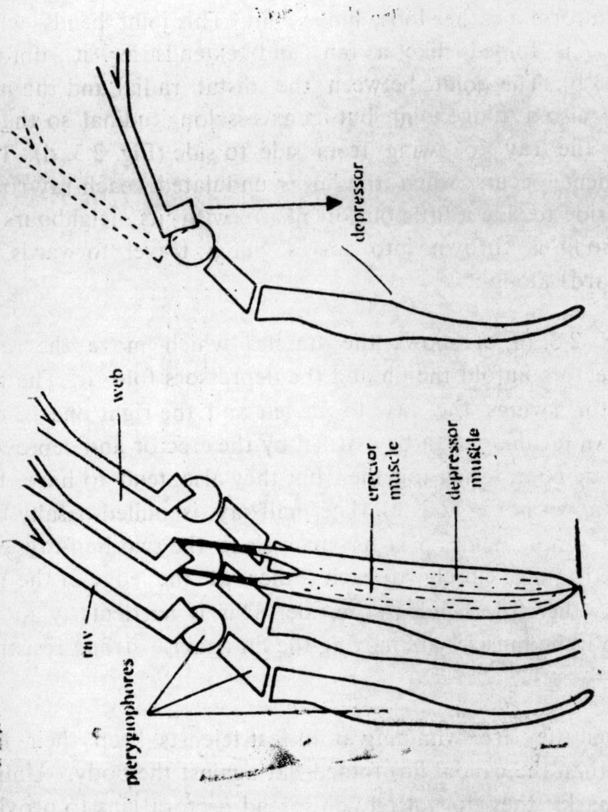
web
ray
erector muscle
depressor muscle
pterygiophores
(a)
depressor
(b)

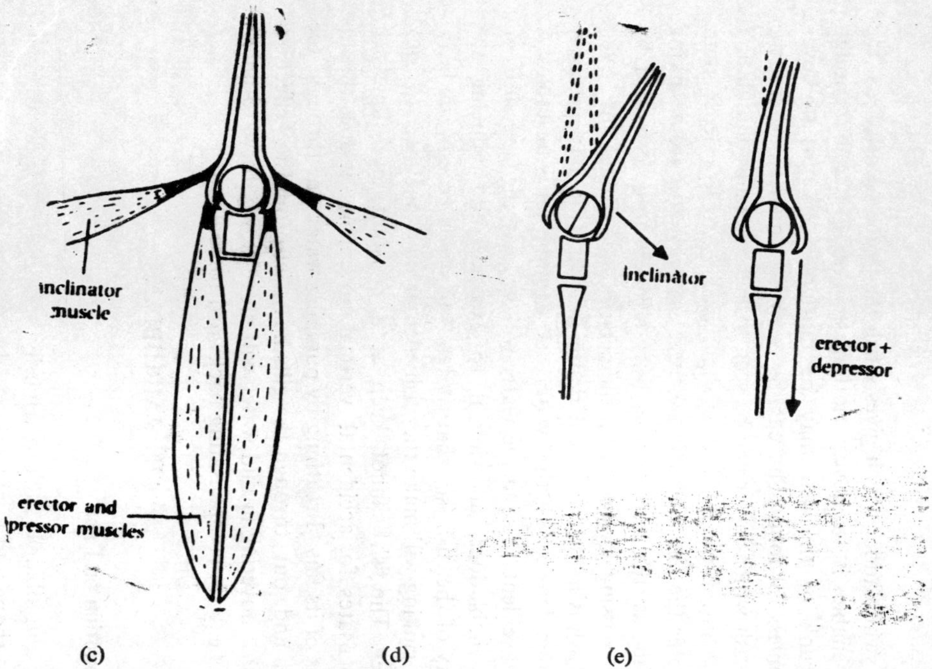

Fig. 2.3. Diagrams of the skeleton and muscles of a teleost fin seen (a), (b) from the side and (c), (d), (e) in transverse section, This is a dorsal or anal fin: the other fins differ in some details of structure.

Fig 2.3. (e). Their edges tend to bend forward, making them more effective as brakes. The forces which can be produced are illustrated by a film sequence of Taurulus (which has large pectoral fins) braking with a deceleration of 15 m s^{-2}. This implies a braking force 1.5 times the weight of the body. Most of this force probably acted on the pectoral fins.

A swimbladder which gives a fish the same density as the water enables it to hover motionless in mid-water, but though the trunk of the fish may be motionless its fins are never motionless for long. This can be seen in aquaria, where the many fish which hover can be seen to be continually undulating various fins. If the fish is even slightly denser or less dense than the water the fins must exert a force to prevent it from sinking or rising, as the case may be. Such a force can be provided by waves passing down or up the pectoral and caudal fins. Even if the fish is at the particular depth at which it has precisely the same density as the water, its equilibrium is unstable. A slight accidental rise in the water will reduce the pressure on the swimbladder, allowing it to expand and reducing the density of the fish so that it tends to go on rising. Similarly, slight sinking will make the fish denser and inclined to sink further. The fish cannot keep at a constant level unless it compensates for accidental vertical movements: it does this by means of its fins. Breathing by pumping water in through the mouth and out through the gills has a jet-propulsion effect, so a teleost hovering in mid water will move slowly forward unless its fins exert a compensating backward force.

IN AMPHIBIAN

Locomotion on Land

Early amphibians such as Ichthyostega were essentially newt-like in shape and might reasonably be expected to have walked on land like newts. Happily we do not have to rely on such comparisons, for there have been preserved in Palaeozoic rocks tracks left by amphibians or reptiles walking over mud. (It is not always possible to decide whether a particular track was made by a reptile or an amphibian.) These indicate newt-like walking with feet well out to the side and generally with the toes pointing forward, though one particularly early track shows the toes pointing laterally. There are seldom marks of the body being dragged along the ground. Apparently early tetrapods normally walked with the belly held clear of the ground, like the newt. Newts do, however, sometimes crawl with the belly sliding on the ground, for instance on a wet surface.

Walking with the belly off the ground poses a problem of balance. Ideally, the forces on the feet should be kept in equilibrium with the body weight at all stages of the stride. If they are not, the animal's trunk will rise and fall as it walks, or pitch or roll like a ship at sea. These perturbations of the animal's forward motion may not matter if they are reasonably small, but if they are too large the belly will hit the ground at some stage of the stride. They are most likely to be large in slow gaits, which allow so that though the sequence is as in Fig. 2.4 and the duty factor may be greater than 0.75, there may be times when only two feet are on the ground.

It has been argued that this may actually give a steadier ride, for animals whose muscles cannot develop tension very rapidly. The animal shown in Fig. 2.4 could maintain perfect equilibrium and move absolutely steadily, if its muscles were capable of exerting forces that changed instantaneouly, when feet were raised or set down. A mathematical model shows that if they can change forces only rather slowly, the animal may wobble less if it moves diagonally opposite feet almost simultaneously. The diagonal pattern also enables the animal to lengthen its stride by bending from side to side as it walks: The newt's bending from side to side may suggest the swimming action of a fish, but fish make travelling waves: each bend travels posteriorly along the body. The bends of the newt are standing waves, like the vibrations of a stretched string.

The pectoral and pelvic fins of Neoceratodus and Latimeria look rather like short legs, but they are not used as legs.

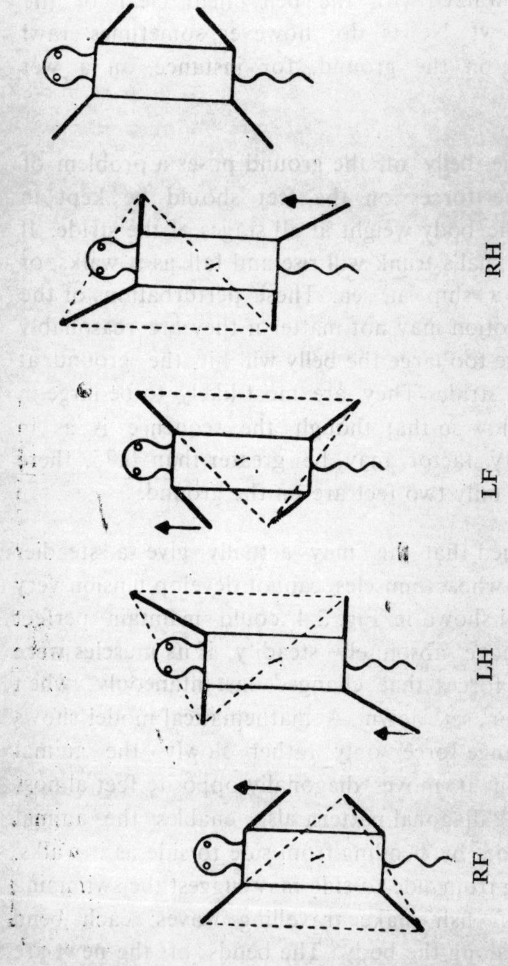

FIG. 2.4. Diagrams showing the order in which a four-footed animal must move its feet if it is to remain stable at all times. Each diagram represents the moment at which the foot marked with the arrow is lifted off the ground. The centre of gravity of the body is marked X. In cases where the centre of gravity lies above one edge of the triangle drawn between the three stationary feet it will move forward overthe centre of the triangle before the foot is set down again.

Neither fish goes onto dry land, and the fins would not be strong enough to support its weight there. Some of the films of Latimeria in its natural environment show it resting on the bottom but none show it using its fins like legs.

No intermediate stage is known between crossopterygian fins and the fully developed limbs of early amphibians.

The form of the primitive amphibian limb is obviously suited to the type of locomotion practised by newts. The large number of bones in the foot enables it to accommodate itself to the shape and angle of the surface the animal is walking on and to take advantage of irregularities to obtain a grip. Fig. 2.5 shows how the radius and ulna can cross over each other

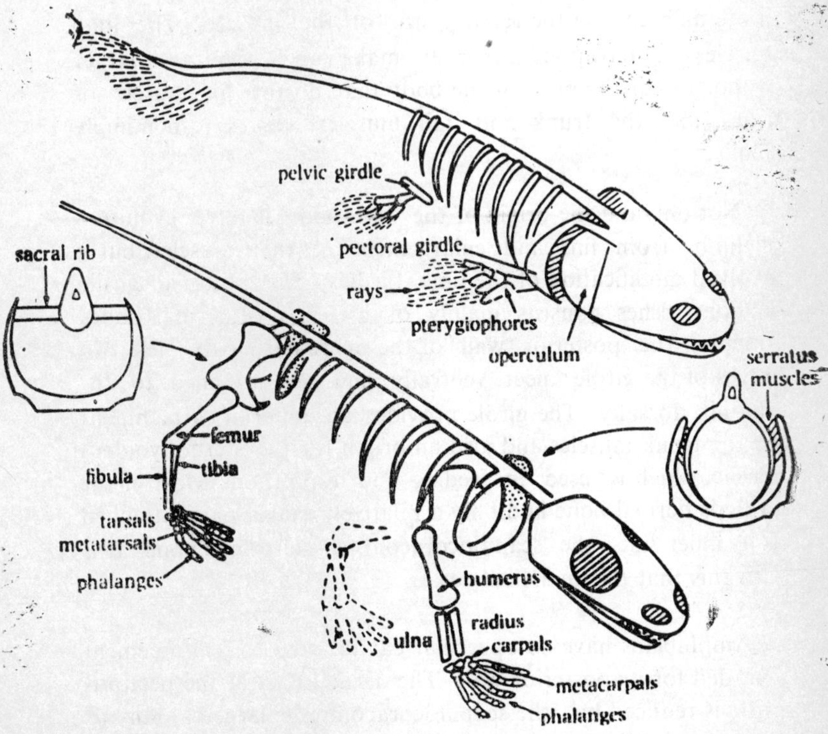

FIG. 2.5. Diagrams of the skeletons of a rhipidistian and an amphibian.

and so keep the toes pointing forward all the time the foot is on the ground, instead of letting the foot swivel round with the humerus (if the foot did swivel round it would be abraded, and work would have to be done against friction to turn it). X-ray cine films of lizards walking confirm that this twisting happens. The tibia and fibula cross similarly in the hindlimb. It is of course possible for an animal without a separate radius and ulna to walk in amphibian fashion and yet not swivel the foot, provided the elbow is not a simple hinge joint but allows sufficient variety of movement: this is the situation in frogs and toads.

The muscles of the trunk and tail do most of the work when a typical fish swims fast, but the muscles of the limbs do most of the work when a tetrapod runs. There is an appropriate difference in the relative sizes of the muscles. The limb muscles of tetrapods generally make up a very much larger proportion of the mass of the body than do the fin muscles of fishes, and the trunk and tail muscles are correspondingly smaller.

Not only did the origin of the tetrapods involve evolution of limbs from fins and enlargement of their muscles, but it involved modification of the limb girdles. The pectoral girdle of bony fishes consists mainly of a series of dermal bones stiffening the posterior wall of the opercular cavity. The two halves of the girdle meet ventrally and are attached to the cranium dorsally. The girdle provides an anterior attachment for the trunk muscles and also an origin for the sternohyoideus muscle, which is used in feeding and respiration. Most of the girdle is dermal bone and only a relatively small part attached to its inner side, the scapulocoracoid, is sclerotome bone. It is with this that the fin articulates.

Amphibians have no opercular cavity, so on reinforcement is needed for its posterior wall. The dermal part of the pectoral girdle is reduced but the scapulocoracoid is enlarged. Many of the limb muscles originate from it. The weight of the animal can best be supported without compressing the viscera and lungs if the pectoral girdle is not simply embedded in the body wall but attached more or less firmly to the exial skeleton. Its attachment to the skull in fishes is not well arranged for this because it is anterior to the fin and linked to it by a series of of flexibly connected bones. It has been replaced in tetrapods by an entirely new connection: the rib cage is suspended from the large scapula by short muscles. Detachment of the girdle from the skull makes the head more freely movable relative to the trunk.

The pelvic girdle in fishes is merely a pair of small plates of sclerotome origin embedded in the ventral body wall. In tetrapods it is larger, giving origin to large muscles and extending dorsally to make a firm attachment with the vertebal column. It is sutured to the ends of short stout ribs, which in turn are sutured to one or more vertebrae (the sacral vertebrae).

IN BIRDS

Flight

Hummingbirds hover in the same way as moths but most birds make quite different movements when they fly. Fig. 2.6 shows at the top the movements of a pigeon flying fairly slowly. The wings beat forward and down, back and up. They are spread more widely for the downstroke than for the upstroke. The primary feathers are bent during the downstroke, showing that large upward forces are acting on them but there is no obvious bending during the upstroke. It seems that in slow flight like this the downstroke is mainly responsible for producing the forces that support the bird's weight.

The pigeons in Fig. 2.6 had electrodes in their wing muscles. In some of the experiments, long (18m) wires trailed behind the birds, connecting the electrodes to recording equipment. In others, the birds carried miniature radio transmitters that transmitted the signals from the electrodes. The figure shows that the pectoralis muscles were active at the end of the upstroke and for some of the downstroke: they stopped the upstroke and pulled the wings down. The supracaracoideus became active late in the downstroke, stopping the downstroke and accelerating the wings upward. The pectoralis has to produce the large forces needed to support the bird's weight but the supracoracoideus has only to decelerate and re-accelerate the wing (and may be helped in that task by aerodynamic lift on the wing). Accordingly, the pectoralis muscles are much larger than the supracoracoideus muscles.

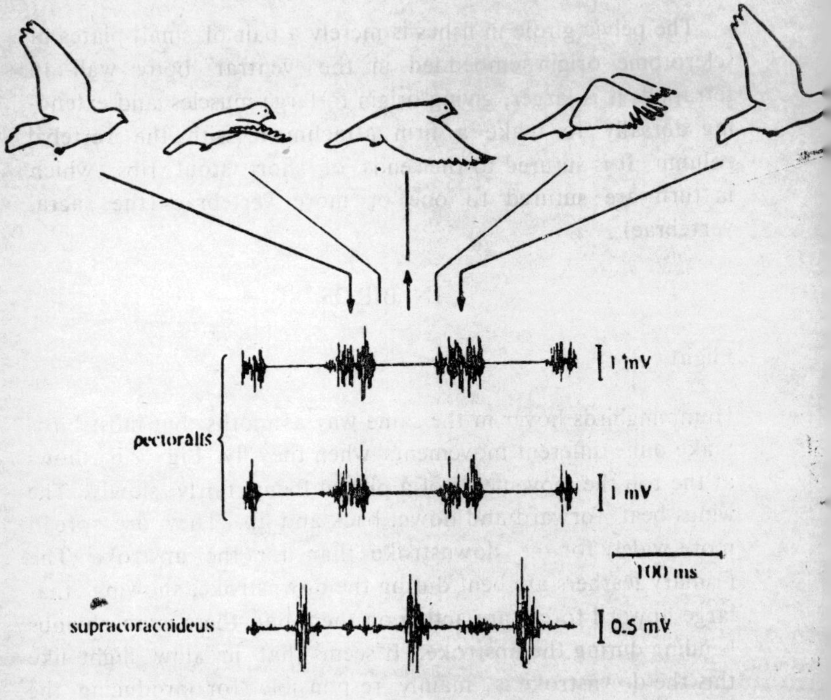

FIG. 2.6. Wing movements of a flying pigeon (Columba livia) and electromyographic records from two parts of the pectoralis muscle and the supracoracoideus muscle.

Flying birds (and aeroplanes) support their weight by pushing air downwards. The air movements involved have been made visible by flying birds through clouds of small soap bubbles. Ordinary soap bubbles are denser than air and sink slowly, but these ones were filled with the low-density gas helium so that the complete bubbles had almost exactly the same density as the air. Thus the movements of the bubbles showed how the air was moving. They were recorded by taking stereoscopic multiple-flash photographs. Each photograph had several images of each bubble, showing where it was when each of the flashes was fired.

Fig. 2.7a shows the pattern of air movement that was seen when a pigeon flew slowly (at 2.4 ms^{-1}) through the bubbles. Each downstroke of the wings drove a puff of air downwards. This moving air set a surrounding ring of air spinning, forming a vortex ring. Such rings are always formed when a discrete puff of air moves through still air; the smoke rings blown by smokers are vortex rings made visible by the smoke trapped inside them. The upstrokes of the wings produced no obvious air movements. This demonstration that air is driven downwards only by the downstroke is consistent with the observation that it is only in the downstroke that the primary feathers are noticeably bent.

Fig. 2.7 (b) shows a different patern of air movements that was seen when a kestrel (Falco tinnunculus) flew at about 8 m s^{-1}. Instead of discrete puffs the bird leaves behind a continuous band of downward-moving air the upstroke drives air downwards, producing lift, as well as the downstroke. Vortices form as continuous lines on either side of the downward-moving air, just as wingtip vortices (that sometimes become visible as vapour trails) are formed by aeroplanes. These vortices undulate up and down with the wing beat. They are closer together in the upstroke, when the wings are less widely spread and the band of air that is being driven downwards is narrower than in the downstroke. The smaller quantity of air driven by the upstroke is indicated by a smaller arrow than the ones used for the downstroke.

The wing must produce mainly an upward force, to support the bird's weight, but a small forward thrust is also needed to counteract the drag on the body. Fig. 2.7 (a′) and (b′) show

how these forces are produced. The evidence from air movements is interpreted in terms of more conventional aerodynamics.

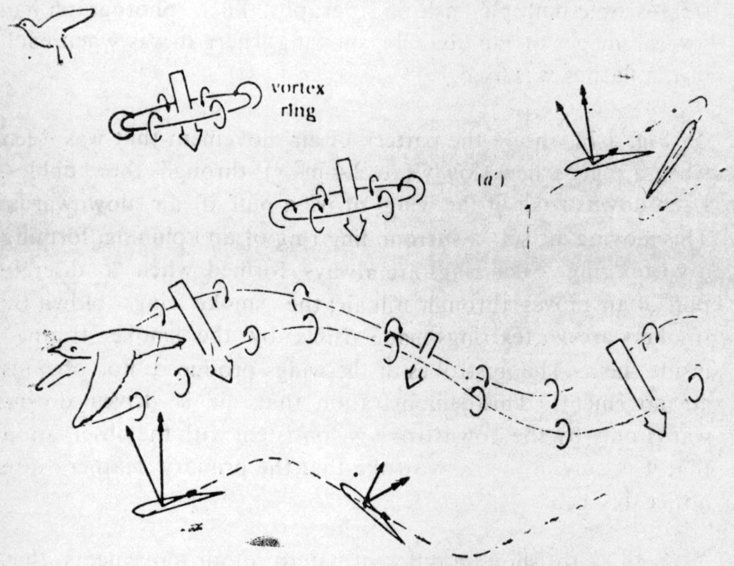

FIG. 2.7. Air movements in the wakes of birds flying (a) slowly and (b) faster, observed in experiments with soap bubles: (a′) and (b′) are diagrams showing the paths through the air of sections of the wings, ond the aerodynamic forces that act on them.

The pigeon flying slowly apparently holds its wings at an angle of attack during the downstroke (Fig. 2.7 a′). The resultant force acts upwards and slightly forwards, as required. In the upstroke, the angle of attack is reduced to zero and the aerodynamic forces on the wing are very small. The kestrel flying faster holds its wings at an angle of attack for the upstroke as well as the downstroke (Fig. 2.7 b′) but the forces must be smaller than in the downstroke: otherwise the backward component of the resultant force produced by the upstroke would concel out the thrust produced by the downstroke.

The pigeons would only fly slowly in the experiments and the kestrel would only fly fast, but noctule bats (Nyctalus) flew at a wider range of speeds. At 3 ms^{-1} they produced air movements like those shown in Fig. 2.7 (a) but at 7-9 ms^{-1} they made the air move as in Fig 2.7 (b). Just as we walk to go slowly and run to go fast, the bats used different gaits at different speeds. It seems likely that birds do the same but so far no bird has demons-trated both gaits in the bubble experiments.

Gliding is an important mode of flight, especially for large birds. A bird gliding in still air inevitably loses height but it can remain airborne or even gain height if the air is rising. Gulls can often be seen soaring on the windy sides of hills, where the wind is deflected upwards by the slope of the ground. Albatrosses and other sea birds use the sloping sides of large waves in the same way and can remain airborne indefinitely, hardly ever beating their wings. Other birds soar in thermals over ground heated by the sun. A patch of hot ground warms the air immediately over it, which rises as a thermal. Vultures and storks gain height by circling in the rising air of a thermal, then glide away (losing height) to another thermal where they circle and rise again. In this way they can cover large distances. Professor Colin Pennycuick, flying a motorized glider in East Africa, once followed a vulture (Gyps ruppellii) for a distance of 75 km. from its feeding ground to its nest on a cliff. It covered the distance in 96 minutes, pausing to circle in five thermals but never beating its wings. Thermals often have cumulus clouds at the top, which may help birds to find them.

Pennycuick used the wind tunnel shown in Fig. 2.8 to investigate the gliding ability of pigeons and a fruit bat. A large fan blew a jet of air of almost one metre diameter, at speeds up to 20 ms^{-1}. (A powerful motor was needed.) The 'honeycomb' and the contraction were designed to make the jet as uniform as possible: similar devices were used in the water tunnel. The tunnel could be tilted to any desired angle θ. The animals were trained to fly in the jet of air so as to remain stationary near the mouth of the tunnel. (The pigeons were persuaded to do this by offering maple peas on a long spoon that they could reach only when flying in the required position.) It the speed and angle of the jet allowed it, the animals saved energy by gliding: otherwise they had to flap their wings. An animal gliding so as to remain stationary in a jet of air of speed u and angle θ has the same motion (relative to the air) as one gliding at the same speed and angle in stationary air. In the latter case it would lose height at a rate u sin θ: this is called the sinking speed.

Fig. 2.8 (a) shows how the lift and drag on the gliding animal balance its weight. The lift acts at right angles to the jet of air, at an angle θ to the vertical, and the drag at angle θ to the horizontal. If the weight is mg the lift must be mg cos θ and the drag mg sin θ, so cot θ is the ratio of left to drag. To glide at shallow angles, an animal must be able to adjust its position so as to get the required lift with very little drag.

In his experiments with the wind tunnel, Pennycuick kept the speed of the jet constant while adjusting its angle, to find the shallowest possible gliding angle for each speed. He calculated the sinking speed and plotted it as in Fig 2.8 (b). The maximum lift that the wings can give is proportional to the square of the air speed so below a certain speed (8 ms^{-1} for the pigeon) the animal cannot glide at all. The smallest sinking speeds are obtained at a higher air speed (14 ms^{-1} for the pigeon). The pigeon glides rather badly, with a minimum sinking speed of more than 2 ms^{-1}, but the falcon and bat do better, about 1 ms^{-1}. Even lower sinking speeds (typically 0.5 ms^{-1}) are obtained with glider aircraft and models but the comparison is unfair: a gliding bird is burdened by the weight of its flight muscles but a glider carries no engine.

The birds that can glide best (at the shallowest angles) have long, narrow (high aspect ratio) wings. Such wings give the highest possible lift: drag ratio, as explained for fish tails. However, excessively long wing might be awkward, especially when taking off from the ground or from trees. Albatrosses have long, narrow wings but vultures have shorter, broader ones.

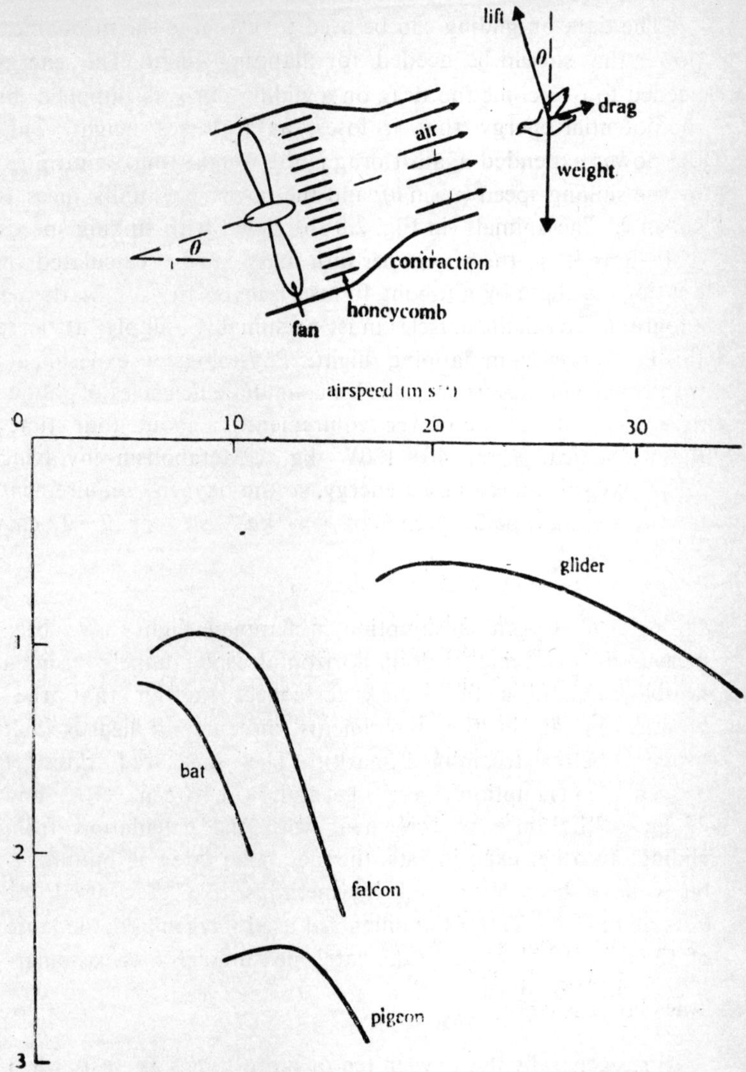

FIG. 2.8. (a) A diagram of a wind tunnel used to investigate the gliding performance of birds and bats. The forces acting on the bird are also shown. (b) Results of gliding experiments in wind tunnels. Sinking speed (u sin θ) is ploted againts air-speed (u). The lines show the performance of a glider aircraft, a pigeon (Columba livia, wing span 0.6 m), a falcon (Falco jugger, 1.1 m) and a large bat (Rousettus aegyptiacus, 0.5 m).

The data on gliding can be used to estimate the mechanical power that should be needed for flapping flight. The energy needed to overcome the drag on a gliding bird is supplied by the potential energy that it loses, as it losses height. Thus the power expended against drag is the weight (mg) multiplied by the sinking speed ($u \sin \theta$), and the power per units mass is $gu \sin \theta$. The animals in Fig. 2.8 (b) glided with sinking speeds of $1-3 ms^{-1}$, so the power per unit mass can be calculated by multiplying these by g (about $10\ ms^{-2}$): it is $10-30$ watts per kilogram. The flight muscles must presumably supply at least this much power, in flapping flight, Physiological experiments with other muscles usually show maximum efficiencies of about 0.25, so the metabolic power requirement is about four times the mechanical power, $40-120W\ kg^{-1}$. Metabolism involving $1\ cm^3$ oxygen releases 20 J energy, so the oxygen requirement for flight should be $2-6\ cm^3$ oxygen $kg^{-1}\ s^{-1}$, or $7-22\ cm^3\ g^{-1}\ h^{-1}$.

Rates of oxygen consumption in flapping flight have been measured for flying birds in horizontal wind tunnels, using a flexible tube and a face mask to collect the air that they breathe out (an in the experiments on running lizards, but using a better-streamlined mask). The measured rates of oxygen consumption were generally between 10 and $25\ cm^3\ g^{-1}\ h^{-1}$, in good agreement with the calculation from gliding. In other experiments, the metabolic rates of humming-birds have been measured by methods like that used for hovering moths. A 3 g hummingbird used oxygen at the rate of $42\ cm^3\ g^{-1}\ h^{-1}$, close to the rate measured for the similar-sized moth Mánduca.

Birds generally use oxygen ten or more times as fast when they fly, as when they are resting. The resting metabolic rate is enough to maintain body temperature, so birds might be expected to overheat on long flights. However, the air moving past the flying bird must help to cool it, and spreading the wings exposes the flanks which are rather sparsely covered with feathers. Birds do not sweat, and heat loss by evaporation of water is much less important than in mammals. In an experiment in which budgerigars flew in a wind tunnel at 20°C and 50 per cent relative humidity, evaporation (mainly in the breath) accounted for only 15 per cent of heat loss. If birds were more dependent on heat loss by evaporation they would have to drink more often on long flights, and migrations across broad deserts or seas might not be practicable.

Many small birds migrate across the Sahara desert and have to fly about 2000 km with no opportunity to feed. They set out with very larg stores of fat, which are used on the journey. For example, yellow wagtails (Motacilla flava) have a mass of 24g. 7.4 g of it fat, when they leave Nigeria. They arrive north of the Sahara with a body mass of only 15g, having lost nearly all of the fat and some water. Oxidation of the fat yields an approximately equal mass of water, which helps to compensate for the evaporative losses.

We will estimate how far a 24 g bird should be able to fly, fuelled 7.4 g fat. The heat of combustion of fat is 40 MJ kg^{-1}, so the energy available is 0.30 MJ. Metabolism using 1 cm^3 oxygen release 20 J, so 15000 cm^3 oxygen would be needed to oxidize the fat, 630 cm^3 per gram of initial body mass. Rates of oxygen consumption of flying wagtails have not been measured so we will have to use data for 35 g budgerigars (Melanopsittacus). In experiments in a wind tunnel, these flew furthest for given oxygen consumption at a speed of 40 km h^{-1} (11 ms^{-1}). At this speed they used 24 cm^3 oxygen g^{-1} h^{-1}, so if they started with the same proportion of fat in their bodies as the wagtails they could have flown for 630/24=26 hours, travelling $26 \times 40 \simeq 1000$ km. This is probably an underestimate, because I have taken no account of their getting lighter (so needing energy for flight) as they used up the fat. Correcting for this would increase our estimate of the range only to about 1200 km so it seems doubtful whether the fat would be sufficient fuel for a budgerigar crossing the Sahara, unless it were helped by a following wind.

IN MAMMALS

Walking and Running

These show that it stands and walks on strongly bent legs, with the humerus and femur more nearly horizontal than vertical. Also, these bones are inclined a little laterally, giving the animal a rather bow-legged stance. The left and right footprints do not form a single line but two parallel lines, a short distance apart.

X-ray cinematography would be little use for studying the leg movements of large animals such as lions and reindeer because the field of vlew is very restricted, but light cine films show that they walk with relatively straight legs, with the humerus and femur more nearly vertical than horizontal. The stance is not bowlegged, and the lines of footprints formed by the left and right feet are relatively close together.

Professor Farish Jenkins called mammals that move like Tupaia 'non-cursorial' and ones that move like lions and deer 'cursorial'. He found that opossums, rats and ferrets (Mustela) were non-cursorial and that dogs and cats were cursorial. Indeed, all mammals of less than 1 kg body mass are non-cursorial and almost all of more than 5 kg are cursorial. (The domestic cat is one of the smallest cursorial mammals.)

Some mammals place more of the foot on the ground than others. Tupaia (Fig. 2.9 c) sets down the whole of each foot, right back to the wrist or heel, and is described as plantigrade. Lions place only the toes on the ground, as far back as the distal ends of the metacarpals or metatarsals, and are described as digitigrade. Deer stand on the tips of their toes with only their hooves on the ground and are described as unguligrade. The unguligrade. The unguligrade stance is peculiar to the ungulate mammals. Most other large mammals are digigrade and small mammals are plantigrade. The only large plantigrade mammals are primates (including ourselves) and bears.

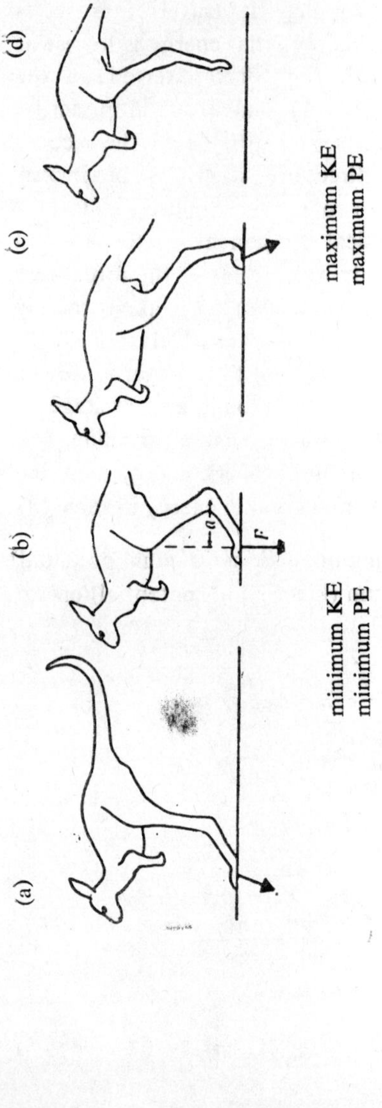

FIG. 2.9. Outlines traced from a film of a kangaroo hopping. The arrows represent forces exerted by the feel on the ground.

The hopping of kangaroos is a conveniently simple gait to study, because only two legs are used and both do the same thing. I will use it to introduce principles that apply also to other mammal gaits, including the bipedal running of humans and the squadrupedal running of most other mammals.

Fig. 2.9 shows a kangaroo hopping. It travels in a series of bounds so it rises and falls: its potential energy is lowest at stage (b) and highest at stage (d). The forces exerted on the ground have been recorded by having kangaroos hop across force plates, force-sensitive plates set into the floor. The records show that the forces act in the directions shown by the arrows, sloping at every stage of the step so as to be more or less in line with the leg. (Only the vertical component of the ground force is needed to support body weight, but if the force were kept vertical it would exert very large moments about the hip joints at some stages of the stride, and much larger forces would be needed in muscles.) At stage (a) the kangaroo decelerates itself by pushing backward on the ground and at stage (c) it accelerates itself by pushing backwards, so it is travelling most slowly at stage (b), and has minimum kinetic energy then, but is travelling fastest and has maximum kinetic energy at stage (d).

Thus the kangaroo has minimum (kinetic plus potential) energy at stage (b) and maximum (kinetic plus potential) energy

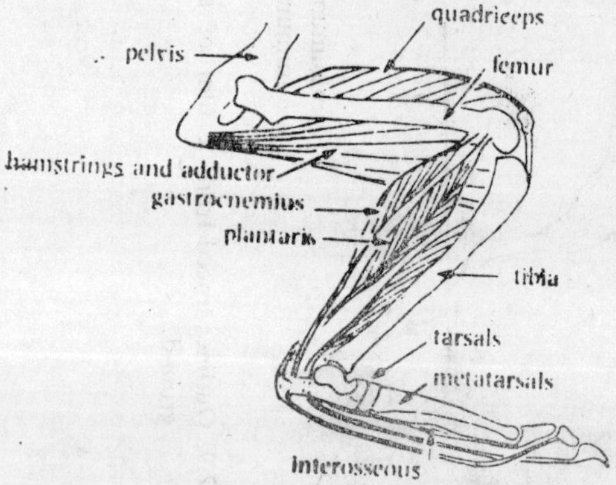

FIG. 2.10. The skeleton, and some of the principal leg muscles, of a typical mammal.

at stage (d). This suggests that the leg muscles must act as brakes at stage (a) of the stride, removing mechanical energy from the body and degrading it to heat: and that they must do work at stage (c), restoring the mechanical energy of the body. The quantities of work involved can be calculated from the force plate records. However, the need to do this work could be avoided if there were springs in the legs that stored up mechanical energy at stage (a) and returned il in an elastic recoil at stage (c): the animal would bounce along like a rubber ball and its leg muscles would have to do only enough work to make up for frictional and viscous losses.

Fig. 2.10 (a) shows how the principal muscles in the legs of kangaroos (and other mammals) are arranged. Notice that the gastrocnemius and plantaris muscles, and some others, have long tendons. The gastrocnemius tendon ends on the heel but the plantaris tendon wraps round the heel and goes right to the toes. These are the springs that are important in hopping.

To assess their importance as springs it was necessary to find out how much energy they stored (as elastic strain energy) the stage of the hop shown in Fig. 2.9 (b). To do that, it was necessary to discover the forces in the tendons at the stage, and to measure their elastic properties. The forces were estimated from force plate records and measurements on films. In Fig. 2.9 (b), if the force plate registers a force 2 F we can assume that force F acts on each foot. If the line of action of the force is at a distance a from the ankle joint, its moment about the joint is Fa and the ankle muscles must exert an equal, opposite moment. This moment must be supplied mainly by the gastrocnemius and plantaris muscles because the other muscles that could assist them run too close to the ankle joint to exert large moments, but it could have been supplied by the gastrocnemius alone, by the plantaris alone or by both acting together. My colleagues and 1, who did the experiment, assumed that both muscles were active, exerting equal stresses (force per unit cross-sectional area). We found out how much the calculated forces would stretch the tendons, and how much elastic strain energy they would store, by dissecting the tendons from carcasses and stretching them in a testing machine of the kind used by engineers. The tests showed that tendon has a fairly high Young's modulus, in the usual range for fibres so quite large stresses stretch it only a little. The forces involved in hopping would have stretched the gastrocnemius and plantaris tendons by about 3 per cent of their lengths, and the graphs of force against extension told us how much strain energy would be stored. The tests also showed that tendon is an excellent elastic material, returning in its elastic recoil 93 per cent of the work done stretching it.

From the results of these experiments we calculated that when an 18 kg kangaroo hopped slowly, it lost and regained 39 J (kinetic plus potential) energy in each step. If its tendons were inextensible it would have had to use its muscles as brakes and then do 49 J work with its muscles, in each step (49 rather than 39 J because during part of the step muscles do work again each other). However, the elastic properties of the tendons enable them to store and return 16 J strain energy, reducing by about one-third the work that the muscles have to do. The work would be reduced by larger fractions in faster hopping.

This presumably saves metabolic energy but it is uncertain how much. Even if the tendons had ideal elastic properties, stretching and shortening by so much that the muscles remained constant in length and so did no work, the muscles would still have to exert tension. (Remember that a force does not work unless it moves its point of application.) Muscles use metabolic energy whenever they develop tension, even if they do no work.

However, a lot of metabolic energy is presumably saved by similar mechanisms in advanced ungulates such as deer and horses. Experiments on them show that the tendons in the lower parts of the legs have near-ideal elastic properties, so that their muscles have to make only very small length changes during running. The muscles have responded by evolving extremely short muscle fibres, so that the forces that are needed can be exerted by much smaller volumes of muscle. (Fig. 2.9 shows how this has happened in the forelegs of antelopes.) Less muscle has to be activated in each step, so less metabolic energy is needed. The extreme adaptation is found in camels, in which the muscle fibres of the plantaris have almost vanished. The plantaris remains as a substantial tendon running all the way from knees to toes, that can only serve as a passive spring. (Though the muscles of the lower part of the leg are small there are large muscles in the thigh, capable of providing the power needed for acceleration.)

Most mammals walk at low speeds and run to go faster. Walking gaits are ones in which the left and right feet of a pair move alternately, and each is on the ground for more than half a stride. Thus there are stages of the stride when both feet are on the ground. In running, each foot is on the ground for less than half a stride and there are stages when both are off the ground. Energy savings by tendon elasticity seem to be important in running, but relatively unimportant in walking.

Quadrupedal mammals walk at low speeds, moving their feet in the order. At moderate speeds most of them trot: the trot is a running gait in which the left forefoot and the right hindfoot move together, and the right forefoot and the left hindfoot together. A trotting quadruped is like two people running one behind the other, half a cycle out of step. At high speeds quadrupedal mammals gallop, setting down the two hindfeet and then the two forefeet (Fig. 2.12).

To understand why mammals gallop, we have to consider a component of the body's kinetic energy that has not so far been mentioned. The forces between the feet and the ground decelerate and re-accelerate the body as a whole, causing fluctuations of kinetic energy due to the legs being decelerated and re-accelerated, relative to the trunk, as they wing backwards and forwards. These fluctuations increase rapidly as speed increases, but galloping seems to use an elastic mechanism to save some of the energy. At frame 360 of the film (Fig, 2.11) the hindlegs have finished their forward swing and are about to

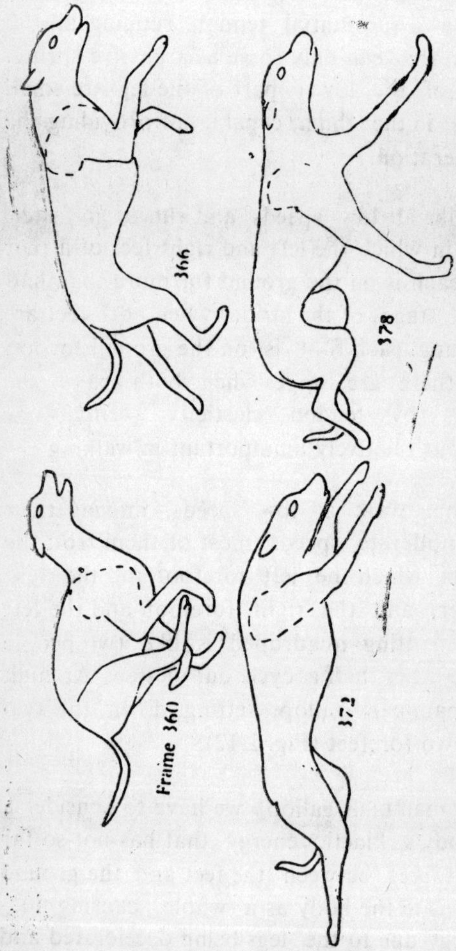

FIG. 2.11. Outlines traced from film, showing four stages of a stride of a greyhound galloping at 15 m/s.

swing back, and the forelegs are finishing their backward swing and are about to swing forward. Both pairs of legs are temporarily halted, so the kinetic energy of the body falls. However, the back is bent and its extensor muscles are taut. A sheet of tendon by which these muscles attach seems to serve as a spring, storing elastic strain energy as the legs are halted and returning it in the elastic recoil to start them moving the other way.

Human walking and running are very different from the gaits of other mammals, but we are not unique in being bipedal. Kangaroos and various other mammals hop bipedally. Among the primates, captive gibbons (Hulobates) walk bipedally when they are on the ground, but wild gibbons seldom leave the trees. Wild chimpanzees (Pan) sometimes travel bipedally in long grass, or when carrying a load. However, we are unique in walking with our trunks erect, and keeping our legs straight while the foot is on the ground. Chimpanzees and other primates walk bipedally with the trunk sloping forward and the knees bent (Fig. 2.12). With the legs bent like this, the line of the ground force must inevitably be a long way from some of the leg joints, so large moments act about some of the joints and large forces are required in muscles. When we walk with straight legs, keeping the ground force more or less in line with the legs, the moments about the leg joints are small and small

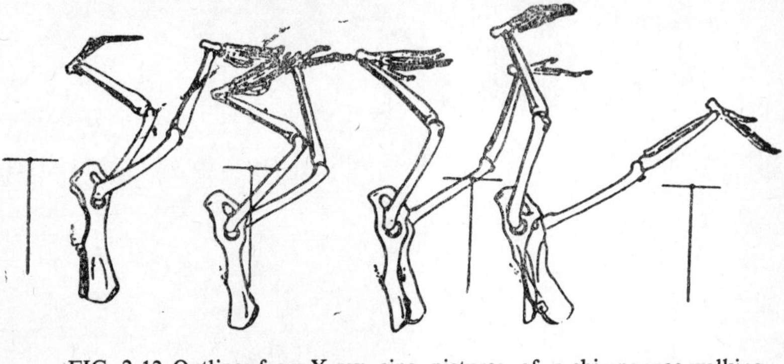

FIG. 2.12 Outlines from X-ray cine pictures of a chimpanzee walking bipedally. The dots with straight lines through them represent a fixed point in the background.

muscle forces suffice. Metabolic energy is needed to generate tension in muscles, so our straight-legged style of walking should reduce metabolic energy costs.

The arguments in this section suggest that horses and antelopes should be able to run very economically, because the elastic properties of their leg tendons have made it possible for some major leg muscles to be small and short-fibred: and that people should be able to walk very economically because we keep our legs straight. Unfortunately, measurements of oxygen consumption have failed to confirm tnese expectations. The metabolic energy cost of locomotion for horses, antelopes and people are little different from the costs for other species of the same body mass, travelling at the same speed.

3

Orientation and Navigation

Most animals spend a great deal of time moving around in their environments, but most of them have some sort of nest or "home base" from which they move out regularly to find resources. Such behaviour require that the animal know where it is in relation to its home and how to get back there. Some animals move long distances from their homes and may have different summer and winter homes that are separated from one another by thousands of kilometres. Migration between different ranges requires very sophisticated navigational abilities, but many animals are able to get about with simple mechanisms.

The simplest orientation movements are known as taxes (singular: taxis) in which the animal assumes a particular spatial relationship to an orienting stimulus. The nature of the stimulus is denoted by adding the appropriate prefix to the word taxis: a Phototaxis is movement guided by a light; a geotaxis is movement guided by gravity; a chemotaxis is movement guided by detection of some chemical substance. A further distinction can be made between a positive taxis (movement toward the stimulu) and a negative taxis (movement away from the stimulut). The flight of a moth toward a light is a positive phototaxis, whereas the retreat of a cockroach from the same light is a negative phototaxis.

Many invertebrates possess a light-compass reaction, generally using the sun, in which the angle between the direction of movement and the direction of the stimulus is kept constant. Ants use this type of orientation in moving to and from their nests, but they cannot compensate for movement of the sun. If an ant is held for 2½ hours in a dark box while it is on its way home, the sun will have traveled approximately 37°. When the ant is released, it shifts its direction approximately 37°, maintaining its original angle to the sun. Orientation to the moon has been demonstrated with inertidal amphipods of the genus Talitrus on beaches in Italy. These crustaceans escape seaward when disturbed; populations on coasts facing different directions have different but appropriate, escape directions even when tested in the laboratory. On moonless nights or under overcast skies, this orientation is lost. Unlike ants, however, the amphipods are able to compensate for the movement of the moon and maintain correct compass orientation of their escape.

More intensive observations of orientation behaviours have been made with been than with any other invertebrate. Bees can be trained to feed at a particular time at a particular spot in a certain direction from the hive. They have internal clocks that have periodicities of approximately 24 hours (circadian rhythms), which they use to tell time. Like amphipods, bees compensate for the movement of the sun. If they are held in lightproof boxes at a feeding tray they still fly directly to their hive when released. If trained bees are transplanted to an unfamiliar environment, they still appear at the appropriate direction at the correct time to day to find food. If they are shifted through great longitudinal distances, they first appear at the feeding sites according to the time of day at their training location. Over a period of about a week they gradually adjust to local time, just as a person does after a transoceanic flight.

Navigation

The most spectacular long-distance migrations are undertaken by vertebrates. Pacific salmon hatch from eggs, laid in gravel beds of rivers and, after varying amount of time in fresh water—intervals that depend on species and location—they migrate to the ocean, where they feed vigorously until they have reached maturity. At this time they migrate back to the coast, find the mouth of their home river, and swim back upstream, often of the very tributary in which they were born. They then spawn and die. The remarkable migration cycle has two components; navigation at sea to the general vicinity of the home river, and location of the correct river and tributaries. Evidence gathered by a number of investigators reveals that salmon and other fishes can orient using a sun compass at sea buy that they use a sense of smell to locate their home rivers. Salmon that are transported back downstream successful return to their home tributaries, but they cannot do so if their nostrils are plugged.

Even in small lakes, fishes may use a sun compass to orient. While bass (Roccus chrysops) in Lake Mendota, Wisconsin, congregate in two small places for spawning. In one study, fish were displaced from the spawning grounds, and floats were attached to them so that they could be followed by observers in a boat. The fish showed good orientation toward the spawning grounds on sunny days, but they swam raddomly on cloudy days.

The first codclusive evidence of navigational abilities in birds was not obtained until 1951, but for centuries it was known that some birds undertake intercontinental migrations and yet return to specificites to breed. The first studies employed homing pigeons displaced from their lofts. The pigeons quickly orient homeward under clear skies but depart randomly under cloudy skies. Since the many species, of birds have been shown to use the sun when navigating. Many small birds migrate at night, however, and they use the stars to orient. Indigo buntings will orient to "stars" in the antificialsky of a planetarium. The physiological condition of buntings may be changed by ex-positing some birds to increasing day lengths (as in spring and others to decreasing day lengths autumn). In a planetarium, buntings that are in spring condition orient toward the north whereas those in fall condition orient toward the south, showing that their responses to the sky depend upon a physiological star that can be induced simply by manipulating daylengths. It is also possible to blot out selected constellations in a planetarium to find which ones and necessary for the birds to maintain correct orientation. The relative positions of just a few constellations, particularly those in the vicinity of the Non Star, are used by the to buntings establish their migratory directions: all others can be removed and the birds can still orient correctly.

Remarkable though the ability to use the sun and stars to establish directions may be, it still cannot account for true homing. In order to home, bird must know where home is and its own position with respect to home, so that it knows which direction take. It is like a person on a raft lost in the middled of an ocean. A compass is of no use unless the person knows the latitude and longitude of both home and the raft. Once a bird has determined its position then it can use the sun or stars to take the right direction. Just how birds determine their positions all unknown, but there is increasing evidence that part of the information they use is derived from the earth's magnetic fleld. Homing pigeons can be disoriented on overcast days by attaching tiny magnets to their necks; the magnets cancel the effects of Earth's magnets field, Also, pigeons ore unusually disoriented in the vicinity of sites or Earth where there are serious magnetic anomalies due to large non-deposits. There is as yet no magnetic receptor nearly identified, but scientists are convinced that here must be one.

Invertebrates

Migrating locusts, flying in swarms, give the appearance of oriented directional movement. Studies of locust swarms have revealed that wind oxerts a displacing effect on individuals and thus on the swarm over time. Within the warm there are groups of locusts that are flying in directions different from that of the swarm as a whole. When these subgroups reach the periphery of the swarm, they turn back inwards toward the main flow of the swarm (Rainey 1959, 1962; Waloff 1958).

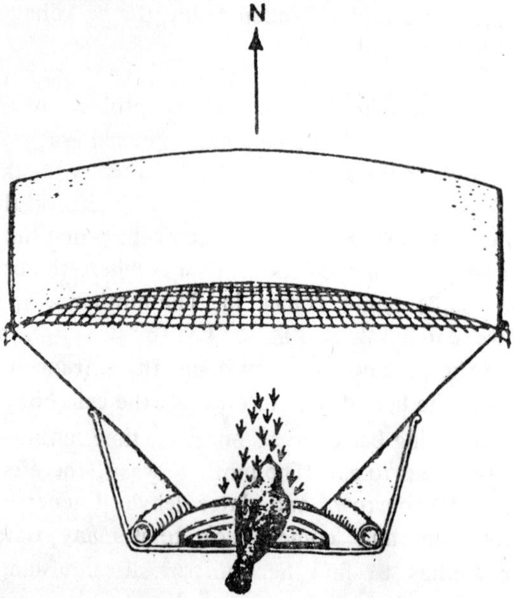

FIG. 3.1. Nightime Restleness in a Planetarium.

The means by which honey been (Apis mellifera) communicate the location of food sources to one another in the hive and then make foraging flights to locate the food have been studied by a number of investigators (von Frisch 1967; Lindauer 1961; Gould 1975, 1976; Wenner 1971, 1974).

Bees and some other insects, including ants, can perceive polarized light (von Frisch 1967). The plane of polarization can provide an axis for orientation for these animals. It is also possible that in some animals the perception of polarized light may enable them to locate the position of the sun when it is obscured by partially overcast skies. The use of polarized light for orientation has also been suggested for a variety of other organisms, including fish (Waterman and Hashimnoto 1974), octopodes (Waterman 1966), and salamanders (Adler and Taylor 1973).

Another example is taken from the be killing digger wasp Philonthus triangulum.

Let us consider first the homing problem of the bee-killing digger wasp Philanthus triangulum. Female wasp of this species dig burrows in the sand, to provide a nursery for their offspring. Each female digs several (about 6 or 7) cells off the side of her burrow, and lays an egg in each cell. When the egg hatches into a larva it will need food. That is where the bee-killing part of the wasp's name comes in. The mother wasp goes of her burrow, catches and kills a bee (by stinging), and brings it back to the burrow. She then opens the entrance to the burrow and takes the bee down to one of the cells. She continues to catch and bring back bees, one at a time, until each larva has about two bees to eat. The mother wasp, therefore, does not merely dig a burrow, and later leave it neuer to return: she departs from and comes back to it many times. On every return she has to find her burrow, distinguishing it from its surroundings. It is not an easy task to find a particular burrow, because these wasps can nest in quite dense groups; there might be more than twenty burrows within a circle of five yard radius. The problem of how the digger wasp locates her home has a special place in the history of ethology, it was one of the first questions about behaviour mechanisms ever to be asked, and experimentally answered. It was studied by Niko Tinbergen in 1929, on the heaths and sand dunes of Hulshort in Holland.

Tinbergen first confirmed, by individually marking all the nests and wasps in a particular area, that each wasp does indeed always return to her own burrow. He then performed some simple experiments to test whether the digger wasp recognized her own entrance by the distinctive array of odd objects (haphazardly fallen sticks, pine cones, stones etc.)

around it, or by some stimulus emanating from the entrance itself. He placed around the entrance of each of several chosen burrows a neat circle of pine cones. He left them therefore a few days, checking that the wasp still kept returning to their burrows. He Hen waited for the wasps to leave on bee-hunting expeditions, and while they were away he moved the cones a yard or so away from the entrances. When the wasps returned they landed where the entrance 'should' have been, in the centre of the circle of pine cones. Evidently, they recognized the entrance by its surrounding landmarks, not by any stimulus from the entrance itself.

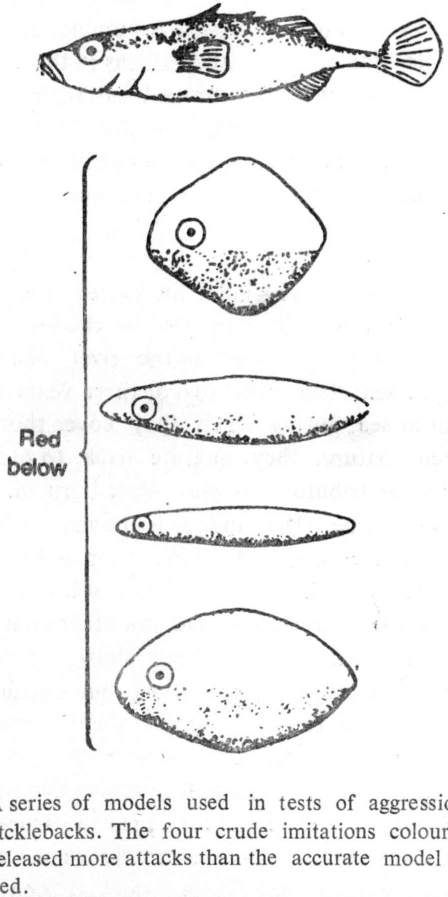

FIG. 3.2. A series of models used in tests of aggression with male sticklebacks. The four crude imitations coloured red below released more attacks than the accurate model which lacked red.

Many of the cues discussed in this chapter with regard to orientation in birds and mammals have also been demonstrated in some invertebrates. Among these are sun-compass orientation in spiders, beetles, and ants; gemoagnetic force cues in bees; and landmarks in a variety of species (Able 1980). In addition, data have been reported which suggest beachhoppers (Talitrus) use the moon (lunar orientation) to guide their nocturnal movements.

Fish

We have defined homing behaviour as the use of landmarks and navigation to return to a homesite. An animal may become displaced from its home by weather, or, as in the case of the Pacific salmon, it may return to a home location in the course of the life cycle. Alternatively, researchers may artificially move an organism from its homesite to another location in order to investigate the cues and means of navigation it employs to return home.

All species of salmon have similar life cycles. They are laid as eggs in river tributaries all over the northern hemisphere. They live their first few months in the river, then migrate downstream to the sea; they spend two or three years in feeding and growing out at sea, during which they cover thousands of miles; and when mature, they migrate back to exactly the same river, and same tributary, as they were born in. How do they find their way home? The answer is thought to be mainly by smell. The attractive odour may come from either (or both) of two sources: the young salmon which are still in the stream, not yet having migrated to the sea, and any other characteristic odours in the stream. Salmons have been shown to be capable of the necessary olfactory discrimination, but the most direct evidence that they use their sense of smell comes from experiments, of the kind first performed by W.J. Wisby and A.D. Hasler, in which the salmon's olfactory sense was impaired. Wisby and Hasler actually plugged the noses of their salmon, which then homed less accurately than untreated controls recent experiments, the same result has been obtained by cutting the salmon's olfactory nerves. Experiments in which young salmons have been transferred from their stream of birth, to be released in another stream, have been used as evidence that salmon imprint on the smell of their river during a sensitive period just before they migrate downstream. In one such experiment, L.R. Donaldson and G.E. Allen took 72,000 young salmon at the 'fingerling' stage when they are about one year old from the Soos Creek Hatchery in Washington and

TABLE 3.1

Numbers of Coho Salmon Released and Later Recaptured in Two Rivers in Washington, With and Without Their Olfactory Sense Impaired. Salmon with their Noses Plugged Homed Less Accurately

Stream of origin		*No. released*	*No. recaptured*	
			Issaquah	*East Fork*
Issaquah	Controls	121	46	0
	Nose plugged	145	39	12
East Fork	Controls	38	8	19
	Nose plugged			

divided them into two groups. In 1952 they released one group (identified by the removal of the right pelvic fin) at Issaquah Hatchery, and the other (which had their left pelvic fin removed) at the University Hatchery. The salmon returned in the winter of 1953/4; they homed on their release sites. None of them returned to Soos Creek, the site of their birth and first year of life. Of 71 marked salmon caught at Issaqua Creek, 70 lacked the right pelvic fins; all 124 marked salmon caught at the University Hatchery lacked their left pelvic fins. Results like that suggest that the young salmon learn the odour of their native stream (or of the young salmon in it), and later find their way home by seeking that scent. But there is another explanation. Salmon migrate in schools. The results of the transplantation experiment may only be due to the transplanted salmon following the lead of the native salmon. Recent commentaries, therefore, such as the review by O.B. Stabell maintain that the case for imprinting is at best not proved. But it is not in doubt that salmon rely critically on their sense of smell to guide them home. Whether they learn their home stream's distinctive smell during a discrete sensitive period is undecided.

AMPHIBIANS

Orientation has been studied experimentally in southern cricket frogs (Acris gryllus) and northern cricket frogs (Acris crepitans) (Ferguson, Landreth and Turnispeed 1965; Ferguson, Landreth, and McKeown, 1967). Experimenters placed captured frogs in a large circular plastic pen in the water, where the frogs could see no landmarks on the horizon. When southern cricket frogs were released in the pen under sunlit skies, they swam in the direction that would have taken them to land had they been released off the shore at their home pond. The results were the same when the experimenters carried the frogs to the pen with or without giving the frogs a view of the sky overhead, when they released the frogs under starlit skies, and when they released the frogs under moonlit skies. When researchers placed the frogs in the pen on a moonless night, with only stars visible in the sky, the frogs oriented themselves in two opposite directions; some headed toward the home shore and others headed directly away from home. When experimenters released the frogs during the twilight period, after sundown but before the stars or moon were visible, or under cloudy, overcast conditions, the frogs oriented in a random fashion.

Cricket frogs held in darkness for periods of thirty hours to seven days exhibited a decline in accuracy of orientation, which culminated in random orientation after seven days, "Dephased" frogs reestablished proper orientation when experimenters exposed them to normal dark/light cycles or to daily fluctuations of temperature and humidity. These frogs appeared to use a learned shore position, celestial cues, and a timing mechanism to orient themselves properly to the home shore. Similar results have been obtained for northern cricket frogs.

Reptiles

Sea turtles perform some of the most remarkable of all migratory feats. They may travel over a thousand miles to a small island to breed. The entire population of the Atlantic ridley turtle, Lepidochelys Kempi apparently breeds on a small stretch of the Gulf Coast of Mexico. So for there is no evidence on the sensory basis of these journeys.

Carr (1967) has summarized a number of investigations on orientation by green turtles. Female turtles (Chelonia mydas) deposit their eggs on sandy beaches on certain islands; they lay about 10 eggs on each trip to the beach, and make from three to seven trips spaced twelve days apart. In non-reproductive periods the turtles live in area, with abundant turtles grass, generally along the coasts of continental land masses. In a specific study (Carr 1967), turtles were banded at Ascension Island. Young turtles that hatches on the island floated on currents that took them to the coast of South America. Adults had to swim against the same current to reach Ascension Island to reproduce. How did they find their way? The current may have certain features that allow the turtles to distinguish it.

Birds

In the last three decades, a variety of studies have been conducted on orientation in and navigation by birds. Many of the hypotheses that have been advanced for birds are also potentially applicable to other animals, such as insects and mammals (Mathews 1968; Griffin 1974; Emlen 1975; Able 1980).

Navigation requires both a compass to provide directional information and a map to provide the animal with information on its position relative to home or some other goal. As the following summaries of evidence for different cues illustrate, considerable evidence exists for the compass, but we still know very little about the "map" (Keeton 1969, 1970).

Topographic Features

Use of familiar topographic features is certainly a factor in navigation, but it is probably of secondary importance compared with other cues. Nonetheless, many diurnal and nocturnal migratory birds are influenced by and may utilize topographic features. For example, birds congregate near and fly along coastlines or river valleys, and they may also use landmarks for piloting, particularly in areas at both ends of the migration where they are more familiar with specific topographic features.

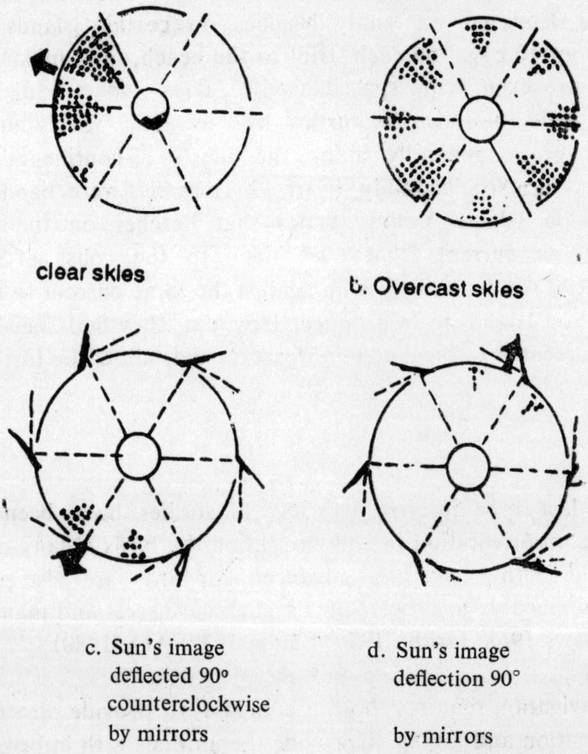

a. clear skies

b. Overcast skies

c. Sun's image deflected 90° counterclockwise by mirrors

d. Sun's image deflection 90° by mirrors

FIG. 3.3. Starling orientation experiment.

However, the use of topography as a guidance system has inherent drawbacks. How do first-time migrants who have never learned the landmarks find their way? What if a storm blows birds of the normal route into areas they have never traversed before? Also, visual landmarks alone cannot provide long distances. Both landmarks and certain topographic features the necessary compass direction for proper orientation over could be used in combination with some type of compass mechanism to provide sustained orientation during longer movements. Also, in familiar terrain, landmarks may suffice without any compass (Able 1980).

Sun

The most prominent regular cue for diurnal migratory birds, and possibly also for those that migrate at night, is the sun. The latter take a bearing at sunset and use that bearing to fly at at night (Able 1982). The early studies of bird navigation devoted considerable attention to the sun (Keeton 1974). Kramer (1950, 1951) first showed that European starlings (Sturnus vulgaris), placed in an outdoor cage with the sun visible, exhibited migratory restlessness in the appropriate direction in the spring and fall. When Kramer used mirrors to alter the apparent position of the sun, the starlings migratory restlessness shifted direction in a predict.

One way to demonstrate the use of sun compass is to examine whether birds maintain a constant orientation at different times of the day even though the sun's position varies as it traverses the sky. Data from some diurnal migrants provides support for this alternative. Clock-shift experiments have provided a second, more rigorous test of sun-compass orientation. First, clock-shifts of six hours, brought about by artificially shifting the birds' internal clock mechanism, resulted in predictable alterations of 90 degrees in the initial bearing of birds heading for the home site. This finding is consistent with the notion that birds use the sun as a simple compass. As a further test birds were clock-shifted six hours birds were using true sun navigation, they should have decided that they were 4,000 miles east of home and headed directly west. However, the birds headed directly east, as would be predicted if they were using the sun only as a compass and not as both a map and compass (Keeton 1974).

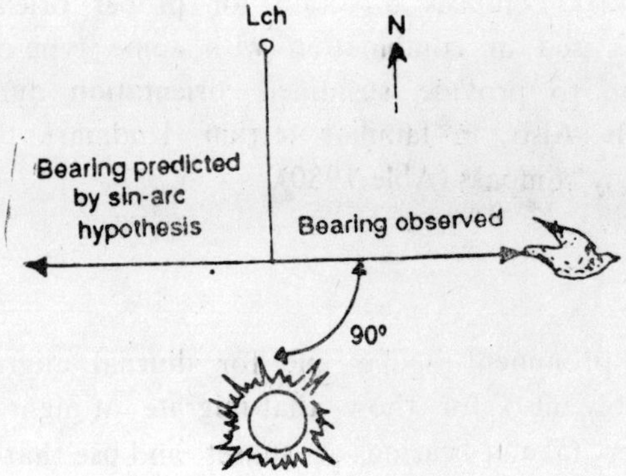

FIG. 3.4. Clock-shifted pigeon experiment.

The sun, then, is an important directional cue for birds that are diurnal migrants and for homing pigeons. Evidence in support of the hypothesis that the sun can be used as both compass and map is lacking.

Stellar Cues

When nocturnally migrating birds are placed in outdoor cages with a view of the clear night sky, they exhibit migratory restlessness in a direction appropriate to the seasonal migration (Kramer 1949, 1951; Sauer 1975). Sauer exposed birds to planetarium skies, which permitted him to manipulate star patterns experimentally. Not only were the results of outdoor studies confirmed, but when he shifted star patterns 180 degrees in the planetarium, the direction of the Zugunruhe exhibited by the birds also shifted. Under cloudy skies or with no stars visible and with only diffuse, dim illumination in the planetarium, the birds exhibited random orientation, Sauer's studies on Warblers (Syviidae have been confirmed in both field and laboratory by Emlen (1967a, 1967b) on indigo buntings (Passerina Cyanea). Emlen's technique is innovative, the bird stands on an ink pad at the bottom of a funnel-shaped cage. The sides of the funnel are covered with paper, and the top is a wire mesh screen that permits the bird to see the overhead sky. When the bird jumps up against the sides of the funnel in its restlessness, it leaves marks on the paper. Investigators can turn the record of these marks into a vector diagram for purposes of analysis.

Meteorological Cues

Migratory activity in a particular region tends to be concentrated on a few days or nights. Birds are capable of responding to favourable weather conditions (Drury and Keith 1962, Kreithen and Keeton 1974). Data support the conclusion that birds often fly downward on their migratory flights (Bellrose 1967; Richardson 1971; Bruderer and Steidinger 1972).

Further experiments (Able 1982; Able *et al.* 1982) monitored the natural noctural migratory flights of individuals of several Passerina species under various weather and wind conditions. When migrants could observe the sun near the time of sunset or the stars, they flew in the appropriate direction regardless of wind direction. If observations were made under totally overcast skies with no view of sun or stars possible, birds often flew downwind and thus sometimes in seasonally inappropriate directions. When migrant white crowned sparrows were fitted with frosted lenses and released from balloons aloft, they headed downwind, even though that direction was sometimes seasonally inappropriate. These results indicate the primacy of visual cues in birds for determining appropriate migratory direction and that, when deprived of visual cues, some species can determine wind direction.

Olfactory Cues

Several investigators have suggested that olfactory cues, primarily odours carried on the wind, may serve as an aid in orientation, particularly in homing pigeons (Papi *et al.* 1972, 1973, 1974). The most direct test of such a hypothesis involve the removal of the olfactory capacity and the testing for expected orientation and homing behaviour. When the olfactory sense is blocked or when nerves are transected, behavioural effects that are not related to the homing ability being tested many result; and, therefore, unequivocal evidence of the possible role of olfaction in pigeon homing is difficult to provide. Indeed, studies by Keeton and Brown (1976), Keeton, Kreithen, and Hermayer (1976), Schmidt-Koenig and Phillips (1978) and Hartwick, Kiepenheuer, and Schmidt-Koenig (1978) have all supplied data that demontrate little or no effect on homing behaviour when the nasal passages are blocked or a local anesthetic is used on the nasal epethelium.

Geomagnetic Cues

Evidence from several sources has accumulated that supports a role for geomagnetic cues in orientation and navigation in some birds. After many failures by experimenters in attempting to determine whether birds could even sense the .05 Gauss geomagnetic field of the earth (*e.g.*, Emlen 1970; Kreithen and Keeton, 1974, Brookman 1978), after a successful laboratory experiment, reported conditioning pigeons to magnetic fields. Measurement of Zugunruhe of European robins (Erithacus rubecula) has provided a second line of evidence. Birds were placed in a cage around which a set of Helmtholtz coils, which provide an artificial magnetic field, was constructed. Wiltschko and Wiltschko (1972; and Wiltschko 1972) have used this apparatus to show that the birds were not responding to the polarity of the horizontal component of the magnetic field; they did not shift the direction of their migratory restlessness when horizontal polarity was shifted. However, they did reverse direction when the vertical component was reversed; and they reversed the direction or orientation in a predictable manner. In an artificial magnetic field with a zero vertical component and strong horizontal component, the activity was random.

Other evidence for the role of magnetic fields in orientation comes from the work of (Southern 1969, 1972) on ring-billed gulls (Larus delawarensis). Gulls in cages exhibited a strong tendency to walk in a southerly direction except when storms produced temporary aberrations in the earth's magnetic field. Also when gulls with small magnets affixed to their backs were taken some distance from home and then released, they displayed a random pattern of dispersion. Control gulls without the magnets exhibited normal and consistent directional headings to the south. Additional confirmation has been reported by Keeton (1971) and Larkin and Keeton (1976), who used bar magnets attached to the wings of homing pigeons.

Walcott, Gould and Kirschvink (1979) have reported the existence of an organ, located along the midline of the pigeon's brain, that contains magnetite granules. This organ may prove to be the source of sensitivity to geomagnetism in the pigeon.

It is clear from much of the foregoing discussion that many bird species possess multiple systems for orientation; the use of available cues apparently follows some type of hierarchical scheme. Which mechanism the organism employs depends on its preferences for using certain types of cues and the prevailing weather conditions at that location.

Mammals

In their studies on orientation and navigation in mammals, researchers have devoted particular attention to small rodents and bats. In many of the studies on rodents, experimenters have displaced the animal for a distance and observed the rate of return in terms of the number of animals that successfully returned home, the speed with which they returned, and aspects of the rodent's orientation as they left the displacement site. In some studies experimenters have attached reflector tape or radio transmitters to the animals to plot the rodents' course during homing.

Rodents have an area, called the home range, in which they carry out their daily functions. When experimenters have determined the home range, they can displace the rodent and monitor its homing behaviour until the rodent's return and recapture within its home tange. Data from several species of mice indicate that, when displaced these animals can find their home range again, some from a considerable distance (up to 30 kilometres) and with speeds ranging up to 300 metres/hour. In most studies the rodents are generally displaced for distances of only up to 500 metres and may take several days to return to the home range and reenter a trap (Joslin 1977). The return rate, or the percentage of mice that successfully return home, varies with the species and the nature of the habitat. Overall figures have indicate return rates ranging from 3 to 5 per cent to over 80 per cent.

Little work has been done on how rodents orient within their home ranges; however, Drickamer and Stuart (1984) investigated the movements of deermice (Peromyscus). In the winter the patterns of tracks left by the deermice on fresh show can be mapped. Drickamer and Stuart used traverses—each segment of the path from tree to tree or from tree to log—to examine the potential use of cues. They found that 86 per cent of all traverses were oriented from one tree to another tree and covered distances ranging from one to thirty metres. The concentration of tracks suggests that the vertical tree trunks are used for orienting. A significant positive relationship (r=0.51) between the length of the segment travelled by the mouse and the diameter of the tree to which it was headed further suggests that larger trees serve as more conspicuous objects on the horizon for orientation. Laboratory investigations (Joslin 1977) have supported the hypothesis that deermice may use conspicuous vertical cues on the horizon for orienting to goals.

The navigation and orientation systems of bats have received considerable attention. Bats taken from caves, barns, or buildings and displaced some distance often return. Sometimes the distances covered are great. A great deal is known about the homing of bats, but the difficulty of immediately recapturing them when they return and the possibility of missing many that do return make interpretations of the actual homing ability equivocal (Cockrum 1956).

Vision and familiar sounds at the home roost may provide terminal information detailing the precise location of the goal. But as in bird homing and migration, mechanisms involved in navigation may be different from those used in piloting. Determination of the initial directional course remains the primary unexplained phenomenon in bat homing. Preliminary experimental tests of the initial orientation of the fruit bat Phyllostomus hastatus seem to demonstrate accurate initial orientation at the release site.

The possibility that search patterns provide an explanation of the orientation of mgratory flights of bats can be dismissed. Direct migrations covering hundreds of miles have been recorded (Griffin 1958), and the seasonally appropriate directional flight of red bats, Lasiurus borealis, is a commonplace observation in the eastern United States. Yet it is remarkable that no author has suggested any explanation for how these travellers find their way! Bats often collide with towers in the same circumstances and often on the same night as migrating birds do (Tordoff and Mengel 1956). Perhaps celestial orienting cues are a source of directional information for them.

Other Navigation Mechanisms

Orientation without Optical Cues: an Artifact? Several authors have reported oriented migratory fluttering in birds in the apparent absence of directional optical cues. Merkel and Fromme (1958) and Fromme (1961) worked with the European robin, Erithacus rubecula, and the whitethroat Sylvia communis. Birds were evenly and faintly illuminated, but they had no view of the sky. Various experimental locations were used, the most restricted of these being the placement of the test apparatus in a cellar without windows. Here directional orientation persisted, even in a steel cylinder with airtight doors, as long as one of the doors remained open. The magnetic field of the earth at the cage was undisturbed. This finding contrasts with Sauer's

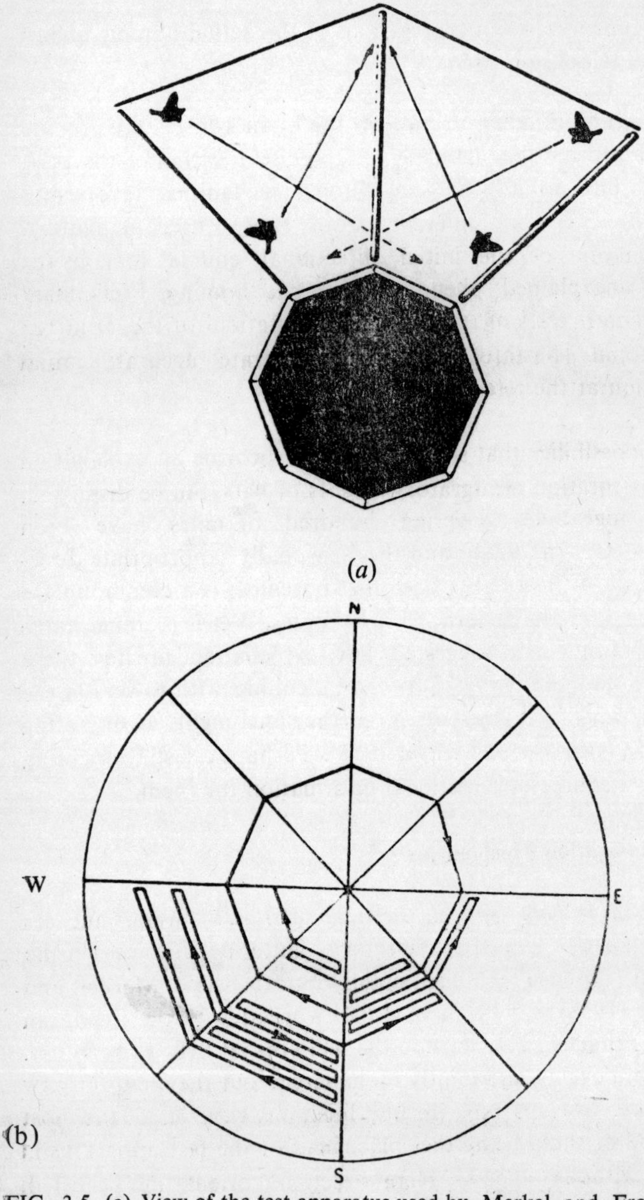

FIG. 3.5. (a) View of the test apparatus used by Merkel and Fromme. The movement from one radial perch to the next is automatically recorded. (b) A sample 30-minute record of the nocturnal movements of robin.

demonstration that garden warblers became disoriented when placed in a diffusely lit planetarium or under natural overeast. The conditions and recording methods used by Merkel and Fromme are different from those reported by Sauer, so that no direct comparison of results is possible.

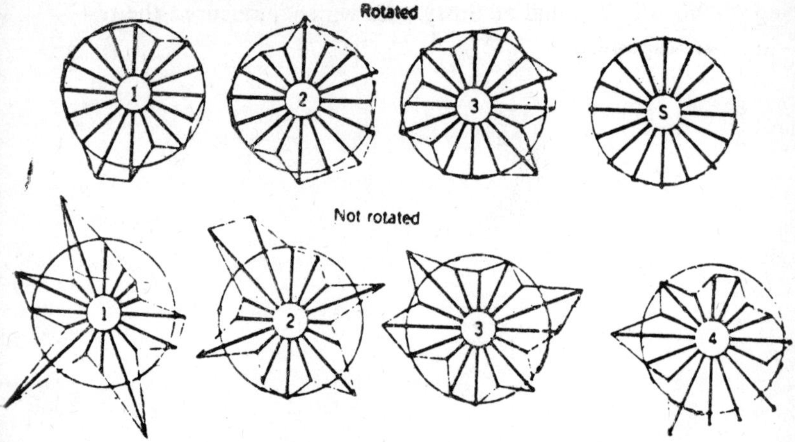

FIG. 3.6. Orientation of the robin, Tilthacus rubecala, in an experimental Chamber lacking directional visual cues. The first row represents, three individual experiments and the summary of thirteen such experiments. In the second row are four Individual experiments which were performed without rotating the apparatus.

Precht (1961) and Gerdes (1961, 1962) recorded the escape behaviour of black-headed gulls, Lerus ridibundus, displaced from their nests. These birds also maintained a significant orientation, heading toward home not only under overcase conditions but also indoor rooms.

In all these experiments in which orientation has been demonstrated under apparently cueless conditions, the birds have not been as highly oriented as birds able to view the everhead sky. The problem was reinvestigated by Perdeck (1963), again with the European robin. Perdeck concentrated on methodological features. He showed that orientation was random under indoor conditions as long as the orientation device was rotated. When the device remained fixed during experiments, directional preferences not related to be migratory orientation of the species were expressed. In addition, Perdeck discovered that the presence of other robins in the same building provided directional auditory cues which influenced the test chamber responses.

4

Buoyancy of Marine Organisms

In oceanic biology one always comes back to the 'water column', which everywhere holds an uppermost layer of planktonic plants and animals that give life to the underlying deep-sea animals. Indeed, '. . . down to the deep-sea floor exist thousands of pelagic organisms, poised, drifting or actively moving at many different levels. Each species lives within a certain range of depth, there to feed, grow and multiply; and, as far as can be seen, the individuals of a species must keep to their particular environment in order to flourish. Study of the means by which plants and animals maintain their level in the ocean is one way of appreciating the latter aspect of living activities.'

During evolution pelagic animals have spread into different deep-sea layers of the ocean: upper mesopelagic (ca. 200-600 metres), lower mesopelagic (600-1,000 metres), upper bathypelagic (1,000-2,000 metres), lower bathypelagic (>2,000 metres) and benthopelagic. There is overlap between these vertical zones, but as we have seen elsewhere, each is occupied by a recognizable community of organisms that exists in stratified but dynamic stability.

Pelagic organisms, which have diverse means of keeping their level in the water column, are alike in one negatively buoyant respect: their living substance, with a range of about 1.03 to 1.11 in specific gravity, is heavier than oceanic sea water (s.g. 1.02-1.028). The proteins that form much of their tissues have a specific gravity of about 1.33. Skeletal materials, as one might expect, are also negatively buoyant. Thus, skeletons containing calcium salts may have a density of about 3. On the positive side, fats (s.g. ca. 0.9), certain body fluids and gas-filled cavities provide the organism with lift. Clearly, in a neutrally buoyant organism, which can stay at one level with virtually no effort, there must be equipoise in the negative and positive factors of buoyancy. The buoyancy balance sheet, to use E.J. Denton's concept, must remain balanced.

Oceanic Plants

For the phytoplankton, light strong enough to energize photosynthesis exists only in the uppermost 150 metres. The continued existence of oceanic life depends on the means by which microscopic plants are suspended in the productive euphotic zone.

Diatoms, whose glassy cell walls make up 20 to 75 per cent of their dry weight, would seem to need appreciable means of positive buoyancy. Most likely, all species are slightly denser than sea water. Their growth and reproduction is not only offset by the grazing of herbivorous zoo-plankters, but also by the fallout below the euphotic zone of living as well as dying cells. Where there are heavy crops of diatoms, as in the Southern Ocean, the Subarctic Pacific and in upwelling regions, samples of diatom ooze may be green and contain living cells.

Diatoms may contain a positively buoyant cell sap, produced by the selective and partial exclusion of heavy sulphate ions, as in large 'bladder-like' species with thin siliceous cell walls. There may also be changes in buoyancy related to nutritional factors and the effect of light. Colony formation by linkages between individual cells of the same species increases density, for smaller colonies sink more slowly than larger ones (Smayda, 1971). The larger the colony the greater the 'need' presumably, for the gelatinous threads, spines, setae and other kinds of projections that increase the surface area and retard the rate of sinking. Moreover, diatoms of the open ocean may not only be smaller but alsohave lighter siliceous skeletrons than their relatives of the neritic province.

Most of the other members of the phytoplankton have active means of movement in the form of one or more flagella. The naked micro-flagellates should easily maintain their level, which must require more effort by the armoured dinoflagellates and calcareous flagellates. The buoyancy and behaviour of these forms is discussed elsewhere (p. 60 and Marshall, 1954).

The Buoyancy of Planktonic Animals

Beside the coccoliths of calcareous flagellates the calcareous oozes of the ocean are formed of the shells of pelagic Foraminifera, while Radiolaria as well as diatoms form siliceous oozes. Little is known of buoyancy in these two groups of pelagic protozoans. The great surface area of their reticular pseudopodial system, and such area in some species is enhanced by projections of the skeleton, must do much to retard their rate of sinking. The latter, according to figures quoted by Parsons and Takahashi (1973), is 30 to 4,800 metres per day in Foraminifera and about 350 metres per day in Radiolaria. If so, how do members of a population remain long enough at one level to complete their life cycle? Actually foraminifera with spiny tests are generally epipelagic, whereas the non-spinose species, which range also into the depths, may have much thickened shells (Be, 1966). The latter mnst have ways of countering their calcareous ballast.

More is known of flotation of radiolarians in the genus Thalassicola, which may reach up to 5 mm in diameter. Thalassicola pelagica has a central capsule encircled by a froth of large cytoplasmic 'bubbles', each filled with a watery solution. The central capsule, even though it contains droplets of oil, skins in sea water: evidently, it is the light fluid contents

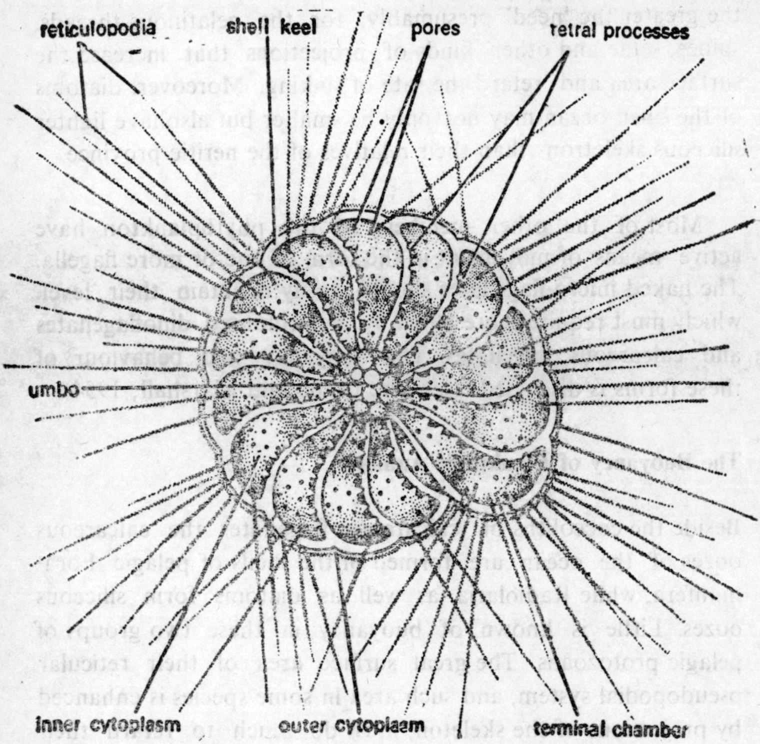

FIG. 4.1. Elphidium crispum (living).

of the 'bubbles' that provide enough uplift (Jacobs, 1935). Observations have shown that the number (and perhaps the contents) of the 'bubbles' can be so regulated that the protozoan can rise or sink in the sea. It is said to descend during stormy weather.

Of the acantharians, species of Acanthometra have other means of flotation. Radiating from the central capsule are a number of symmetrically arranged spicules of strontium sulphate provided distally with contractile threads (myonemes), which are inserted also on the gelatinous extracapsular cytoplasm. When the myonemes contract the gelatinous matrix is greatly expanded and the animal eventually rises in the water: on relaxation of the myonemes the matrix shrinks and is followed by sinking (Jacobs, 1953). The expansion is presumably linked to inner changes that provide uplift.

The many larger gelatinous forms of the zooplankton, whether herbivorous pelagic tunicates or carnivorous cnidarians and ctenophores, have many common adaptations, including their means of flotation, to a pelagic mode of life. Their very watery tissues, which contain very small quantities of sinking materials, require just the uplift provided by body fluids that are isotonic with sea water, but in which the heavy sulphate ions have been replaced in part by lighter chloride ions. This insight came from the investigations of Denton and Shaw (1961) on the jellyfish Pelagia noctiluca, the ctenophores Beroe and Cestus, the tunicates Salpa maxima and Thalia democratica and the heteropod mollusc Pterotrachea coronata. Except for Beroe, which was somewhat overbuoyant, these animals were close to neutral buoyancy. Their body fluids were not only less dense than sea water, because of the partial exclusion of sulphate ions, but also the uplift (mg/m) varied directly with the degree of such exclusion.

The attainment of neutral buoyancy by single planktonic animals is fitting, but the flotation of siphonophores evidently depends largely on over-buoyant members of the colony. When the bracts (hydrophyllia) of a colony are detached from the central axis they float to the surface. Moreover, in physonects there is also uplift from the gas-filled float (pneumatophore) at the top of the colony. The float supports the upper part containing the swimming-bells (nectophores), but the heavy feeding and reproductive members that form most of the colony are buoyed by the gelatinous bracts (Jacobs, 1937).

Siphonophores without a pneumatophore (Calycophorae) also consist in part of buoyant members. In the diphyids the feeding and reproducing members are propelled and buoyed by two gelatinous swimming-bells. One may contain an oily cavity (somatocyst), but in Sulculeolaria quadrivalvis at least, the oil can have little buoyant effect, for the bell containing it floats underneath the other; the upper swimming-bell together with gelatinous bracts on the stem are the main supports of the colony. In Hippopodius hippopus, a representative of another family, there is a cluster of gelatinous swimming-bells, all lighter than sea water. There are also means for altering the specific gravity in Hippopodius and Diphyes: in a half-hour to an hour these forms can become heavier or lighter than sea water and so vertical movements may be facilitated (Jacobs, 1937). Presumably the changes of specific gravity are effected by the sulphate-excluding mechanism in the buoyant members of the colony.

Arrow-worms (Chaetognatha), which are also dominant forms of the zooplankton, have much more muscle relative to their size than has a siphonophore, a jellyfish, a comb-jelly, or a pelagic tunicate. Indeed, Sagitta setosa (2.1-2.2 cm in length) from the Black Sea have been timed to swim at speeds of about 0.5 to 0.6 metres per second, which is striking proof of their propulsive muscles. Such individuals had a mean density (gm/c.c.) of 1.02, whereas 1.01 was the density of the water in which they were caught (Aleyev, 1977). Their slightly negative buoyancy and tendency to sink can be countered by propulsive movements and, as Aleyev says, the flattened body with its horizontal lateral fins will improve its properties as a lifting surface. Aleyev's figure of 1.02 for the density of Sagitta setosa is intriguing in that an arrow-worm of this density would be neutrally buoyant in oceanic waters, But some species are more muscular and thus carry more sinking material than S. setosa. Clearly, further investigations are needed.

There is also much uncertainty regarding the buoyancy of copepods and ostracods. When the copepod Calanus finmarchicus is anaesthetized it sinks in sea water. Individuals 2 mm in length took 3 minutes to sink through a 25 cm column of water (of temperature 18.5°C and salinity to 35.01 per cent): Calanus of twice this length, thus with less surface area relative to their bulk, took only half a minute to sink the same distance (Gardiner, 1933). It would seem that Calanus would need to keep swimming upwards to stay at one level. But underwater

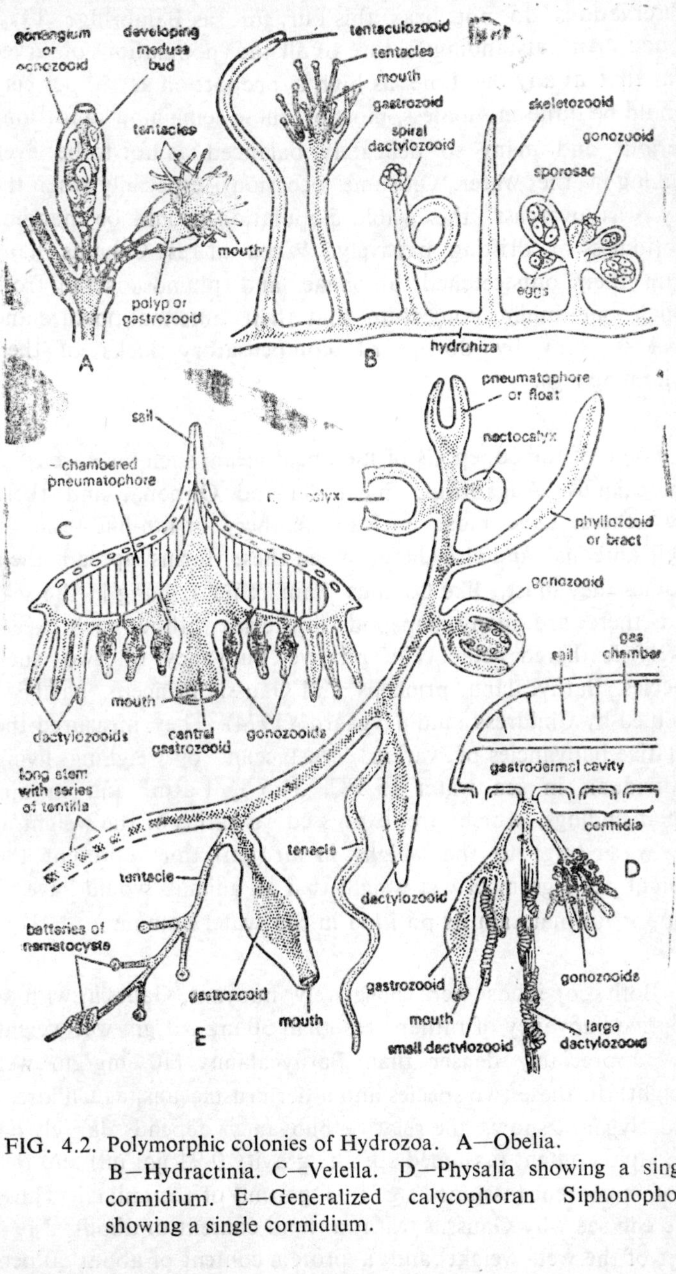

FIG. 4.2. Polymorphic colonies of Hydrozoa. A—Obelia. B—Hydractinia. C—Velella. D—Physalia showing a single cormidium. E—Generalized calycophoran Siphonophora showing a single cormidium.

observations do not bear this our, for, as Bainbridge (1952) found, 'An outstanding feature of all the populations observed was that, at any one time, as high a proportion as 50 per cent would be quite motionless, most of these remaining so for long periods and many so delicately balanced as not to be even sinking in the water. On some occasions, especially when the sky was overcast, the whole population would be hanging motionless or drifting passively.' When at rest Calanus hang from their outstretched antennae and plumose setae, from which one might conjecture that their antennal posture and level is kept by downward compensatory kicks of their swimming legs.

Some of the copepods of the open ocean, such as species of Calocalanus, Augaptilus, Euchirella and Oitbona, and there are others, have most intricate feathery or fan-like setae on their antennae and elsewhere. When these forms spread their plumes they must, like Calanus finmarchicus, hang in the sea. But there are many copepods in the sea without such excessively feathered setae. The relative buoyancy of two such species. Bathycalanus princeps and Gaussia princeps, has been studied by Childress and Nygaard (1974). They measured the relative buoyancies of various crustaceans by weighing living individuals in sea water (5.5°C, 34.5 ‰, 1 atm) and in air. Their findings, which are expressed as ratios of the weight in sea water (mg) to the weight in air (gm), thus represent the weight per 1 gram of wet weight that an animal would have to support to maintain its position in the water column.

Both copepods were negatively buoyant, Gaussia, with a relative buoyancy of rather less than 30 mg per gm wet weight was appreciably denser than Bathycalanus (10 mg/gm wet weight) .In these two species and other crustaceans, as Childress and Nygaard show, the relative buoyancy depends largely on the lapid content (assumed specific gravity 0-91 gm/ml) and the protein content (assumed s.g. 1.33 gm/ml) of the animal. Thus, one can see why Gaussia with a lipid content of about 3 per cent of the wet weight and a protein content of about 50 per cent of the we weight is more negatively buoyant than Bathycalanus, in which these two figures are about 6 per cent and 40 per cent.

By contrast, the mesopelagic ostracod Gigantocypris agassizii gave quite different results. It was neutrally buoyant, and combined with an extremely high water content (95.6%) it had very low lipid and protein comments (0.1% and 0.6% of the wet weight). As Childress and Nygaard observe, its chemical composition is more like that of a jellyfish than a crustacean. Moreover, according to Macdonald (1975) Gigantocypris mulleri has body fluids that are isosmotic but lighter than sea water due to the presence of ammonium ions.

Evidently, Gigantocypris can float at one level, but Angel (1971-72) has described how individuals of another deep-sea ostracod, Conchoecia spinirostris, sank head first at about 1 cm/sec, with the valves gaping and the limbs moving, after swimming movements. Several times individuals hung, as though neutrally buoyant, in midwater with the valve gape closed and only occasional bursts of activity by the vibratory plates.

Buoyancy of the Micronekton

Most members of the micronekton are crustaceans, and we will recall that they are muscular animals with complex skeletons. Euphausiids, which were called the 'copepods' of the micronekton, concern us first. When observed in an aquarium they sink when they stop swimming. Indeed, the common epipelagic species, Euphausia pacifica, proved to be the most negatively buoyant of the midwater crustaceans studied by Childress and Nagaard (1974). Its relative buoyancy of 40 mg/gm wet weight may be set beside its low lipid and high protein contents, which were about 2 per cent and 60 per cent of the wet weight. Even more than arrow-worms, euphausiids are muscular animals without special means of uplift. Evidently they have muscles and legs enough to keep their position in the water column.

Sergestid prawns are also appreciably heavier than sea water. Thus, the relative buoyancy, lipid and protein contents of two mesopelagic species, Sergestes phorcus and S. similis are: 32 and 24 mg/gm wet weight, 4.5 and 3.3 per cent and 53 and 53 per cent wet weight (Childress and Nygaard, 1974). Like most sergestids these are active, muscular animals that undertake daily vertical migrations. But species from lower mesopelagic to upper bathypelagic levels, whose vertical migrations, if any, are small in extent, have reduced muscles. Two such species are Sergestes robustus and S. Japonicus, and Donaldson (1976) concludes that they have adapted to the low food supply of their living spaces by reducing their muscular tissues and building up their lipid reserves. Even so, their chemical composition suggests that they are still somewhat negatively buoyant.

By contrast, three of the four oplophorid prawns that Childress and Nygaard (1974) studied were neutrally buoyant. The negatively buoyant exception, Acanthephyra curtirostris, had a relative buoyancy of nearly 20 mg/gm wet weight and lipid/protein contents of 7.5/35 per cent of the wet weight. Two of the just buoyant species, Hymenodora frontalis and Systellaspis cristata, had lipid/protein contents of 19/30 and 12/35 per cent of the wet weight, which would suggest that the former is more buoyant, but it had a lower water content (63.8%) than the other (72.8%). In the third just buoyant prawn, a bathypelagic species of Notostomus, the water content was very high (91.3%) and the lipid protein contents (1.5/22%) low. But buoyancy in Notostomus, with its 'inflated' cephalothorax, largely resides in ammoniacal body fluids (Herring, 1973), which, as we shall see, is a common means of buoyancy in deep-see squids.

Buoyancy of the Nekton

We begin with the cephalopods, and must note first that active forms of epipelagic squid with much muscle in the mantle have no special buoyancy devices. They are heavier than their environment. For instance, the hooked squid, Onychoteuthis banksi and Ommastrephes pteropus, which have a density (gm/c.c.) of 1.05, came from sea water with a density of 1.03

gm/c.c. (Aleyev, 1977). Mesopelagic squids, such as Pterygioteuthis, Abraliopsis and Lycoteuthis, with no obvious buoyancy mechanisms, are likely to be negatively buoyant. But Clarke (1977b) estimates that 53 to 78 per cent of the midwater squid that are consumed by sperm whales belong to species with light ammoniacal body fluids.

At least five species of cranchiid squid, which can hang almost motionless in the sea, have a large, fluid-filled buoyancy chamber. The fluid has a relative density of about 1.010 (sea water is about 1.026) and its principal ions are ammonium, sodium and chloride (in concentrations of about 470,83 and 630 m mol/1 respectively). It is isotonic with sea water but nearly all of the cations have been replaced by ammonium ions. Without its buoyancy chamber, the relative density of the animal is 1.046 and thus a large volume of (moderately buoyant) ammoniacal fluid is required to make it neutrally buoyant (Denton, 1971; Denton and Gilpin-Brown, 1973). Moreover, such high concentration of ammonium, which stop the conduction of squid nerves, must clearly be segregated from vital tissues (Denton, 1971).

The buoyancy chamber of cranchiid squid, which takes up so much of the body volume (about two-thirds), hardly leaves space for the development of a capacious mantle cavity. Indeed, such forms, though capable of rather quick escape reactions, are small-finned, unhurried swimmers. In larger and more active kinds of squid there is no single buoyancy chamber: instead there are many small vacuolar chambers over the body, particularly in the arms and mantle, that are filled with ammoniacal fluids very similar in composition to those of cranchiid squids. In Histioteuthis these chambers are so disposed over the body that the animal may assume any attitude with very little effort. The arms are the most buoyant parts of the giant squid (Architeuthis), Mastigoteuthis and enoploteuthids, which may thus hang vertically in the sea, suspended by them (Denton and Gilpin-Brown, 1973). Representatives of at least 11 of the 25 families of squid are buoyed by ammoniacal body fluids. As in species of the families Chiroteuthidae, Octopoteuthidae and Histioteuthidae, they are approximately weightless in water. Their buoyancy balance sheet has been analysed by Denton (1971), who writes. 'Now if the only components denser than sea water in a marine animal were protein and amino-acids, and the only component giving lift a body fluid isosmotic with sea water but in which sodium ions were replaced by ammonium ions, then the animal would be neutrally buoyant if the ratio of ammonium nitrogen to total protein and amino-acid nitrogen was about two or three. We can, therefore, by analysing whole animals determine whether or not the accumulation of ammonium is a principal buoyancy mechanism. For a specimen of Octopoteuthis danae this ratio was found to be 0.69 while for a specimen of Calliteuthis reversa it was 1.3. Clearly animals like Calliteuthis and Octopoteuthis derive the lift giving them neutral buoyancy mostly from the ammonium ions which they contain.'

Lastly, as least one squid (Bathyteuthis), like numerous sharks, has a buoyant liver. In this squid, which swims slowly and often hangs motionless in the water, the liver consists of two parts, the forward part consisting of two large chambers filled with reddish-orange oil (C.P.E. Roper, quoted by Denton and Gilpin-Brown, 1973).

Ammoniacal devices are not known in midwater octopods, most of which, like Vampyroteuthis, have the look and consistency of a jellyfish. Indeed, at least one species, Japetella diaphana, has jellyfish ways of providing uplift. This octopod, which weighs only 0.12 gm in sea water, excludes sulphate ions to the extent that it is half way towards neutral buoyancy (Denton and Gilpin-Brown, 1973).

A gas-filled buoyancy chamber, which is so common among teleost fishes, is found in but one deep-sea cephalopod, the sepioid. Spirula spirula. Like cuttlefishes (Sepia) and Nautilus, Spirula has means of pumping liquid out of chambers that become filled with gas.

Spirula is a mesopelagic cephalopod, which off the Canary Islands is found at depths of 600-700 metres by day, and after vertical migrations, between 300 and 100 metres by night (Clarke, 1966). Living individuals float gently upward in a vertical position with the buoyant, shell-containing part of the body uppermost, the head and tentacles pointing downwards. (Fig. 4.3). In adults the spiral shell is formed of thirty or more calcareous chambers through which runs a tubular siphuncle, differentiated in part as a fluid-pumping device. Whatever, the depth of capture, the pressure of gas in the shell chambers is less than atmospheric (0.8 atm.). Under pressure the shell implodes at pressures corresponding to a depth range of 1,300 to 2,300 metres, well below the centre of concentration of Spirula (Denton, 1971). As Spirula grows new chambers are added to the shell. Both the newest and oldest chambers contain liquid. A newly forming chamber is first full of liquid isosmotic with sea water, then after some removal of solutes followed by with-drawal of liquid, the chamber contains a sizeable gas bubble,

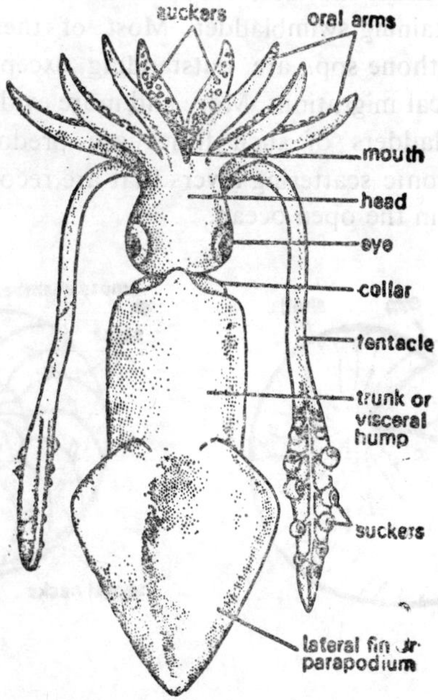

FIG. 4.3. Loligo. Dorsal view.

the liquid around it now being very hyposmotic to sea water. By this time a still newer chamber is partly formed, and when the penultimate one is nearly full of gas the last chamber is not only sealed but also contains a fluid isosmotic with sea water (secreted into it by surrounding tissue) and now ready for evacuation. The liquid is removed by a special epithelial section of the siphuncular tube, but how this pump works against high hydrostatic pressures has yet to be determined (Denton and Gilpin-Brown, 1973).

The Buoyancy of Midwater Fishes

There are some 1,000 species of midwater fishes, and those containing a gas-filled swimbladder live at mesopelagic levels. These species, which are mainly lantern-fishes and small stomiatoids, form about 40 per cent of the mesopelagic fauna (ca. 850 spp.). In numbers of individuals, as judged by the catches of large midwater nets, well over 75 per cent of mesopelagic populations consist of fishes that derive most or some of their buoyancy from a gas-containing swimbladder. Most of these individuals (those of Cyclothone spp. are outstanding exceptions) undertake daily vertical migrations. More and more evidence indicates that the swimbladders of such fishes are predominantly the cause of the sonic scattering layers that are recorded on ships' echo-sounders in the open ocean.

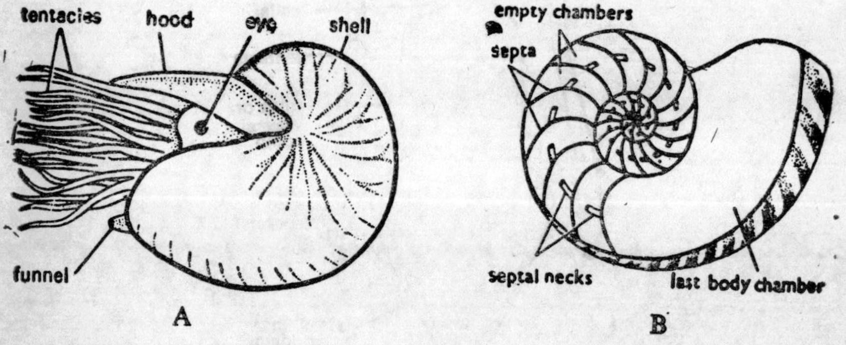

FIG. 4.4. Nautilus. A—Animal within shell. B—Shell in section.

Many of the mesopelagic fishes lacking a swimbladder are predatory forms. Some such as Chauliodus, Stomias and various melanostomiatids, are migrators: others, like the alepisauroids and giganturids, are non-migrators. There are also non-migrators with small jaws (bathylagids and moderately long jaws (searsiids). Even though lacking a swimbladder, such fisher, as we shall see, are only slightly heavier than sea water.

Bathypelagic fisher either have no trace of a swimbladder, as in ceratioid angler-fishes and gulper-eels, or a regressed swimbladder (Cyclothone spp.). More than mesopelagic fishes that lack a swimbladder, bathypelagic species approach a neutrally buoyant condition.

To make its user weightless in water, the swimbladder of a marine fish should have a volume equal to about 5 per cent of the animal's total volume (Jones and Marshall, 1953). In mesopelagic fishes this percentage is more likely to range between 4 and 1, but species with the smallest swimbladders may derive buoyancy from reserves of lipid substances. A marine teleost also gets uplift from dilute body fluids that provide about one-sixth the lift exerted by a 5 per cent swimbladder. The main heavier-than-sea-water components are proteins, especially in its lateral muscles, and skeletal materials.

In deep-sea fishes, as in other teleosts with a closed swimbladder, gas is secreted into the swimbladder by a gland that receives blood through closely parallel systems of arterial and venous capillaries (retia mirabilia). In lantern-fishes an artery-vein pair that enters the fore part of the swimbladder subdivides to form three retia mirabilia, each made of hundreds of contiguous arterial and venous capillaries that run in parallel to a corresponding lobe of the gas gland*. Blood flows to the gland through the arterial capillaries, and after circulating

*Gas glands, which are rich in glycogen, are specialized parts of the inner epithelial lining of the swimbladder. Between this epithelium and the tough outer fibrous coat of the swimbladder wall is a loose gelatinous connective tissue layer (submucosa) that may hold a layer of smooth muscles.

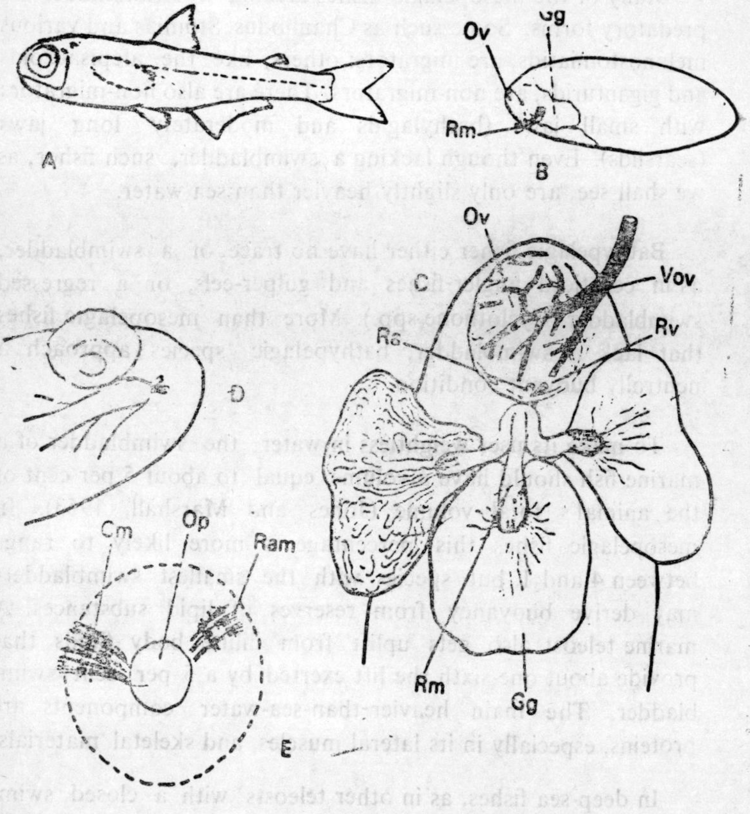

FIG. 4.5. Swimbladder of a lantern-fish, Myctophum punctatum; A, in position in body cavity of fish; B, laterally; C. ventrally (anterior part of organ). D and E, the gas-resorbing part (oval) of the swimbladder of another lantern-fish, Diaphus rafinesquei. Cm, circular muscle of oval; Ram, radial muscle of oval; Gg, gas gland; Op, opening from oval into swimbladder chamber; Ov oval; Ra, retial artery; Rv retial vein; Rm, rete mirabile; Vov vein to oval. Veins shown black, arteries white.

among the gland cells in capillary loops, returns to the venous capillaries of the rete. There is a single posterior rete mirabile in the swimbladder of stomiatoid fishes but the capillaries do not pass directly into the gas gland: instead they reunite to form larger vessels that run into the gland and then form the looping circulation.

The countercurrent system of capillaries in a rete mirabile serves both to keep gas within the swimbladder and to build the requisite gas tensions for secretion through the gas gland. The first function is by way of gas exchange between the blood in the two sets of capillaries. Blood leaving the swimbladder will enter the venous capillaries bearing dissolved gases at a pressure equal to that in the swimbladder (which in atmospheres at deep-sea levels is about one-tenth of the depth in metres). As there are merely two thin capillary walls between the venous blood and the blood in the arterial capillaries, gas will diffuse freely into the latter and continue to do so as the blood flows down the venous capillaries. When the blood is about to leave the rete it flows by ingoing arterial blood that contains gas tensions (notably oxygen at about 0.2 atm.) acquired during circulation through the gills. Thus the venous blood in the rete can lose gases until it is in diffusion equilibrium with the inflowing arterial blood (Fig. 4.5).

While gaseous exchange is facilitated by the short diffusion distance (<1 μm) between the blood in the two sets of capillaries, the efficiency of the rete as a gas retainer depends largely on the total capillary surface available for countercurrent exchange, which in turn depends on the number and length of the retial capillaries. Indeed, these two dimensions in the rete of a deep-sea fish ensure inter alia that it is a very effective system for maintaining high gas pressures (even in excess of 4,000 atm.) in the swimbladder (Scholander, 1954).

How a rete mirabile can build gas tensions through countercurrent multiplication was first clearly shown by Kuhn and his colleagues (1963). As they assumed, and later study has revealed, the gas gland produces lactic acid, which on entering the circulation, has two effects on the venous blood that flows into the retial capillaries; it releases oxygen from haemoglobin (Root effect) and decreases (by a 'salting-out' process) the solubility of all gases in the blood. The gas tension

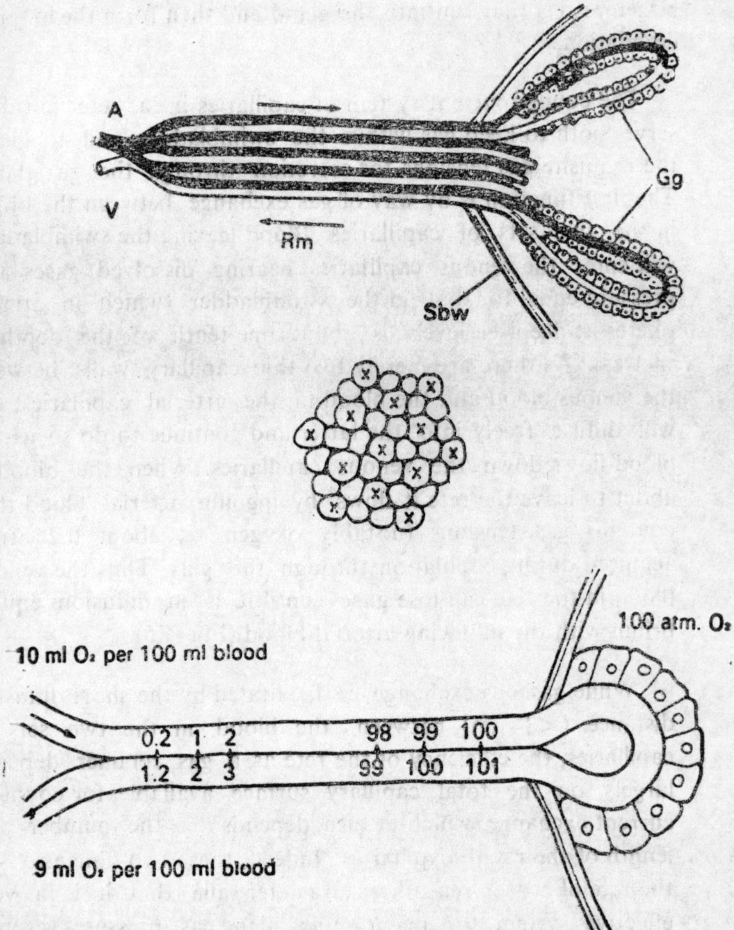

FIG. 4.6. Top, schematic representation of the formation of a rete mirabile (Rm) from an attery (A) and a vein (V). The arterial capillaries are shown in black. Within the swimbladder (Sbw is the swimbladder wall) are shown two capillary loops and associated gas gland cells (Gg).
Centre, a few of the capillaries of a rete mirabile as seen in cross-section. (The larger venous capillaries are labelled X.)
Bottom, schematic representation of the countercurrent multiplier system of the teleost fish swimbladder. The numerous capillaries of the rete mirabile are represented by a single loop. As the gas gland produces lactic acid, the oxygen

tension increases (above that in the swimbladder) in the venous capillary (below), and gas diffuses across to the arterial capillary (above), thus remaining in the loop. As it leaves the rete, the venous blood contains less oxygen than the inflowing arterial blood.

generated by both processes causes the released gases to diffuse into the blood of the ingoing arterial capillaries and gradually accumulate at the gas-gland end of the rate: and here, as the blood continues to circulate, the pressure will steadily increase (Fig. 4.5).

Considering the salting-out process alone, the multiplying-gas-generating power of a rete depends directly on its length and the degree of permeability between the two blood systems The inverse factors are the speed of flow in the capillaries, the bore of the capillaries and the solubility coefficients of the gases. In theoretical equations based on these relationships, Kuhn and his colleagues have shown that a rete 20 mm in length can easily build the gas tensions that exist in the swim-bladders of deep-sea fish. For oxygen itself—the main gas in the swimbladder of a deep-sea fish—a small part (about 1 to 2 mm) of the retial capillary length is enough to generate, by both the salting-out and Root effects, pressures between 10 and 100 atmospheres. At pressures above 100 atmospheres further increase in the tension of oxygen is solely by way of the salting-out effect. High tensions of gas may thus be presented to the gas gland, but just how gas is secreted is still unclear.

The organization of the swimbladder in deep-sea fishes may now be set against these physiological sketches of the maintenance and generation of gas tensions. After survey of the mesopelagic fauna it will be logical to continue with the benthopelagic fauna, numerous species of which live at depth ranges well beyond mesopelagic levels.

Species with populations centred between lower mesopelagic levels (600-1,000 metres) have longer retia mirabilia (3-7 mm) in the swimbladder than are found in upper mesopelagic

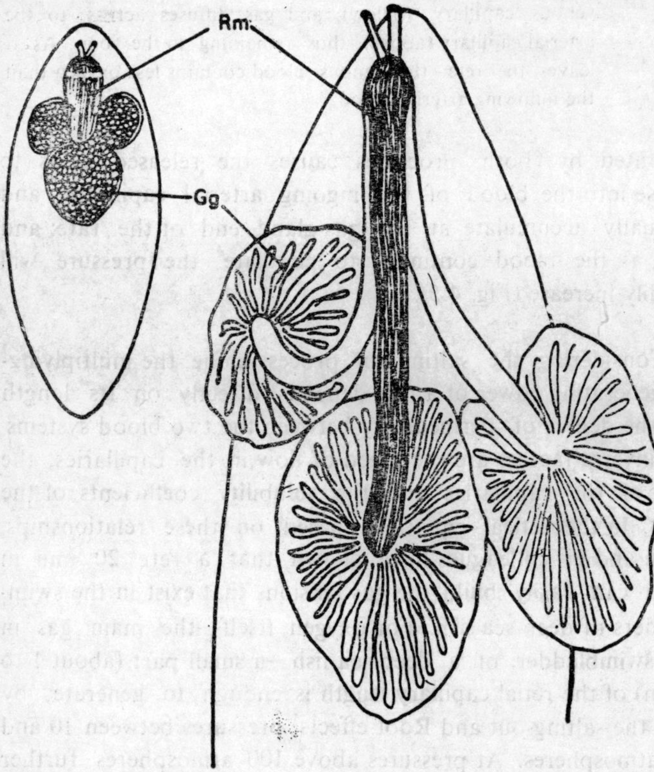

FIG. 4.7. Swimbladders of mesopelagic fishes. Left, Valenciennellus tripunctulatus, an upper mesopelagic species, right, Lampadena speculigera, a lower mesopelagic species, the lantern-fish. Rm, rete mirabile; Gg, gas gland.

species (200-600 metres. As the former have to maintain and generate higher gas tensions than do the latter, and as the efficiency of both processes depends on the length and interchange surface area of the retial capillaries, the difference in the length of the countercurrent system is not surprising. But first we must note that while the length of the retia mirabila is between 1 and 2 mm. in most upper mesopelagic species with a well-developed swim bladder (*e.g.*, numerous lantern-fishes, such as Myctophum punctatum (Fig. 4.5), M. affine, Diaphus garmani and Lobianchia garmani, and hatchet-fishes of the genus Argyropelecus), the length of the rete is about 0.75 mm in the stomiatoid fishes Maurolicus, Valenciennellus and Vinciguerria (Fig. 4.6). The retia in the latter three genera are unusual also in having narrower capillaries (arterial 2-4 μm; venous 5-10 (μm) than those of the majority (−18 μm). Moreover, to fit the narrow-bore capillaries the red cells (erythrocytes) of the blood are unusually small and lack nuclei. (Maurolicus muelleri has rod-shaped erythrocytes 5×2.5 μm; Valenciennellus has globular erythrocytes 7×7 μm). The erythrocytes of other deep-sea teleosts (Bathylagus, various stomiatoids, lantern-fishes, rat-tails, tripod fishes, etc.) are larger nucleated and flattened, the largest dimension ranging from 7 to 20 μm (Hansen and Wingstrand, 1960). These authors suggest that the functional significance of small, non-nucleated erythrocytes is in their high surface to volume ratio, which will ensure rapid exchange of gases, and in their increased oxygen-carrying capacity through a high content of haemoglobin, owing to loss of nuclear ballast. Moreover, despite their shortness the small bore of retial capillaries still provides, relative to the contained volume of blood, a large surface area for countercurrent functions (Marshall, 1972). There is also a high number of capillaries in the rete, which in Vinciguerria is about 5,000 a figure that may be compared to the 2,000 odd larger-sized capillaries in the longer retia (by 2.5) of the latern-fish, Myctophum punctatum (Marshall, 1960).

In lantern-fishes from lower mesopelagic levels instances of length of the retia mirabila are: Lepidophanes guentheri 3.0 mm, Hygophum benoiti 3.0 mm, Lampanyctus pusillus 3.0 mm, L. alatus 4-5 mm, Lampadena chavesi 6.0 mm, L. speculigera 7.0 mm (Fig. 4.6), Taanigichthys bathyphilus 7.0 mm. The greater retial dimensions of the last three species, whose depth ranges extend into the upper bathypelagic zone, should be noted.

The depth ranges of benthopelagic fishes with a gas-filled swimbladder extend from about 200 to 7,000 metres. Such species, as we will recall, are mainly rat-tails (Macrouridae), deep-sea cods (Moridae), brotulids, halosaurs, notacanths and

Depth ranges of benthopelagic deep-sea fishes and lengths of swimbladder retia mirabilia

Initial letter of family name to the left of each specific name: S. Synaphobranchidae; H. Halosauridae; Mo. Moridae; M. Macrouridae; B, Brotulidae; T. Trachichthyidae.)

Species	*No. of retia*	*Length of retial capillaries (mm)*	*Depth range where most (m)*	*depth range common*
1	*2*	*3*	*4*	*5*
M. Malacocephalus laevis	2	3.4	150-1315	(150-600)
M. Ventrifossa occidentalis	2	5-6	150-585	(300-500)
M. Coelorhynchus carminatus	4	6	90-850	(300-600)
M. Macrourus ber glax	4	6	100-1350	(200-600)
M. Hymenocephalus italicus	2	6	200-2080	(300-800)
T. Hoplostethus mediterraneus	2	5	235-780	
M. Coryphaenoides rupestris	4	8	160-2260	(500-1400)
M. Coelorhynchus occa	4	8-10	460-2220	(700-1100)
M. Nezumia aequalis	2	11	200-2320	(400-1000)
Mo. Antimora rostrata	4	8-9	400-29000	(800-1800)

1	*2*	*3*	*4*	*5*
S. Synaphobrachus kaupi	2	10	200-3600	(800-2000)
M. Bathygadus melanobranchus	2	10	640-1590	(800-1400)
M. Gadomus longifilis	4	12	630-2170	(800-1500)
M. Cetonurus globiceps	2	15	860-4260	(1100-1830)
M. Coryphaenoides guntheri	4	20	830-2600	(1000-2000)
B. Basssogigas profundissiumus	1	15	5610-7160	
B. Bassozetus compressus	1	17.5	1920-2750	
H. Halosaurus ovenil	2	c. 20	440-1620	(800-1300)
H.H. parvipennis	2	c. 20	780-1610	
H. Aldrovandia affinis	2	c. 20	840-2615	(1000-2500)
B. Bassozetus taenia	2	25+	4570-5610	
M. Nematonurus armatus	5	25+	280-5700	(2600-3600)
M. Lionurus carapinus	6	25+	1210-5300	(mainly below 2000 m)

synaphobranchid eels. The number of retia mirabilia, length of the retial capillaries and depth ranges of representatives from these families is shown in the table (Marshall, 1972).

It will be seen that the first four species, whose commonest depth ranges most closely correspond with those of upper mesopelagic fishes, have retia mirabilia of 4-6 mm in length (cf. mesopelagic species 1-2 mm). Species such as the rat-tails Coryphaenoides rupestris and Coelorhynchus occa, with commonest depth ranges at lower mesopelagic levels, have retia of 8-10 mm (cf. mesopelagic species 3-7). These contrasts in retial dimensions '. . . are most probably related to differences in the vertical spread of mesopelagic and benthopelagic fishes. The former, as echosounder records suggest, and as proper surveys of vertical distribution are revealing (see Badcock, 1970), have relatively narrow depth ranges. Such vertical disposition is presumably related to the coactive place of each species in a complex ecosystem, and is clearly of essential biological value in species that undertake daily vertical migrations. Benthopelagic fishes tend to range downward well beyond the lowest depth of their most common occurrence (see the table). Hence, it is suggested, their built-in safety factor in the shape of comparatively long retia mirabilia'.

The deepest living benthopelagic fishes, centred at levels below 1,500 metres have swimbladder retia from 15 to 25 mm in length. Thus in a series of mesopelagic and benthopelagic fishes ranging between depths of 200 metres and 7,000 metres there is a direct relationship between depth range and the length of the retia mirabilia in the swimbladder (Fig. 4.8). To recapitulate, the efficiency of the swimbladder, whether in maintaining or generating gas tension, depends directly on the length and interchange surface area of the retial capillaries. Both functions become harder with increasing depth and pressure; hence, the graded increase in retial capillary length in a series of species that range from shallow to deep reaches of the ocean. Lastly, we should note how well the theoretical maximum length (20 mm) of a rete (Kuhn *et al.*, 1963) fits the lengths (25 mm) found in the deepest living benthopelagic fishes.

Returning now to the mesopelagic fish fauna, some species when adult contain a relatively small swimbladder. In adult

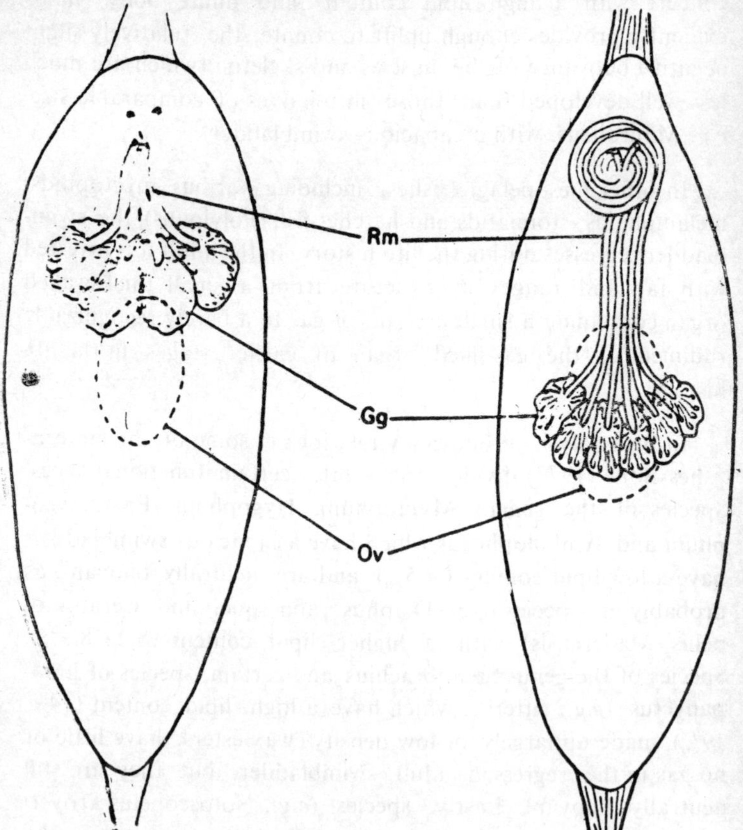

FIG. 4.8. Swimbladders of rat-tail fishes. Left, Coelorhynchus coelorhynchus, an upper slope-dwelling species; right, Lionurus carapinus, an abyssal species. Note the much longer retia mirabilia (Rm) of the latter. Gg, gas gland; Ov, outline of oval.

species of Cyclothone (braueri, alba, pseudopallida) the swimbladder has much the same position and relative size that it once had in the postlarval stages. It still contains gas; indeed, the length of the rete (3—4 mm) is comparable to that in fishes with well-developed swimbladders from the same levels. The Cyclothone are neurally buoyant and the swimbladder in concert with a high lipid content and dilute body fluids evidently provides enough uplift to counter the relatively slight negative buoyance of the muscles and skeleton (which are much less well developed than those in relatives of comparable size, *e.g.* Maurolicus, with a capacious swimbladder).

In other mesopelagic fishes, including various myctophids, melamphaids, stomiatids and hatchet-fish (Polyipnus), the swimbladder regresses during the life history. In the adult it is invested with far and ranges in structure from a small thick-walled organ containing a small amount of gas to a barely recognizable rudiment of the gas-filled organ of earlier stages in the life history.

In his analysis of buoyancy relations in some of the lantern-fishes Bone (1973) divides them into certain functional types. Species of the genera Myctophum, Hygophum, Protomycophum and Symbolophorus which have a capacious swimbladder, have a low lipid content (<5%) and are neutrally buoyant, as probably are species (*e.g.*, Diaphus rafinesquei and Ceratoscopelus Maderensis) with a higher lipid content (8.2—8.9%). Species of the genus Stenobrachius and certain species of Lampanyctus (*e.g.*, ritteri), which have a high lipid content (14—19%), made up largely of low density wax esters, have little or no gas in the regressed adult swimbladder, but they are still neutrally buoyant. Lastly, species (*e.g.*, Notoscopelus kroyeri and Tarletonbeania crenularis) also with little or no gas in the adult swimbladder have a low lipid content (3—5%) and, not surprisingly, are negatively buoyant.

Vertical Migration and Buoyancy

In numbers of individuals most of the fishes that commute daily between midwater and higher levels have some form of swimbladder. In both numbers of species and individuals the majority of the migrators are lantern-fishes, which are now considered.

If a neutrally buoyant species with a capacious gas-filled swimbladder (say, 4 per cent of the body volume) is to maintain its buoyancy at all levels the volume of the swimbladder must be kept more or less constant. When the lantern-fish descends after feeding in the surface layer the swimbladder will be compressed by the increasing pressure. Thus, during a descent from 100 to 500 metres the volume of the swimbladder, will decrease by about a fifth (if no gas is secreted). During an upward migration from 500 to 100 metres the swimbladder will undergo a five-fold increase in volume. Clearly, gas must be lost if the fish is not to become dangerously over-buoyant.

In lantern-fishes gas is lost through a thin-walled fore part of the swimbladder wall known as the oval. When fully exposed through the contration of its radial muscles the highly developed capillary circulation of the oval comes very close to the gases in the swimbladder. These will diffuse into the oval at a rate depending on the area of the capillary bed, the rate of its circulation, the pressure (the difference in tension between the swimbladder gases and those in the blood) and the temperature. Apart from the first two factors, there is no way of increasing the rate of gas resorption; in fact, the area of the capillary bed is not especially large compared to the volume of the swimbladder (in a lantern-fish the ratio between these two dimensions is about twice that in a freshwater perch Perca), but judging by the size of the venous vessels that drain the oval the capillary circulation must be very well developed (Marshall, 1960).

While deep-sea fishes that make extensive migrations have both a very well-developed gas-producing and gas-resorbing system, it seems unlikely that they can secrete and reabsorb gas quickly enough to stay neutrally buoyant at all depths. In most sound scattering layers Hersey and Backms (1962) found that the peak frequency of reverberation increases with depth during the course of a diurnal migration, and they continue, 'The relationship that would obtain were the swim bladder and its gas passively responding to changes in ambient pressure ($F/F_0=(P/P_1)^{5/6}$) has been observed during both a sunrise and sunset migration. This seems to imply that the retention of the swimbladder is worthwhile even though the fish is at neutral buoyancy only at the top of its depth range.' Certainly lantern-fishes when netted at the surface are neutrally buoyant.

Taking as a model a free spherical bubble rising at a rate such that its expansion due to decrease of pressure is exactly opposed by the decrease in volume due to gas resorption, D'Aoust (1970) has shown that for a spherical bubble of a size equal to or less than the volume of swimbladder normally found in mesopelagic fishes, the times theoretically needed to rise 200 and 300 metres are up to an order of magnitude less than those actually got from diurnal records of sound scattering layers. Thus, one secreted there need be no stress in removing gas at rates relevant to observed rates of vertical migration. The more difficult physiological task, as perhaps reflected by the reduction in size or regression of the swimbladder in numerous species, may well be the secretion of gas during and after descent. Moreover, even though the deepest living lantern-fishes (*e.g.*, Taaningichthys spp.) have the longest retia mirabilia in the swimbladder, they are non-migrators.

If the swimbladder of a migrating deep-sea fish falls short of its ideal hydrostatic function, a better alternative would be to replace gaseous lift by lipid lift. As implied already, certain lantern-fishes with a regressed swimbladder but with a lipid content high enough to give them neutral buoyancy should retain this advantage at all depths. This seems an ideal arrangement and makes one wonder why so many species retain a swimbladder. Evidently the provision of high quantities of wax esters is not for all species.

In any event, it is likely that most mesopelagic fishes without a gas-filled swimbladder are somewhat negatively buoyant. Concerning two common species, Gonostoma elongatum is often within 0.5 per cent and Xenodermichthys copei within 1.2 per cent of neutral buoyancy. Their fat content (about 3% of the wet weight) is not high, but this in concert with the greater lift given by their dilute body fluids, comes close to matching the relatively small negative buoyancy of their muscular and skeltal system (Denton and Marshall, 1958) (see also Fig. 4.9).

Bathypelagic fishes, as already stated, lack a gas-filled swimbladder. During their postlarval life in the surface layer Cyclothone spp. have a small gas-filled organ, but after metamorphosis and descent to the adult habitat the swimbladder regresses and is entirely without gas (Marshall, 1960). In a catch that includes both mesopelagic and bathypelagic species of Cyclothone many individuals of the former float whereas the latter sink to the bottom of the container. As stated elsewhere (Marshall, 1960), a fish with a capacious gas-filled swimbladder can carry a well-developed muscular system and a firm skeleton at neutral buoyancy. The simple organization of bathypelagic fishes, which fits so well with their food-poor living space, has necessarily involved the loss of the swimbladder. Again the

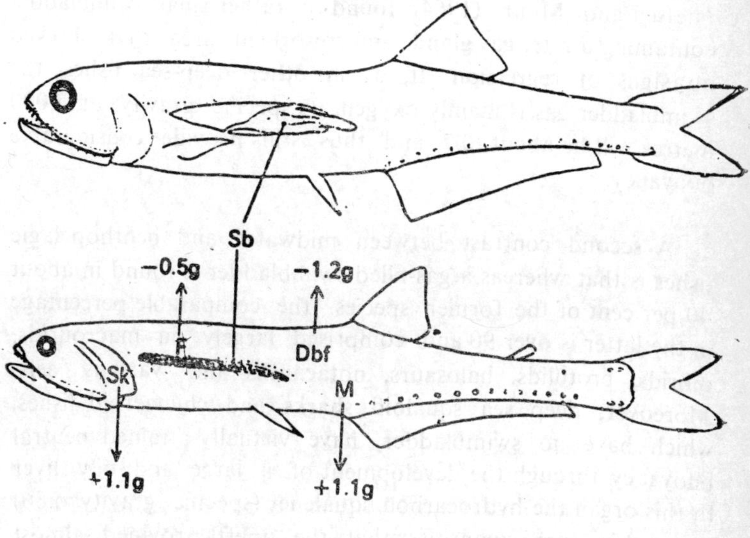

FIG. 4.9, Buoyancy balance sheets of two species of Gonostoma. G. denudatum (above) has a large gas-filled swimbladder (Sb) and is neutrally buoyant. In G. elongatum (below) the swimbladder is regressed but the uplift given by fats (F), and dilute body fluids (DBF) almost balances the two main heavier than seawater components, comprised by the skeleton (Sk) and proteins.

reduction of the muscular and skeleton system means that relatively little lift will bring them close to neutral buoyancy. For instance, while a female angler-fish (Melanocetus murrayi) was somewhat heavier than water one of comparable size in another species (Ceratias bolboelli) was very close to neutral buoyancy (personal observation).

The Buoyancy of Benthopelagic Animals

To recapitulate; while at midwater levels a gas-filled swimbladder occurs only in mesopelagic fishes, such an organ is present, and well developed, in benthopelagic species that range from slope to abyssal levels. In an individual of the brotulid fish Bassogigas profundissimus, which was trawled by the Galathea in the Sunda Trench at a depth of 7,160 metres, Nielsen and Munk (1964) found a rather small swimbladder containing a rete, gas gland and resorbent area that showed no signs of regression. If, as in other deep-sea fishes, the swimbladder gas is mainly oxygen, its specific gravity at 7,000 metres will be about 0.7 and thus still provide considerable buoyancy.

A second contrast between midwater and benthopelagic fishes is that whereas a gas-filled swimbladder is found in about 40 per cent of the former species, the comparable percentage in the latter is over 90 and comprised largely on macrourids, morids, brotulids, halosaurs, notacanths and various eels. Moreover, deep sea squaloid sharks and chimaeroid fishes, which have no swimbladder, have virtually attained neutral buoyancy through the development of a large and oily liver In this organ the hydrocarbon squalene (specific gravity o.86) is stored in such quantities that the uplift provided almost eliminates the fishes' weight in water (Corner, Denton and Forester, 1969) For instance, the shark Centroscymnus coelolepis, which weighed 5,260 gm in air, contrained a liver of 1,550 gm, most of which (82%) consisted of oily squalene. The buoyancy balance sheet, as calculated by Corner and his colleagues, showed that at the depth of capture (1,500 metres) the shark is likely to weigh a mere 9.1 grams. Lastly, benthopelagic teleost fishes lacking a smimbladder, notably alepocephalids and ateleopids, must also be close to neutral buoyancy, judging by their rather fragile skeletons and watery, loose-textured muscles.

It is clear that neutral buoyancy, or a close approach to such conditions, is biologically valuable in fishes that live near the deep-sea floor. Concerning the species with gaseous uplift, it is not simply that their ancestors endowed them with a swimbladder, but rather that they live in surroundings that provide sustenance enough to support the heavier tissues notably muscle and bone, that can be carried at neutral buoyancy by a gas-filled float. The best support for this conclusion involves midwater macrourids (*e.g.*, Cynomarurus and Odontomacrurus) and ophidioid fishes (some of the aphyonids that have evolved from benthopelagic ancestors. Both kinds, like many of the primary midwater fishes, have a regressed (or absent) swimbladder and reduced muscular and skeletal systems (Marshall, 1960, 1964).

Neutral buoyance is apparent also in benthopelagic cephalopods. After study of the literature and a fine series of deep-sea photographs. Roper and Brundage (1972) concluded that Cirroteunthis spp. and probably some of the other cirrate octopods as well are not only benthopelagic in habit but also close to being weightless in water. Perhaps in parallel with the fishes, the midwater Cirrothauma murrayi is more reduced than its presented benthopelagic ancestors.

Buoyancy Relations of Benthic Animals

Nearly all of the larger invetebrates that move over the continental shelf are plainly heavier than their environment. But the same is by no means true of some of their relatives on the deep-sea floor. As we saw elsewhere many of the elasipod sea-cucumbers are not much heavier than sea water. Thus, little effort is needed for them to rise above their resting place and be carried or move elsewhere. There is a parallel difference between shelf-dwelling and deep-sea rays. The latter, which have a large oily liver, are probably near to neutral buoyancy (Bone and Roberts, 1969). Even Bathysaurus, which from observations and photographs seems a classic type of benthic teleost, may not be much heavier than sea water. In a Bathysaurus agassizi (length 495 mm) the larger liver, which weighed about one-fifth of the total weight, consisted largely (70% by weight) of lipids with a specific gravity of 0.92 (R.J. Morris, quoted by Marshall and Merret, 1977). One's impression is that diverse members of the deep-sea benthos have acquired the means to rise easily above the bottom and thus move on to new feeding grounds.

The Biological Significance of Neutral Buoyancy

Our survey of buoyancy relations in deep-sea animals poses an outstanding question: what is the biological advantage of neutral buoyancy or its near attainment 'sought' not only by simply organized planktonic forms but also by muscular cephalopods and fishes? Indeed, there is evidence that the sperm whale, which spends part of its in deep-sea feeding grounds, has means of buoyancy control that involve the assumption of weightlessness.

Even so, we must first recall that many midwater reprerentatives of successful groups of animals, notably copepods, euphausiids, segestid prawns and arrow-worms, are without special buoyancy devices. Evidently, they are not so negatively buoyant that the effort needed to maintain height in the water column is a substantial drain on their total energy budget. The biological advantages of adequate buoyancy in the phytoplankton have been considered elsewhere. Until more is known of their buoyancy relations and life cycles little can be said of the midwater protozoans. Indeed, not enough is known of the slender energy budget of neutrally buoyant gelatinous forms, whether cnidarian or tunicates, to consider how much energy is saved by weightlessness in water. Perhaps the latter condition enables them to feed efficiently. Certainly, the proper setting of the deadly drift lines of siphonophores depends on appropriate assemblage of the over-buoyant members of the colony.

It is easier to consider the advantage of neutral (or near-neutral) buoyancy in nekronic animals. In a midwater fish one may consider the energy needed to maintain a level if the fish is deprived of its swimbladder. A marine fish with a well-developed gas-filled swimbladder will have a reduced weight in water or about 5 per cent of its weight in air. If it lacked such a swimbladder 'to stay at a constant level it would need to exert a downward force equal to its reduced weight (and unlike terrestrial animals, pelagic creatures can only exert forces on the surrouning medium by making movements), When it is remembered that even an active pelagic fish seldom exerts a force of more than 25—50 per cent of its weight in air for more than a very brief period, the force necessary for a fish to maintain its level appears to be considerable.'

During swimming a just buoyant nektonic animal saves energy in that virtually all the power developed by its locomotor muscles is concentrated in its forward motion. If an animal is heavier than water a considerable part of its propulsive power must go to maintaining its level. In most sharks, for instance, lift during swimming is generated mainly by the pectoral fins in concert with the asymmerical tail fin. But in buoyant deep-sea sharks the pectoral fins no longer look like hydroplanes and the tail fin approaches a symmetrical form (Marshall, 1971).

In midwater fishes and squids the energy needed to secrete gases or ammoniacal fluids is evidently much less than that saved through the uplift derived by either of these processes. Even though a gas-filled swimbladder may provide neutral buoyancy only during their main feeding period, it is surely striking that in numbers of individuals the majority of migrating midwater fishes contain such a float. These migrators, notably lantern fishes, may often have to work hard to obtain enough planktonic food. As they search for prey, neutral buoyancy will not only enable them to concentrate propulsive energy in forward motion, but give them easy manoeuvrability in the seizure of prey. Turning to numbers of species, it is equally striking that the majority have dispensed with a gas-filled swimbladder. But the reduction of muscle and skeleton associated with near neutral buoyancy is fitting also in surroundings that contain limited supplies of food. Such reduction, as we saw, is most marked in bathypelagic fishes. These fishes, like many mesopelagic species that also lack a swimbladder, do not undertake daily vertical migrations.

Midwater cephalopods, which are largely confined to mesopelagic levels, differ in their bouyancy relations. The octopods, unlike so many squids, have not developed ammoniacal means of uplift. They are rather fragile, gelatinous creatures with relatively little muscle, and one at least (Japetella) has ionic means of approaching neutral buoyancy. In their more active the ammoniacal squids, one may argue again that neutral buoyancy not only saves energy as they hover and move but also gives them fluid manoeuvrability in the capture of prey.

Near the deep-sea floor neutral buoyancy is more or less obligatory among the fishes. Those with a gas-filled swimbladder are predominantly elongated or rat-tail-like fishes with long, many-rayed median fins. They have slow deliberate ways, and as films also show, they are neutrally buoyant. At low and moderate speeds, according to Alexander's (1966) estimations, the energy saved through such buoyancy is considerable. Yet again, one thinks also of the postural versatility afforded by weightlessness in water—a versatility put to best use by the many species that feed on both benthic and free-swimming prey. After watching the unhurried ease and varied motions of rat-tails on films taken by baited cameras, it is easy to forget that neutral buoyancy is at the centre of their existence.

5

Migration of Insects

A few insects migrate seasonally over great distances rather like birds. Recently it has been learned that most other species also migrate by simply beating their wings while the wind carries them.

The literature of entomology, beginning with the Bible, records many astonishing accounts of the migration behaviour of insects. Locusts in the desert, mosquitoes in the Arctic and the Tropics and butterflies, moths, beetles, bugs and dragonflies almost everywhere have been seen in sudden mass flights, often involving millions of insects all travelling in the same general direction at the same time. Particularly in the case of the desert locust of Africa and the monarch butterfly of North America-insects that can make seasonal flights of more than 2,000 miles—this behaviour has been likened to that of migratory birds. The spectacular character of these flights has invited explanation in terms of a migration instinct that would cause the individual insects to congregate in response to overcrowding or to some other unfavourable change in the environment, and has inspired speculation about sensory mechanisms that would enable the individuals to orient themselves in space, navigate and hold course toward distant destinations. Migratory flight thus more or less elaborately defined became one hallm ark of a few species and much controversy has flamed from efforts to distinguish "true" migrators from mere drifters whose populations were dispersed by the wind.

It was closer acquaintance with the desert locust—established with the aid of photographic film interpreted by R.C. Rainey and Z.V. Waloff of the Anti-Locust Research Centre in London—that helped to set modern studies of insect migration on an entirely new and more fruitful course. Locusts are insect migrants per excellence. Vigorous fliers, they give an observer the strong subjective impression of concerted, purposeful progress in a direction under their own control. The camera too will show whole groups of insects headed in one direction. In successive images registered on the same film, however, individuals headed in one direction are seen to be carried in another direction by the wind. Within a swarm there will be many groups of similarly oriented insects flying in different directions, but the swarm as a whole will move with the wind, albeit more slowly.

The discovery that locust swarms are wind-borne eliminated the need to postulate mechanisms of orientation and navigation as essential features of insect migration. What is common to all migrants, as J.D. Kennedy of the Agricultural Research Council in Cambridge, England, has insisted, is an intense "locomotory drive" that sets the insect off on a relatively prolonged and undistracted flight compared with other kinds of flight associated with egg-laying or feeding. The drive arises from the internal physiology of the individual in a way not yet understood, but it is associated somehow, at least in females, with sexual immaturity. With the behaviour of individuals synchronized by the simultaneity of their development, the migratory flight seems to have evolved as an adaptation designed primarily to relinquish habitats destined to become unsuitable and to secure new ones. This often leads to dispersal of the species. In this view migratory behaviour appears as a key phase in the life history not of merely a few species of insects but of a large proportion of winged species. To the universal cycle of birth, reproduction and death must be added the process of migration or dispersal, with the winged adult as the essential, mobile element in the system.

For an understanding of migratory behaviour in a generalized form, the lowly plant-sucking aphid provides a most typical and instructive example. Aphids move on the wind. In spring and summer some species called the migratory aphids feed and deposit their larvae on herbs (these aphids are viviparous, delivering small larvae rather than eggs), and in fall and winter they transfer their activities to woody shrubs. These insects have been under intensive study by our group at the Rothamsted Experimental Station and by other workers at the Agricultural Research Council in Cambridge.

At the end of the larval stage the aphid molts into either a wingless or a winged adult. The wingless form stays where it is, but every new winged adult, after waiting for some hours until its cuticle hardens, takes off on a strong vertical flight, attracted by the blue light of the sky. It is caught by the wind and, still flying actively, may be carried hundreds or thousands of feet aloft and often many miles overland. Typically the locomotory drive begins to fail within an hour or two and the aphid starts its fortuitous descent. If the plant on which it lands is one to which it is adapted, the insect feeds for the first time since leaving its birthplace. Then it may begin to deposit larvae. If the flight has been short or if the plant is unsuitable, the aphid may take off again. It may make several flights and cover many miles in the few days before it is finally grounded by the dissolution of its wing muscles. The migratory flight thus spreads the species to new habitats and effects the seasonal transfer of the population from green herbs to woody shrubs, even though the direction and distance of travel depend in part on the wind.

All other migrants share some elements of this behavioural routine with the aphid. Characteristically, most insects make their migratory flight soon after completion of their metamorphosis and before they enter the reproductive phase of their life history. The mosquito Aedes taeniorhynchus, studied in

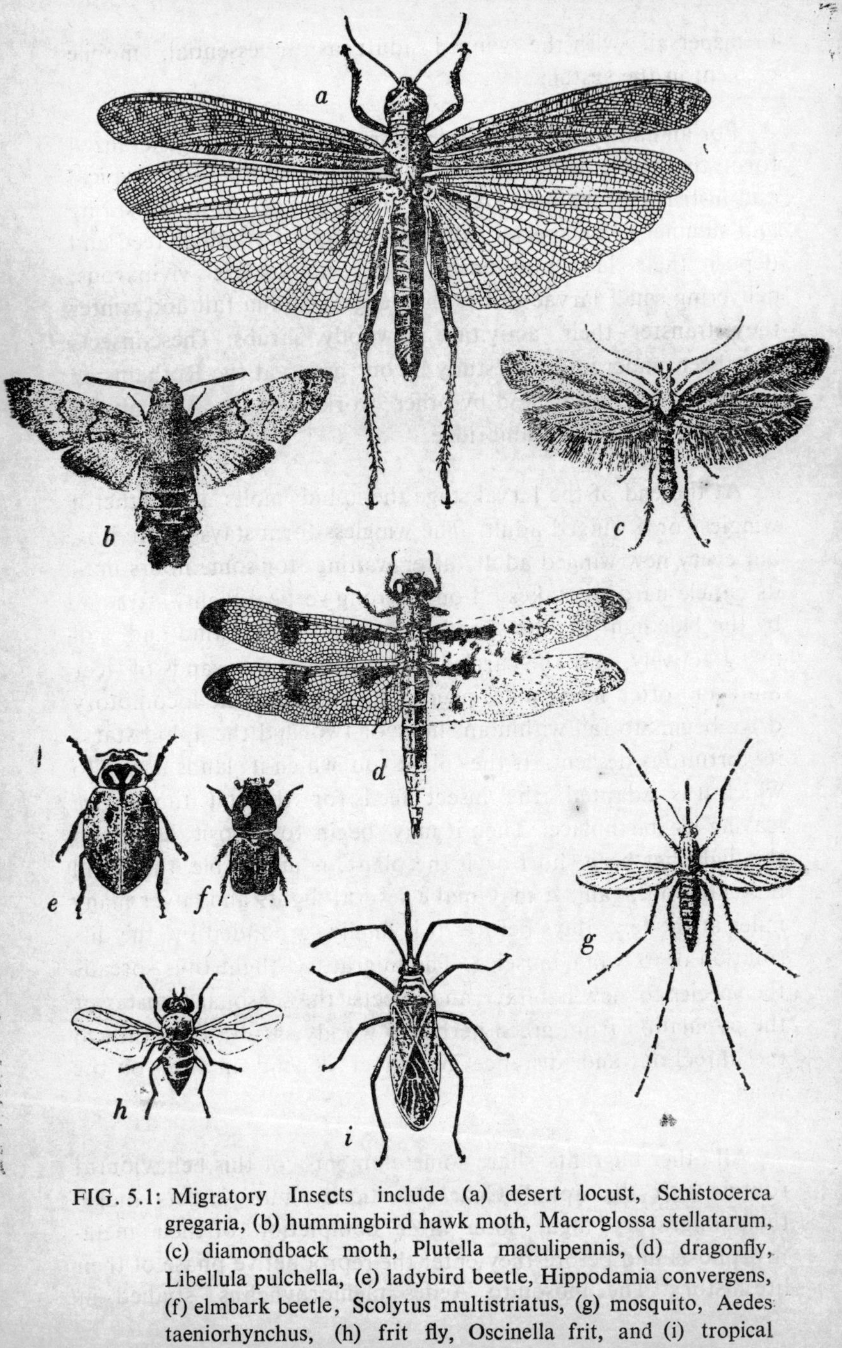

FIG. 5.1: Migratory Insects include (a) desert locust, Schistocerca gregaria, (b) hummingbird hawk moth, Macroglossa stellatarum, (c) diamondback moth, Plutella maculipennis, (d) dragonfly, Libellula pulchella, (e) ladybird beetle, Hippodamia convergens, (f) elmbark beetle, Scolytus multistriatus, (g) mosquito, Aedes taeniorhynchus, (h) frit fly, Oscinella frit, and (i) tropical cotton stainer, Dysdercus suturellus.

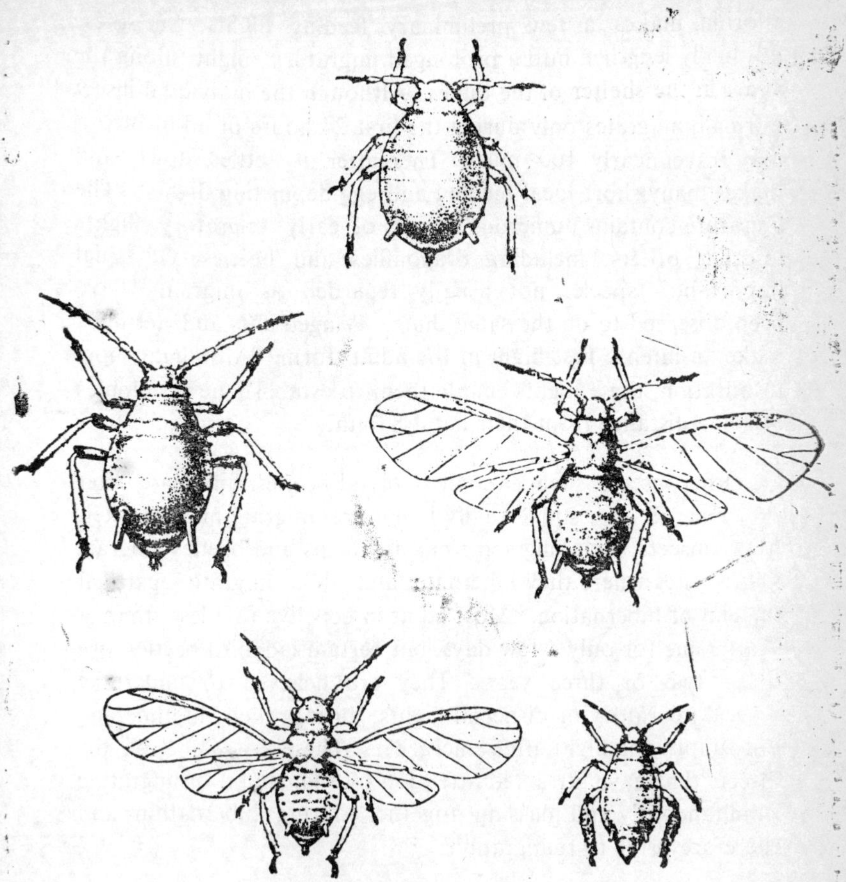

FIG. 5.2: APHIS FABAE, a migratory species of aphid, occurs in many forms during a year. The five shown here are the fundatrix (top), which hatches in spring from an overwintering egg and founds a new line; a wingless female that gives live birth to new larvae (centre left); a winged female that migrates and gives live birth (centre right); a male (bottom left) that mates with an egg-laying female (bottom right). Species changes host plants with season.

Florida by Maurice W. Provost, E.T. Nielsen and James S. Haeger, begins its migration with its first flight after emergence and may travel up to 20 miles in a direction strongly influenced by the wind. The butterfly Ascia monuste, also observed in

Florida, makes a few preliminary feeding flights, but these gradually lengthen into a prolonged migratory flight along the coast in the shelter of the dunes. Although the individual insect normally migrates only during the first 24 hours of adult life, it may travel nearly 100 miles. Thereafter it settles down and makes many short local feeding and egg-depositing flights. The literature contains numerous reports of early migratory flights in other orders, including dragonflies and beetles. Of equal importance, species not usually regarded as migrants have been observed to do the same thing. Winged ants and termites make an intense first flight in the adult form. Although short in duration, these flights enable them to establish new colonies at some distance from their natal habitat.

Other insects, such as some scolytid beetles, hibernate after they emerge and, still sexually immature, migrate months later. Many insects, including some moths, thrips and beetles, migrate first to sites where they hibernate and then migrate again at the end of hibernation. Most adult insects live for less than a year, some for only a few days, but certain ladybird beetles live up to two or three years. They are believed to undertake several migratory or dispersal flights during their lifetime and not simply one flight after emergence. These are some of the insects that have attracted attention by their habit of migrating simultaneously and massing together during hibernation and therefore prior to remigration.

In their migratory flight aphids do not excite as much wonder as the same behaviour does in the case of larger insects. They appear against the sky as tiny elements in the aerial plankton, part of the busy swarm of "gnats" that may lend visual discomfort to the heat of a summer's day. At Rothamsted we have developed suction traps to plot the diurnal rise and fall of the aphid population in the air. Suspended at measured heights from steel towers or from barrage ballons, these traps suck in the air and collect the insects in a tube. Disks are dropped into the collecting tube at preset intervals in order to divide the catch into fractions over periods of time. A count of the insects on each disk reveals the shifts and changes in the concentration of the air-borne population of aphids (and other insects) throughout the day.

By this technique we have found that aphid flights have a daily rhythm that frequently shows two peaks. We have also established the underlying mechanism of this cycle and thereby the explanation of the massing of the aphids in their migratory flight. The aphids that molt from the larval stage during the night are inhibited from flying by the cold and darkness. When the light and temperature rise above flight thresholds in the morning, the new adults that have accumulated all take off at the same time. Their exodus is recorded by our suction traps as a midmorning peak. Meanwhile the individuals whose maturation has been slowed by the low temperatures of the last hours before dawn mature more quickly as the day grows warmer. Their simultaneous departure yields another peak of migration in the early afternoon. By evening the fall in temperature and intensity of light from the sky brings a corresponding drop in the number of air-borne aphids. Since the temperature threshold for maturation is lower than that for flight, however, the next morning's crop of aphids now begins to accumulate. Trevor Lewis of Rothamsted has found that the mass migrations of another insect, the thrip, also occur in the warmup after a period of low temperature. Thus many mass insects flights must be regarded as gushes of newly emerged and waiting winged adults, released by a rise in temperature.

These observations can be matched to the recorded behaviour or many classical and well-recognized migrants. As long ago as 1880 S.B.J. Skertchly noted an early-morning emergence of thousands of painted butterflies (Vanessa cardui); within an hour they all flew together in the same direction. It is now known that such migrants as locusts, monarch butterflies, Aedes and Ascia similarly emerge together at their breeding sites and then depart on the simultaneous flight that is featured in the literature on insect migration. The analogy so often drawn with bird migration in these cases fails to take into account the fact that all the individuals are of the same age; in contrast, flocks of birds may contain adults of all ages, and they may have congregated, unlike insects, from different places.

The strength and direction of the wind play an important but not necessarily crucial role in determining how far wind-borne migrants will go. The contribution of the insect—the duration of its locomotory drive—constitutes an equally important variable, for most insects must beat their wings if they are not to fall out of the air. The desert locust, geared to travel great distances, refueling as it goes, spends most of its adult life migrating, after which it lays eggs and dies. In contrast, termites migrate for only a few minutes but spend months reproducing. The travel of aphias (at least of those with temporary wing muscles) is limited by their initial fuel (fat) supply; they cannot refuel during migration.

In laboratory studies of flight capacity some insects turn in extraordinary performances. For example, although aphids are weak fliers, A.J. Cockburn of Rothamsted has shown that some can continue to beat their wings for 16 hours nonstop, until all their fuel is exhausted. These intrinsic performances, however, bear little relation to distances travelled in nature. An analysis of daily flight rhythms of aphids at different altitudes at Cardington in England shows that the average flight lasts for only one to three hours. In elegant flight-chamber experiments by J.S. Kennedy, aphids permitted to fly freely stay up for about the same length of time. Height-density studies of aphids and other insects reveal that the general aerial population, at least over Britain, resembles a daily explosion, with millions of insects thrown up, often to great heights, followed by an almost complete settling-out by nightfall.

In contrast with the short, daylight-only flights of the insects in the aerial plankton over Britain, desert locusts have been known to travel across the sea for 24 to 60 hours, going as many as 1,400 miles in one hop. This greatly exceeds their records for tethered flights in the laboratory. R.A. French and J.H. White have reported an observation by a ship 1,000 miles at sea of millions of tiny diamondback moths. Apparently they had been going for at least two days non-stop, carried irresistibly by the wind at 20 miles per hour, beating their wings all the time in order to stay up.

The distance travelled overland does not necessarily reflect the duration of the insects' flight for another reason: the air currents may carry them upward as well as horizontally. In vigorus convection currents locusts form towering cumuli-form swarms thousands of feet high. When convection is low, the swarms become flat and remain near the ground. As would be expected, smaller insects are even more sensitive to air currents. In still, stable air crowds concentrate near the ground; this accounts for dense swarms observed at eye level on quiet summer evenings. In rising currents they go up for thousands of feet. There is an approximately linear inverse relation between the logarithms of the density of the air-borne population and altitude, as would be expected in plotting the turbulent diffusion of small particles. The slope of the curve, when density is plotted against height, correlates well with the degree of atmospheric stability and makes it possible to calculate from measurements of air currents the density of insects in particular altitude zones. These calculations show, for example, that in Britain in the middle of a typical summer day half of the population of frit flies (Oscinella frit, a cereal pest hitherto thought to fly only near the crop) is above 1,300 feet!

Travel in air currents is haphazard and many insects are lost in unfavourable places. Perhaps only a few reach a good site, but these are enough to perpetuate the species. Even this seemingly fortuitous system of movement can be highly adapted for survival, as Rainey has shown for the desert locust. Their migrations are associated with the movement of the convective air currents of the intertropical convergence zone, a region in which winds from the north and from the south converge, tend to ascend and so produce rain. As this zone moves across Africa and the Middle East with the change of seasons, the winds coming into it from the desert carry locusts from dry areas into wet areas. Locust eggs need moisture to survive and the larvae need vegetation when they hatch. Thus the apparent haphazard movements of the swarms are geared to their basic ecological requirements.

In all other species it can be anticipated that migratory behaviour will prove to be correspondingly adapted to placing them in the right habitat at the right time in their life cycles. Some insects, such as the monarch butterfly, migrate long distances to overwintering sites and so escape adverse seasonal changes in climate and food. The migratory aphid, by moving a short distance from herbs to woody plants as the season changes, is doing the same thing. Adults of other species merely scatter in successive generations, without a regular change of host or complete change of habitat. In every case, however, the species engages in migration or adapted dispersal (sometimes in both, since some classical migrants disperse en route) in order to relinquish a habitat that would eventually become untenable. T.R.E. Southwood of the Imperial College of Science and Technology in London has recently shown that species occurring in relatively temporary habitats produce more migrant adults than species living in more permanent situations. Many migratory dragonflies, for example, live in ponds that dry up periodically, whereas non-migratory species tend to inhabit rivers. Agricultural pests are wide-ranging and mobile because nearly all crops are temporary, annual plants. Hence pests are here today, gone tomorrow and where they come from often seems to be a mystery. It is not so mysterious, however, when one realizes that millions of individual insects of the aerial plankton rain down almost everywhere on the earth day after day.

Whereas adaptive advantages of migration are found in study of the ecological relations of each species, the cause of the behaviour must be sought in the individual insects. In some cases a currently unfavourable environment induces the insects to leave; froghoppers have been known to move from cut vegetation and locusts have been thought to respond to dryness. Nevertheless, most mass migrations do not seem to be in direct response to any current adversity. Monarch

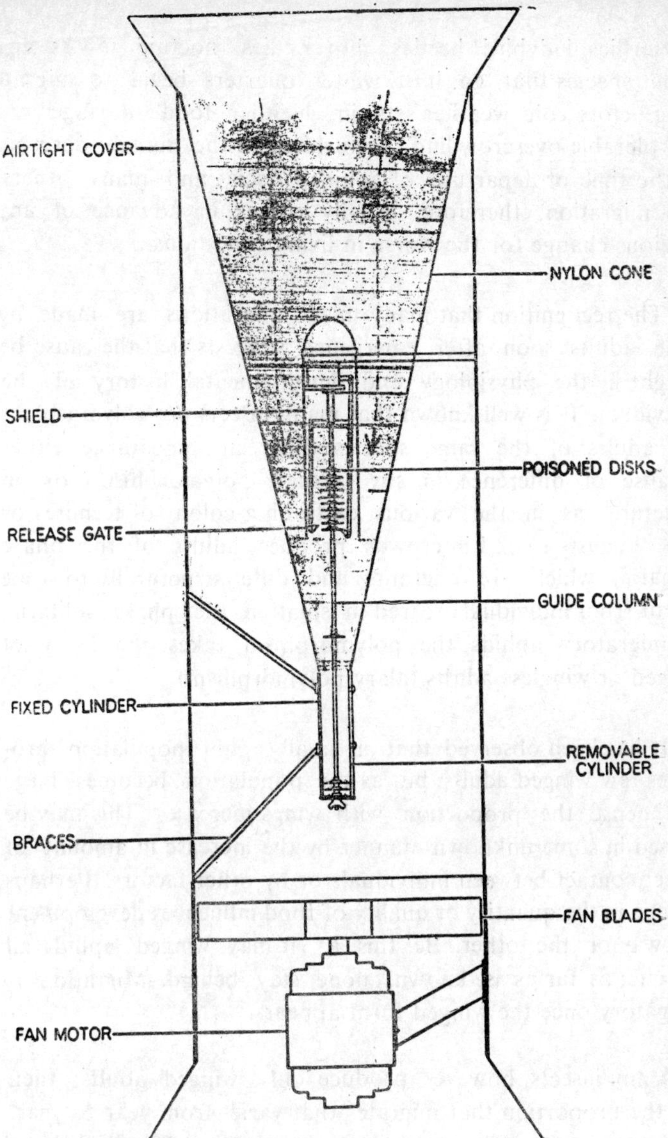

FIG. 5.3: Segregating Suction Trap pulls in air and insects (arrows). At preset intervals, poisoned disks drop, dividing catch. This gives a directmeasure of aerial density of insects at a given time. Various types of trap are suspended from posts, towers and balloons.

butterflies, ladybird beetles, hover flies, noctuid moths and other species that go into winter quarters begin to migrate long before cold weather sets in. Neither food shortage nor "intolerable overcrowding" are evident to the human obserber at the time of departure of aphids, Ascia and many others. The migration, therefore, usually comes in advance of any obvious change for the worse in living conditions.

The recognition that many mass migrations are made by fresh adults soon after emergence suggests that the cause be sought in the physiology and developmental history of the individual. It is well known that many insects are polymorphic: the adults of the same species differ in appearance either because of difference in sex (sexual polymorphism) or in structure, as in the various castes in a colony of termites or ants. Locusts reared in crowds produce adults of the phase gregaria, which are migrants and differ structurally to some extent from individuals reared in isolation, the phase solitaria. In migratory aphids the polymorphism takes the form of winged or wingless adults (alary polymorphism).

It has been observed that a small aphid population produces few winged adults, but as the population becomes large and dense the proportion with wings increases. This may be caused in some unknown manner by the increase in amount of direct contact between individuals or by other factors. Perhaps change in the quantity or quality of food influences development one way or the other. Be this as it may, winged aphids all migrate; as far as is known, none stay behind. Migration is obligatory once the winged form appears.

Many insects, however, produce only winged adults; then it is the proportion that migrates that varies from year to year, and some generations may have no migrants. This has been another mystery of migration. Yet perhaps the mystery is no deeper than that associated with aphid migrations. It is certainly not unreasonable to assume that if aphids produced winged individuals in response to the environmental pressures that induce migration, the insects that always produce winged adults can generate some individuals with a "behavioural polymorphism" that endows them with the locomotory drive necessary to migration. These polymorphs would appear when the ecological situation of the species demanded it, just as the winged aphids appear.

The principal physiological clue to the triggering of the locomotory drive is the finding that it is associated with delay in the development of ovaries and of the endocrines that control ovary development. Most migratory flights are made by sexually immature females. Among migratory insects it is always the females that engage in migration, because it is they that spread the species. Males present different and distinct problems since even in some migratory species the males never migrate.

It is known that several factors, such as crowding, a short day or poor food (particularly protein deficiency) all delay development of the ovaries. It is also known that many insects cease migrating when ovaries mature. It should be possible, therefore, to make experimental tests of whether or not changes in such factors as length of day, food and temperature produce the postulated behavioural polymorphism. If they do, this would bring us back to the old idea that migration and adaptive dispersal are induced by overcrowding, changing seasons or poor food, but that the mechanism is in the biochemical rather than the sensory responses of the individual insect.

Still awaiting investigation by entomologists are some of the real puzzles of nature that surround such highly specialized migrants as the monarch butterfly. How, for example, do vast numbers of different species make their way seasonally along well-defined routes, often in mountain passes? What are the factors that determine the orientation of migrants that fly at a certain height when it is relatively calm and so control their own direction?

It will be difficult to find the means of attacking such problems. On the other hand, recent progress in the study of migratory behaviour in insects has opened up equally interesting new questions to investigation. It remains to be determined, for example, if there is any general connection between the length of the period before egg-laying begins and the tendency to undistracted migration in different species. We must study the effect of length of day, larval crowding and quantity and quality of food on the length of the migratory period and also find out how these factors relate to the attraction of the sky that takes the migrating insect into the windy air. Finally, we should like to know how many species are in the aerial plankton by accident and how many by adaptation. I suspect that far more are there by adaptation than is commonly believed.

6

The Swimming Energetics of Salmon

Man has far surpassed the flight of birds in the air, but he has yet to match the hydrodynamic feats of a fish in the water. The swimming of fishes and whales shows an efficiency of propulsion, in terms of the energy expended, that is much beyond anything human designers have been able to create. Indeed, the efficiency of the aquatic vertebrates is so remarkable that it has been called "Gray's paradox," after the British zoologist who has stimulated an intensive study of it [see "How Fishes Swim," by Sir James Gray; Scientific American, August, 1957].

Among the most spectacular fish performances are those of the homing salmon of the Pacific Ocean. These renowned swimmers are born in the head-waters of rivers flowing into the Pacific, migrate down to the ocean to spend the major portion of their lives and eventually swim back to their birth-place to spawn. They frequently make their way upriver on journeys of 500 to 1,000 miles, against strong currents, over rapids and waterfalls and past artificial obstacles, such as dams and locks, that man has erected in these rivers. Except when they are blocked by a difficult passage the salmon swim steadily day and night without rest and without food. With their magnificent sense for hydrodynamic advantage, they complete the journey by the thousands with dramatic success.

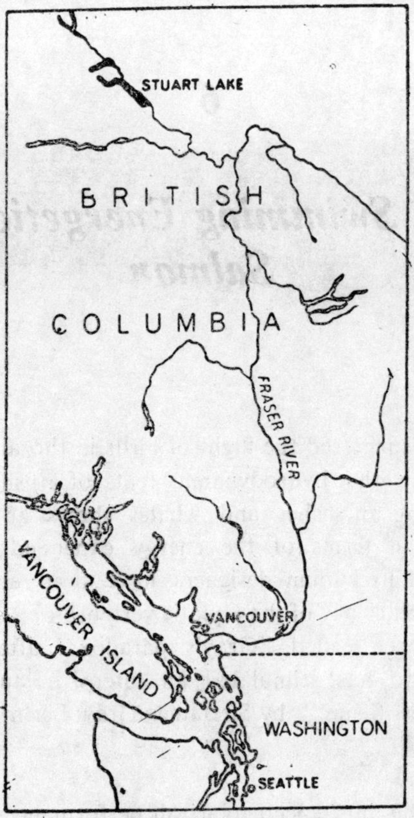

FIG. 6.1: Migration of sockeye salmon up the Fraser River from the Pacific to the spawning stream above Stuart Lake covers some 700 miles. It is one of the longer migrations.

Part of the story of how the salmon accomplish this can now be told. Because industrial development along the rivers threatens extermination of the salmon, a food resource that mancannot afford to sacrifice, systematic studies have been undertaken to determine what conditions are necessary to ensure the animal's survival. Exactly how much energy does the migrating salmon expend? What is the metabolic basis of its extraordinary performance and endurance? How is its ability to complete its marathon affected by temperature, its size and other factors? How fast can it swim? What is the secret of its hydrodynamic efficiency? Some of these questions, incidentally, are exactly those a naval architect must consider in designing a ship: the cruising range, the top speed, the most economical speed, maintenance problems, the nature of the waters to be traversed and so on.

In pursuit of answers to the questions about the salmon my colleagues and I made field studies of the fish in rivers and also measurements of its performance under strictly controlled conditions in the laboratory. The laboratory experiments were conducted at the Biological Station of the Fisheries Research Board of Canada at Nanaimo in British Columbia. For the experiments we developed two special kinds of apparatus. One is a respirometer: a water tunnel in which the fish's consumption of oxygen, indicating its rate of metabolism, can be measured while it is forced to swim at a given speed and at a given temperature of the water [see top illustration on pages 1946 and 1947]. Swimming speed is controlled by the water flow through the tube, against which the fish has to swim upstream; if it fails to match the velocity of the flow, it is forced back against a charged screen and receives a slight electric shock. The salmon quickly learned to avoid the shock and would swim at the enforced speed until they were exhausted. Measurements of a fish's rate of consumption of oxygen during a test can be obtained at intervals by shutting off new water temporarily and recirculating the enclosed water so that the depletion of its oxygen can be ascertained.

The second apparatus is an exercise cage, placed in a funnel-shaped structure, in which the fish can be made to swim at speeds up to 5.4 feet per second, or 3.7 miles per hour [see lower illustration on opposite page]. This device was employed mainly to observe the effects on the fish of sustained swimming for long periods under various conditions.

FIG. 6.2: ADULT MALE SOCKEYE is portrayed as it looks at about the time it begins its upriver migration. Such a be about two feet long and weigh about five pounds. In its almost ceaseless effort to migrate from the ocean to its fish may birthplace to spawn an adult bound for Stuart Lake progresses upstream at an average rate of about 1.33 miles per hour; its sustained speed must be about 2.65 m.p.h. because of the current. Sockeye salmon were the principal subjects of experiments conducted by author and his colleagues.

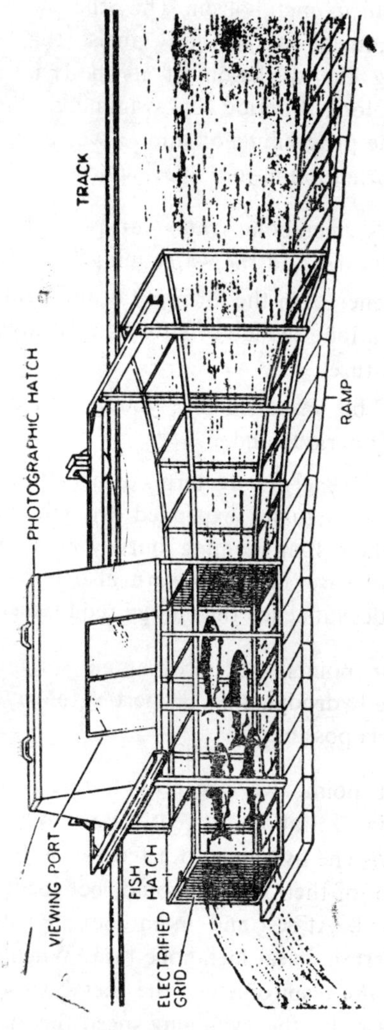

FIG. 6.3: EXERCISE CASE was devised by the author and his colleagues to observe the effect of long periods of swimming, comparable to the upriver migration for spawning. The fish's effective speed is determined by the velocity of the water which can be controlled by moving the entire apparatus forward or backward on the ramp. As depicted here the cage is in the position for maximum velocity. Salmon are induced to swim by the electrically charged screen behind them; they receivea mild shock if they touch it.

The experiments gave two independent measurements of the energy expended by salmon in swimming. One measure was oxygen consumption, showing the rate at which the fish generated energy through its metabolism. The other measure was the fish's consumption of stored body fuels; this was determined by measuring the depletion of its body fat and protein at the end of a prolonged period of swimming during which it ate no food. The programme of studies was designed to obtain quantitative information in four specific areas:

1. how the salmon's metabolic rate varies with its swimming speed, its size and the water temperature;
2. how much stored energy, in the form of fat and protein, a fish burns up in a long session of swimming at a given speed and temperature;
3. what speeds can be sustained by young and adult salmon over a long period; and
4. the total amount of energy, in terms of depletion of body fuels, that is actually expended by salmon of various sizes in their long homing journey up a river, the object in this case being to learn also how the river's characteristics influence the energy requirement.

In addition to these four points, our programme sought to learn something about the hydrodynamic properties that make the salmon's swimming feets possible.

To investigate the first point, we began by testing young sockeye salmon of a certain size (about 20 centimeters, or eight inches) in the respirometer. The fish were not fed for 36 hours beforehand so that none of their energy would be spent in digesting or absorbing food. At rest and in an unexcited state the young salmon had a certain basal metabolic rate. When the fish was forced to swim in the respirometer, the metabolic rate rose; with further increase in the swimming speed the metabolic rate mounted like a compound-interest curve, each gain in speed calling for a greater proportional increase in output of energy. Ultimately the curve peaked where the fish reached its maximum rate of oxygen consumption. The peak, called the active metabolic rate, represents the top speed the fish can sustain. It can swim faster by going into oxygen debt, but any such higher speed is limited to short bursts. The experiments showed that at low swimming speeds an excited or overactive fish uses more oxygen than a fish not under stress, but at high speeds the energy output of the excited fish is about the same as that of the unexcited one.

In these first tests the temperature of the water was set at 15 degrees centigrade (59 degrees Fahrenheit), which is about the average summer temperature of many of the rivers the salmon ascend. To what extent is the fish's expenditure of energy affected by temperature? Clearly there must be an effect. Rising temperature causes the chemical machinery of a cold-blooded vertebrate to turn faster, either at rest or under a work load. Moreover, a fish is compelled to put forth more effect to obtain oxygen at elevated temperatures, because water then holds less dissolved oxygen. To obtain enough oxygen the fish must flush more water through its gills, just as a mountain climber must breathe harder to obtain sufficient oxygen from the rarefied air at high altitudes.

Our young sockeye salmon were subjected in the respirometer to various temperatures from near zero degrees C. to 24 degrees (75 degrees F.), the highest the salmon can tolerate. Their basal metabolic rate was six times higher at 24 degrees C. then at five degrees. To put it another way, at 24 degrees a young salmon at rest burned as much oxygen as it did while swimming at about a mile an hour at five degrees. The effort required for swimming at the salmon's top sustained speed was a great deal higher at 15 degrees than at five degrees. It is apparent on the basis of temperature alone that the salmon's migration from the ocean up comparatively warm rivers must be a costly business.

When one considers the combined effect of temperature and the speed required to swim up these rivers, the salmon's accomplishment becomes impressive indeed. The young salmon proved

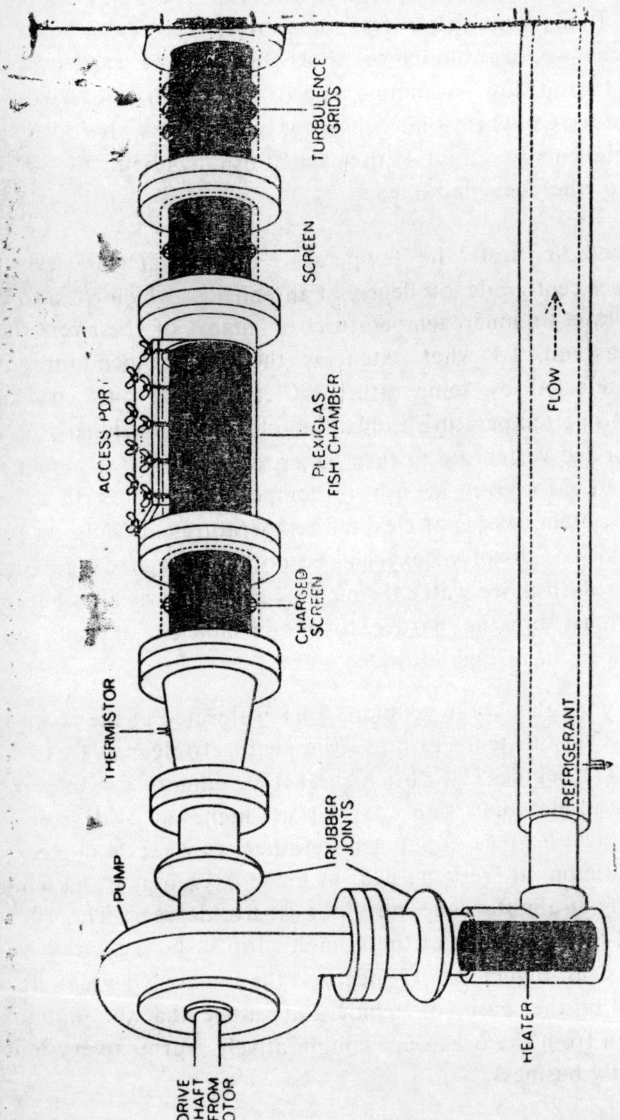
TURBULENCE GRIDS
SCREEN
ACCESS POR
PLEXIGLAS FISH CHAMBER
FLOW
CHARGED SCREEN
THERMISTOR
REFRIGERANT
RUBBER JOINTS
PUMP
DRIVE SHAFT FROM MOTOR
HEATER

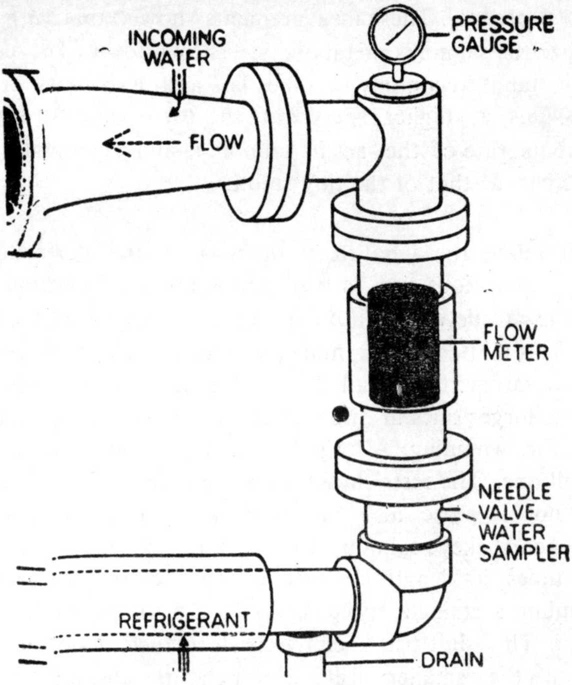

FIG. 6.4: RESPIROMETER APPARATUS is used to determine the metabolic rate of a salmon by measuring its oxygen consumption. The amount of energy required by the fish depends on the velocity of the water through the fish chamber. Water temperature, which aflects metabolism, can be controlled. Use of oxygen is found by recirculating water.

to be capable of increasing their energy output more than twentyfold in attaining their top sustained speed at optimum temperatures. The performance is comparable to that of some mammals; it had not been realized that a fish is capable of such metabolic increase. The salmon can truly be placed in a class with the racehorse.

In order to determine the relation of the metabolic cost to the size of the salmon, sockeye salmon of a wide range of sizes, from five-gram smolts to 1,500-gram (three-pound) adults, were tested in the respirometer with the temperature held constant at 15 degrees C. These measurements showed that with increasing size the salmon's metabolic rate slows down. The comparative metabolic economy becomes less and less, however, as the fish swims at higher speeds; at the top sustained speed the metabolic rate cf the adult salmon is almost as high (per unit of weight) as that of the tiny smolt.

To relate the laboratory findings to the performance of salmon of various sizes in their actual upriver migrations it was necessary to determine how the top sustained speed varies with size. This matter was examined by means of endurance tests in the laboratory, the small fish being tested in the respirometer and the lorger ones in the exercise cage, where they had more room for swimming. A large fish can of course swim faster than a small one. The tests showed, however, that the performance does not increase in proportion to the increase in size. An eight-inch sockeye salmon, for example, can maintain a speed of4.1 times its length per second, but a 24-inch salmon has a maximum sustained speed of only 2.5 times its length per second. The adult fish's relative performance is not as good as that of the smaller fish, although its absolute speed is considerably higher.

What is a salmon's total energy expenditure in its journey from the ocean upriver to its spawning ground? For an accurate answer to this question the most reasonable direct approach was to measure the fish's consumption of its store of body fuels. Two methods of measurement were pursued, one in the laboratory, the other in the field.

For the laboratory study adult sockeye salmon beginning their upriver migration were trapped in the river and taken to our laboratory. They were divided into an experimental group, which was to be put through a long swim in the exercise cage, and a control group, which was killed and quick-frozen immediately so that its content of stored fuel (fat and protein) could later be compared with that of the salmon depleted by swimming. The living were made to swim in the cage at 1.83

miles per hour for 12.7 days, the equivalent of covering a distance of 560 miles. At the end of this "journey" their remaining content of fat and protein was measured; the difference between this and the original amount, as indicated by the control group, gave a measure of the metabolic price the swimmers had paid for their swim. How much energy did this consumption of fuel represent? Calculations were made on the basis of the known energy yields from oxidation of fats and proteins by mammals. Although differences occur in the fish's metabolic products, the use of these assumptions proved to be valid when the results were compared with the rates of energy expenditure by salmon in the respirometer.

The field data on the same question came from intensive studies of the migration of sockeye salmon up the Fraser River in British Columbia, made by the Fisheries Research Board of Canada and the International Pacific Salmon Fisheries Commission. In 1959 D.R. Idler and W.A. Clemens followed a "wave" of sockeye salmon that journeyed up the river to Stuart Lake, about 640 miles from the ocean. These fish, most of them four years old, make the 640-mile trip in approximately 20 days. Their average rate of progress for the total journey therefore amounts to 1.33 miles per hour. To achieve that upriver movement they obviously must swim a great deal faster than 1.33 m.p.h. By conservative estimates the mean velocity of the Fraser River is about three miles an hour. If the salmon breasted this river velocity, to cover the 640 miles to Stuart Lake in 20 days they would have to swim at an average speed of 4.4 m.p.h.! A sustained speed of that order is impossible for the salmon. It must seek out paths in the turbulent river where the downstream flow offers less resistance. Considering the devious route the salmon must follow, there is no way to estimate its true speed or swimming effort from the physical statistics of its journey. The biological data collected by Idler and Clemens, however, make it possible to do so.

By the end of its journey, after it has laid its eggs in a stream above Stuart Lake, the female sockeye salmon has consumed 96 per cent of its body fat and 53 per cent of its protein reserves. Part of the female's energy has been spent in development of its large ovary. Idler and Clemens were able to establish, however, that the average daily output of energy by male and female salmon in swimming up the Fraser was 43 calories per kilogram of body weight. When we plot this output on the chart showing the results of tests of a full-grown salmon's capabilities, it becomes apparent that the Fraser River migrants were spending energy at nearly 80 per cent of the maximum rate they could maintain, leaving little margin for any emergency demands on them. Their rate of energy expenditure corresponds to swimming at an average speed of 2.65 m.p.h. Since swimming straight against the current in the Fraser would call for a speed of 4.4 m.p.h., this means that by skillful maneuvering the salmon reduced the requirement by about 1.7 m.p.h.

According to computations based on the performance of salmon in the respirometer, the most efficient swimming speed for a typical adult sockeye salmon is 1.1 m.p.h.; at this velocity the ratio of the speed obtained to the energy spent is optimal. At sea tagged salmon migrating toward shore have been found to travel an average of 30 miles per day. Assuming that they swim in the same general direction for most of the 24 hours, this indicates that in the ocean they would be swimming at somewhere near the most efficient speed. To make their way upriver against the river currents, however, the salmon must swim at higher, less efficient speeds. At an average of 2.65 m.p.h.—the speed maintained by the migrants up the Fraser River—the adult salmon must sacrifice nearly half of the possible efficiency in use of energy.

Having explored the metabolic aspects, we went on to look into the hydrodynamic attributes that help to account for the salmon's remarkable swimming feats and endurance. The first step was to measure the drag force on the streamlined shape of the fish in the water. For this measurement the model was a freshly killed salmon that was held at the end of a thin rod in the flowing water in the respirometer, like a model in a wind tunnel. The drag force, measured by a spring connected to the rod, was found to vary with velocity according to standard hydrodynamic relations, This procedure made it possible to express the drag in terms of the power required to overcome it at a given velocity, and therefore the drag on the dead model could be compared with that on a live fish swimming at the same speed.

A swimming fish, propelling itself by undulating movements of its body, of course is a different object from a fixed, motionless model as far as drag is concerned. The swimmer leaves remarkly little turbulence in its wake, which indicates that the flow around its body is more laminar, or smoother, than that around a nonoscillating body of the same streamlined shape. Our measurements showed that the drag on the swimming salmon was indeed considerably less than that on the dead model, except at low velocities. The difference increased right up to the maximum sustained speed, where the power required for a swimming fish to overcome drag was less than half that for the dead model. At speeds beyond the maximum that the fish could sustain—that is, at speeds attainable only in short bursts—the swimming fish lost its hydrodynamic advantage.

In short, the salmon's hydrodynamic efficiency is greatest at its top sustainable speed. This helps to counteract the high metabolic cost of swimming at that speed. Evidently the evolution of the salmon has endowed it with a hydrodynamic perfection that tends to minimize the energy it must spend for its extraordinary swimming performance. It appears that the best prospects for further improvement of the performance lie in evolution of increased efficiency in the salmon's metabolic yield of energy.

Evolution has carried the salmon very far. It remains for man to take pains to avoid destroying the fish's chances of surviving and flourishing. By unravelling the secrets of the salmon's hydrodynamic efficiency man may also learn some lessons of benefit to his marine technology.

7

How Fishes Swim

The submarine and the airplane obviously owe their existence in part to the inspiration of Nature's smaller but not less attractive prototypes—the fish and the bird. It cannot be said that study of the living models has contributed much to the actual design of the machines; indeed, the boot is on the other foot, for it is rather the machines that have helped us to understand how birds fly and fish swim [see 'Bird Aerodynamics," by John H. Storer; Scientific American Offprint 1115]. But engineers may nevertheless have something to learn from intensive study of the locomotion of these animals. Some of their performances are spectacular almost beyond belief, and raise remarkably interesting questions for both the biologist and the engineer. In this article we shall consider the swimming achievements of fishes and whales.

Looking at the propulsive mechanism of a fish, or any other animal, we must note at once a basic differencer in mechanical principle between animals and inanimate machines. Nearly all machines apply power by means of wheels or shafts rotating about a fixed axis, normally at a constant speed of rotation. This plan is ruled out for animals because all parts of the body must be connected by blood vessels and nerves: there is no part which can rotate freely about a fixed axis. Debarred from the use of the wheel and axle, animals must employ levers, whipping back and forth, to produce motion. The levers are the bones of its skeleton, hinged together by smooth joints, and the source of power is the muscles, which pull and push the levers by contraction.

The chain of levers comprising a vertebrate's propulsive machine appears in its simplest form in aquatic animals. Each vertebra (lever) is so hinged to its neighbours that it can turn in a single plane. In fishes the backone whips from side to side (like a snake slithering along the ground), in whales the backbone undulates up and down. A swimming fish drives itself forward by sweeping its tail sidewise; as the tail and caudal fin are bent by the resistance of the water, the forward component of the resultant force propels it. As the tail sweeps in one direction, the front end of the body must tend to swing in the opposite direction, since it is on the opposite side of the hinge, but this movement is usually small—partly because of the high moment of inertia of the front end of the body and partly on account of the resistance which the body offers to the flow of water at right angles to its surface. Thus the head end of the fish acts as a fulcrum for the tail, operating as a flexible lever.

At the moment when the tail fin sweeps across the axis of propulsion, it is travelling rapidly but at a constant speed. During other phases of its motion the speed changes, accelerating as the tail approaches the axis and decelerating after it passes the axis. The whole cycle can be regarded as comparable to that of a variable propeller blade which periodically reverses its direction of rotation and changes pitch as it does so.

How efficient is this propulsion system? Can the oscillating tail of a fish approach in efficiency the steadily running screw propellers that drive a submarine, in terms of the ratio of speed to applied power? To attempt an answer to this question we must first know how fast a fish can swim. Here the biologist finds himself in an embarrassing position, for our information on the subject is far from precise.

As in the case of the flight of birds, the speed of fish is a good deal slower than most people think. When a stationary trout is startled, it appears to move off at an extremely high speed. But the human eye is a very unreliable instrument for judging the rate of this sudden movement. There are, in fact, very few reliable observations concerning the maximum speed of fish of known size and weight. Almost all the data we have are derived from studies of fish under laboratory conditions. These are only small fish, and in addition there is always some question whether the animals are in as good athletic condition as fish in their native environment.

With the assistance of a camera, a number of such measurements have been made by Richard Bainbridge and others in our zoological laboratory at the University of Cambridge. They indicate that ordinarily the maximum speed of a small freshwater fish is about 10 times the length of the fish's body per second; these speeds are attained only briefly at moments of great stress, when the fish is frightened by a sudden stimulus. A trout eight inches long had a maximum speed of about four miles per hour. The larger the fish, the greater the speed: Bainbridge found, for instance, that a trout one foot long maintained a speed of 6.5 miles per hour for a considerable period.

It is by no means easy to establish a fair basis of comparison between fish of different species or between different-sized members of the same species. Individual fish—like individual human beings—probably vary in their degree of athletic fitness. Only very extensive observations could distinguish between average and "record-breaking" performances. On general grounds one would expect the speed of a fish to increase with size and with the rapidity of the tail beat. Bainbridge's data suggest that there may be a fairly simple relationship between these values: the speed of various sizes of fish belonging to the same species seems to be directly proportional to the length of the body and to the frequency of beat of the tail—so long as the frequency of beat is not too low. In all the species examined

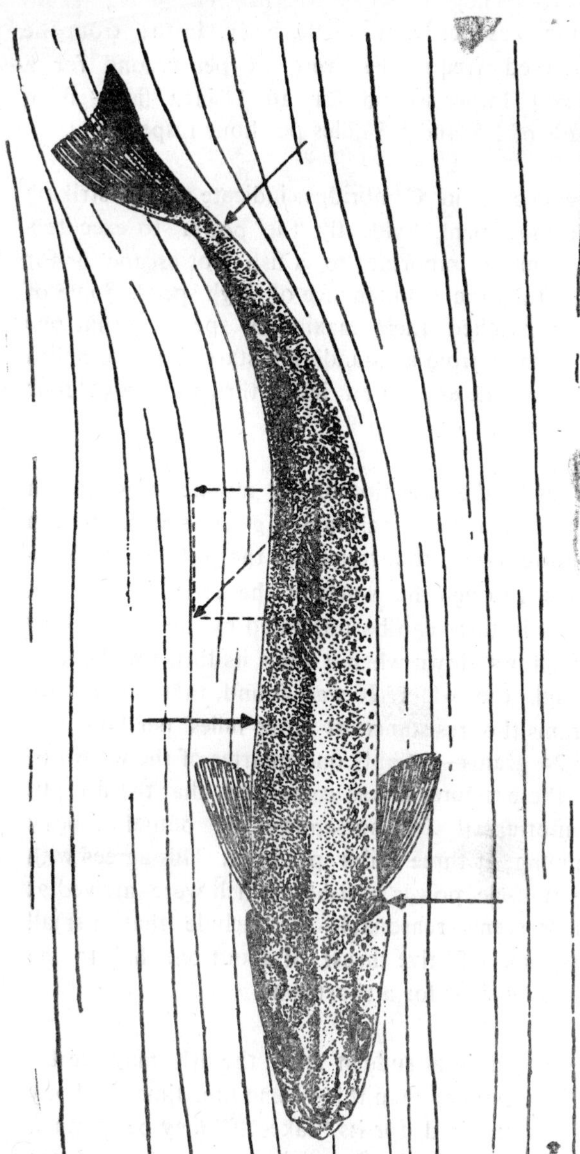

FIG. 7.1: How Medium Exerts Force on the fish is indicated by the arrows on this drawing of a trout. As the tail of the fishmoves from right to left the water exerts a force upon it (two diagonal arrows). The forward component of this force (heavy vertical arrow) drives the fish forward. The lateral component (broken horizontal arrow) tends to turn the fish to the side. This motion is opposed by the force exerted by the water on sides of the fish.

the maximum frequency at which the fish can move its tail decreases with increase of length of the body. In the trout the maximum observed frequencies were 24 per second for a 1.5-inch fish and 16 per second for an 11-inch fish—giving maximum speeds of 1.5 and 6.5 miles per hour respectively.

The data collected in Cambridge indicate a very striking feature of fish movement. Evidently the power to execute a sudden spurt is more important to a fish (for escape or for capturing prey) than the maintenance of high speed. Some of these small fish reached their maximum speed within one twentieth of a second from a "standing" start. To accomplish this they must have developed an initial thrust of about four times their own weight.

This brings us to the question of the muscle power a fish must put forth to reach or maintain a given speed. We can calculate the power from the resistance the fish has to overcome as it moves through the water at the speed in question, and the resistance in turn can be estimated by observing how rapidly the fish shows down when it stops its thrust and coasts passively through the water. It was found that for a trout weighing 84 grams the resistance at three miles per hour was approximately 24 grams—roughly one quarter of the weight of the fish. From these figures it was calculated that the fish put out a maximum of about .002 horsepower per pound of body weight in swimming at three miles per hour. This agrees with estimates of the muscle power of fishes which were arrived at in other ways. It seems reasonable to conclude that a small fish can maintain an effective thrust of about one half to one quarter of its body weight for a short time.

As we have noted, in a sudden start the fish may exert a thrust several times greater than this—some four times its body weight. The power required for its "take-off" may be as much as .014 horsepower per pound of total body weight, or .03 per pound of muscle. The fish achieves this extra force by a much more violent maneuver than in ordinary swimming. It turns its head end sharply to one side and with its markedly flexed tail executes a wide and powerful sweep against the water—in short, the fish takes off by arching its back.

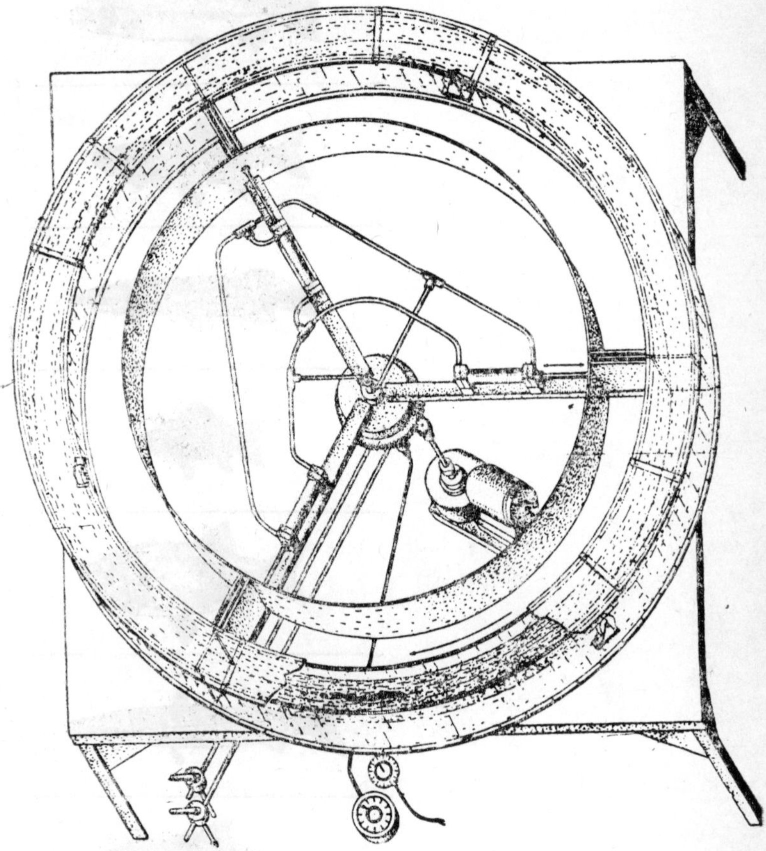

FIG. 7.2: Speed of Fishes was Measured in this apparatus at the University of Cambridge. The fish swims in a circular trough which is rotated by the motor at right centre. The speed of the trough is adjusted so that the swimming fish is stationary with respect to the observer. The speed of the fish is then indicated on the speedometer at bottom. Above the speedometer is a clock. When the apparatus is started up, the water is made to move at the same speed as the trough by doors which open to let the fish pass.

FIG. 7.3: Speed of Fishes is Listed in this table. The speed of the first five fishes from the top was measured in the laboratory; that

Species	*Length (feet)*	*Maximum Observed Speed (feet Per Second)*	*(Miles Per Hour)*	*Ratio of Maximum Speed to Length*
Trout	.656	5.552	3.8	8.5
	.957	10.427	6.5	11
	.301	5.229	3.6	17.8
Dace	.594	5.552	3.8	9
	.656	8.812	5.5	13.5
Pike	.529	6.850	4.7	13
	.656	4.896	3.3	7.5
Goldfish	.229	2.301	1.5	10.3
	.427	5.552	3.8	13
Rudd	.730	4.240	2.9	6
Barracuda	3.937	39.125	27.3	10
Dolphin	6.529	32.604	22.4	5
Whale	90	33	20	.33

of the barracuda, dolphin and whale in nature. The barracuda is the fastest known swimmer. Whale in the table is the blue whale.

This sort of study of fishes' swimming performances may seem at first sight to be of little more than academic interest. But in fact it has considerable practical importance. The problem of the salmon industry is a case in point. The seagoing salmon will lay eggs only if it can get back to its native stream. To reach its spawning bed it must journey upriver in the face of swift currents and sometimes hydroelectric dams. In designing fish-passes to get them past these obstacles it is important to know precisely what the salmon's swimming capacities are.

Contrary to popular belief, there is little evidence that salmon generally surmount falls by leaping over them. Most of the fish almost certainly climb the falls by swimming up a continuous sheet of water. Very likely the objective of their leap at the bottom of the fall is to pass through the fast-running water on the surface of the torrent and reach a region of the fall where the velocity of flow can be negotiated without undue difficulty. The brave and prodigious leaps into the air at which spectators marvel may well be badly aimed attempts of the salmon to get into the "solid" water!

A salmon is capable of leaping about six feet up and 12 feet forward in the air; to accomplish this it must leave the water with a velocity of about 14 miles per hour. The swimming speed it can maintain for any appreciable time is probably no more than about eight miles per hour. Accurate measurements of the swimming behaviour of salmon in the neighbourhood of falls are badly needed—and should be possible to obtain with electronic equipment.

At this point it may be useful to summarize the three main conclusions that have been reached from our study of the small fish. Firstly, a typical fish can exert a very powerful initial thrust when starting from rest, producing an acceleration about four times greater than gravity. Secondly, at times of stress it can exert for a limited period a sustained propulsive thrust equal to about one quarter or one half the weight of its body. Thirdly, the resistance exerted by the water against the surface of the moving fish (*i.e.*, the drag) appears to be of the same order as that exerted upon a flat, rigid plate of similar area and speed. Fourthly, the maximum effective power of a fish's muscles is equivalent to about .002 horsepower per pound of body weight.

Such is the picture drawn from studies of small fishes in tanks. It has its points of interest, and its possible applications to the design of fish-passes, but it poses no particularly intriguing or baffling hydrodynamic problems. Recently, however, the whole matter of the swimming performance of fishes was given a fresh slant by a discovery which led to some very puzzling questions indeed. D.R. Gero, a U.S. aircraft engineer, announced some startling figures for the speed of the barracuda. He found that a four-foot, 20-pound barracuda was capable of a maximum speed of 27 miles per hour! This figure not only established the barracuda's claim to be the world's fastest swimmer but also prompted a new look into the horsepower of aquatic animals.

A more convenient subject for such an examination is the dolphin, whose attributes are somewhat better known than those of the barracuda. (The only essential difference between the propulsive machinery of a fish and that of a dolphin, small relative of the whale, is that the dolphin's tail flaps up and down instead of from side to side.) The dolphin is, of course, a proverbially fast swimmer. More than 20 years ago a dolphin swimming close to the side of a ship was timed at better than 22 miles per hour, and this speed has been confirmed in more recent observations. Now assuming that the drag of the animal's body in the water is comparable to that of a flat plate of comparable area and speed, a six-foot dolphin travelling at 22 miles per hour would require 2.6 horsepower, and its work output would be equivalent to a man—of the same weight as the dolphin—climbing 28,600 feet in one hour! This conclusion is so clearly fantastic that we are forced to look for some error in our assumptions.

Bearing in mind the limitations of animal muscle, it is difficult to endow the dolphin with much more than three

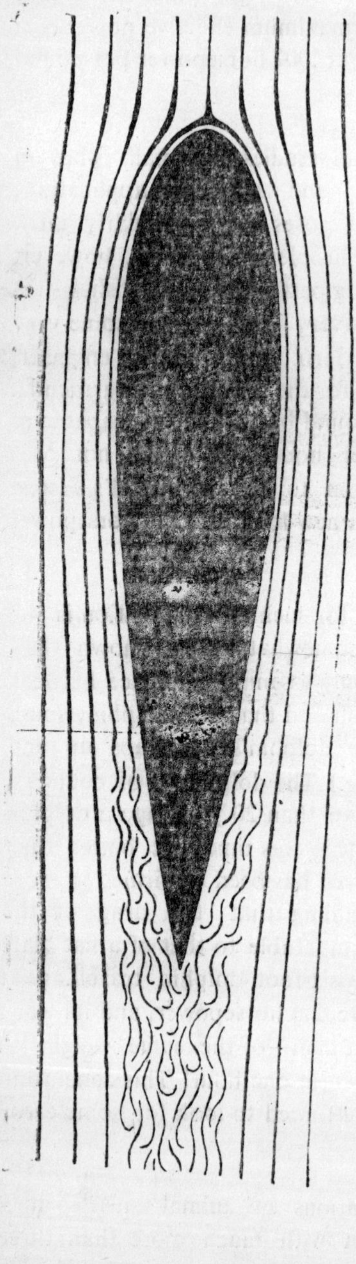

FIG. 7.4: Laminar and Turbulent Flow are depicted in this diagram of a streamlined body passing through the water. The smooth lines passing around the body indicate laminar flow; the wavy lines, turbulent flow.

tenths of one horsepower of effective output. If this figure is correct, there must be something wrong with the assumption about the drag of the animal's surface in the water: it cannot be more than about one-tenth of the assumed value. Yet the resistance could have this low value only if the flow of water were laminar (smooth) over practically the whole of the animal's surface—which an aerodynamic or hydrodynamic engineer must consider altogether unlikely.

The situation is further complicated when we consider the dolphin's larger relatives. The blue whale, largest of all the whales, may weigh some 100 tons. If we suppose that the muscles of a whale are similar to those of a dolphin, a 100-ton whale would develop 448 horsepower. This increase in power over the dolphin is far greater than the increase in surface area (*i.e.*, drag). We should therefore expect the whale to be much faster than the dolphin, yet its top speed appears to be no more than that of the dolphin—about 22 miles per hour. There is another reason to doubt that the whale can put forth anything like 448 horsepower. Physiologists estimate that an exertion beyond about 60 or 70 horsepower would put an intolerable strain on the whale's heart. Now 60 horsepower would not suffice to drive a whale through the water at 20 miles per hour if the flow over its body were turbulent, but it would be sufficient if the flow were laminar.

Thus we reach an impasse. Biologists are extremely unwilling to believe that fishes or whales can exert enough power to drive themselves through the water at the recorded speeds against the resistance that would be produced by turbulent flow over their bodies, while engineers are probably equally loath to believe that laminar flow can be maintained over a huge body, even a streamlined body, travelling through the waterat 20 miles per hour.

Lacking direct data on these questions, we can only speculate on possible explanations which might resolve the contradiction. One point that seems well worth re-examining is our assumption about the hydrodynamic form of the swimming animal. We assumed that the resistance which the animal (say a dolphin) has to overcome is the same as that of a rigid body of the same size and shape moving forward under a steady propulsive force. But the fact of the matter is that the swimming dolphin is not a rigid body: its tail and flukes are continually moving and bending during each propulsive stroke. It seems reasonable to assume, therefore, that the flow of water over the hind end of the dolphin is not the same as it would be over a rigid structure. In the case of a rigid model towed through the water, much of the resistance is due to slowing down of the water as it flows past the rear end of the model. But the oscillating movement of a swimming animal's tail accelerates water in contact with the tail; this may well reduce or prevent turbulence of flow. There is also another possibility which might be worth investigation. When a rigid body starts from rest, it takes a little time for turbulence to develop. It is conceivable that in the case of a swimming animal the turbulence never materializes, because the flukes reverse their direction of motion before it has an opportunity to do so.

It would be foolish to urge these speculative suggestions as serious contributions to the problem: they can only be justified insofar as they stimulate engineers to examine the hydrodynamic properties of oscillating bodies. Few, if any, biologists have either the knowledge or the facilities for handling such problems. The questions need to be studied by biologists and engineers working together. Such a cooperative effort could not fail to produce facts of great intrinsic, and possibly of great applied, interest.

Movements of Adult Fish

Adult plaice fitted with 300-kHz transponding acoustic tags (Mitson & Storeton-West, 1971) and tracked in the open sea by sector-scanning sonar (Voglis & Cook, 1966; Mitson & Cook, 1971) show a pattern of semi-diurnal vertical movement related to the tides, which we have called selective tidal stream transport (Greer Walker, Harden Jones & Arnold, 1978).

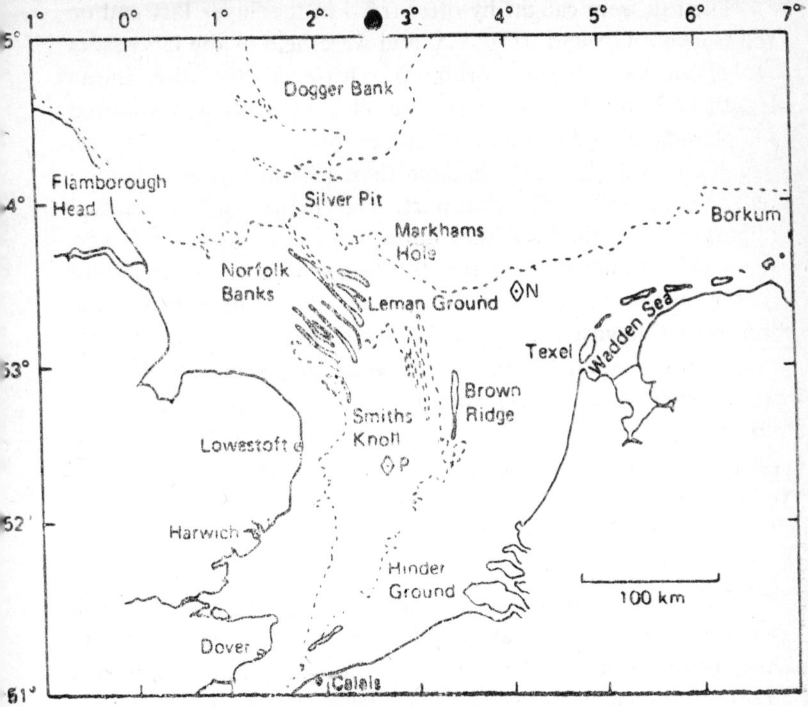

FIG. 7.5. Place names in the Southern North Sea of interest in connection with plaice migrations. The positions of tidal stations N and P are indicated by diamonds. The 30-m depth contour is indicated.

The fish were tracked within an area bounded by latitudes 51°40′ N and 52°40′ N, longitudes 1°40′ E and 2°30 E (Fig. 8.4). The tides in this area are fast, with speeds up to 1.4 m s^{-1} at springs, and highly directional, flowing approximately north-northeast and south-southwest. The tide floods on the south-going stream and the tidal range is 3 m in water of 20-40 m depth. The bottom deposits are mainly fine to medium sand with beds of gravel in some places and patches of sand waves (wavelength 2-3 m) in others. The water is generally turbid and visibility is limited to a few metres.

The fish were caught by otter trawl in the Silver Pits and on the Borkum Ground (Fig. 7.5) and were held in the laboratory for about two months prior to release. Twelve fish (mean length 41.3 cm; S.D. 4.6 cm) were released singly and followed for periods of 25-55 h and distances of 9-61 km. Eight fish which covered distances greater than 15 km showed a consistent pattern of tidal transport. The made regular vertical migrations, coming off the bottom at one slackwater and remaining in midwater for the following 5-6 h before returning to the bottom at the next slack water. When in midwater the fish moved downsteam with the tide for distances of 12-16 km, more or less equal to that of the tidal displacement. Activity was limited when the fish were on the bottom and some fish did not move for 2-3 h. The regular cycle of movements on and off the bottom at slack water was very striking and this pattern of semi-diurnal vertical migration was characteristic of all fish that moved appreciable distances.

The track of plaice no. 7 was typical (Fig. 7.6). The fish came up into midwater after high water slack, gained ground to the north during the north-going tide and returned to the bottom at low water slack. The speed of the fish over the ground was 80, 68 and 105 cm s^{-1} for the three northerly tides compared with only 8 and 9 cm s^{-1} for the two south-going tides.

During this, and some other tracks, water-current measurements were made from a second research ship or with moored current meters. From these measurements it was possible to estimate the speed of the fish through the water. It rarely exceeded one body length per second (1 L s^{-1}) and, while fish no. 7 appeared to swim downstream with the tide, three other fish swam through the water heading consistently southeast as they were being carried northwards over the ground. The orientation was maintained for periods up to one hour and may well have been determined with reference to sandwaves before the fish left the bottom. More recent observations, where the heading of the fish is returned to the research ship by telemetry, are discussed by Dr. Harden Jones in his chapter of this book.

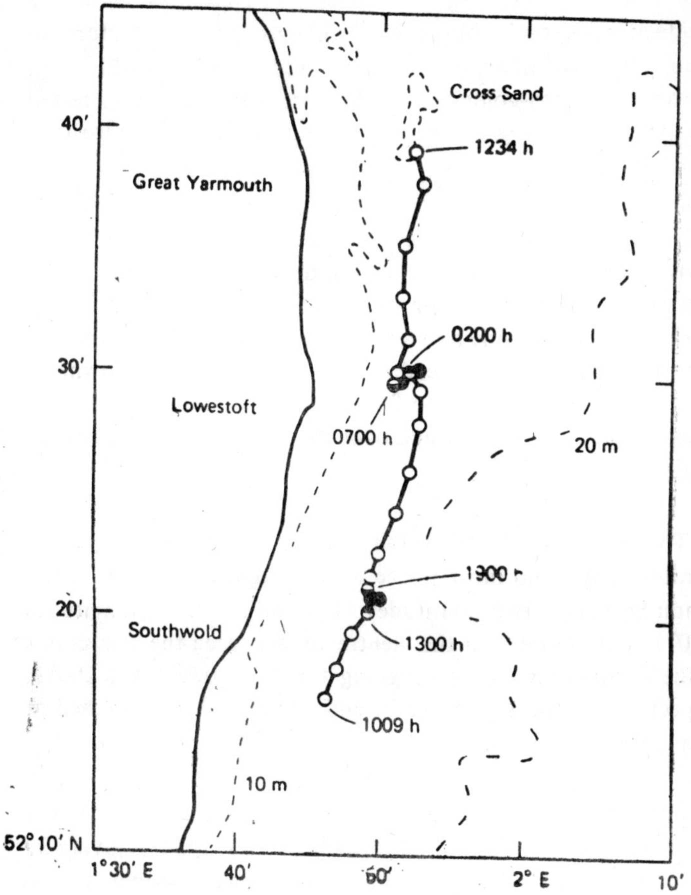

FIG. 7.6. Track chart of plaice no. 7, released at 1009 h, 12 December 1971. Hourly positions of the fish are indicated and the times of slackwater are given. Oven circle, north-going tide; circle with upper segment filled, low water slack; filled circle, south-going tide; circle with lower segment filled, high water slack. The 10 m and 20 m depth contours are shown.

The Spawning Migrations

If plaice used selective tidal stream transport when on migration, ripening fish moving towards their spawning grounds in the Southern Bight during November and December should be in midwater on south-going tides, whereas spent fish returning to their feeding areas in February and March should be in midwater on north-going tides. This prediction has been tested by midwater trawling experiments along the line of the western migration route.

The experiments were carried out with large midwater trawls. Paired tows, usually of 3 h duration, were made on successive tides and the gear was shot so as to start fishing at the required depth one and a half hours before the expected time of maximum tidal speed. Each haul was made downtide along the tidal axis and, so far as was possible, the research vessel covered the same track over the ground on each tow of a pair.

The work was carried out on the western side of the Southern Bight and most plaice were caught in the line of the 'Knoll Stream' between latitudes 52°N and 52°40′ N, longitudes 2°00′ E and 2°30′ E. At the centre of this area the directions of the north-going and south-going tides are 020°T and 200°T, respectively, peak speeds of 70 and 120 cm s^{-1} are reached at neap and spring tides.

Fifty poured tows were made in November and December ('autumn' pairs) and 22 in February (winter' pairs) giving 39 and 21 pairs, respectively, in which plaice were caught in one or both tows. More plaice were caught by night than by day but this diurnal effect was removed by restricting the comparison to the 17 'autumn' and 8 'winter' pairs in which both tows were completed during the night. There were no pairs in which both tows were completed during the day. During the prespawning migration significantly more fish were caught on the south-going tide, during the postspawning migration significantly more fish were caught on the north-going tide (Table 7.1).

In the 'autumn' most of the female plaice were still maturing and none was yet ripe, whilst half the males were already 'running ripe' and extruding sperm. No spent fish were caught in 'autumn' but three-quarters of the females caught in February had completed spawning. Virtually none of these fish had been feeding.

These midwater trawling experiments provide direct evidence that in November and December ripening plaice migrating into the Southern Bight by the western route use tidal steam transport from at least as far north as the Leman Ground. The paired tows made in February show that spent plaice return by the same transport system, at least as far as Smiths Knoll.

Diurnal Vertical Movements on Spawning and Feeding Grounds

Plaice kept in aquaria under natural illumination show a clear rhythm of diurnal vertical migration, feeding on the bottom during the day and swimming freely of the bottom at night (Verheijen & de Groot, 1967). During most off the year catches of plaice in bottom trawls in the Southern Bight are lower by night than by day with an average day night catch ratio of 1.3 : 1.0. But on the spawning grounds, in January and February, this ratio is reversed (de Groot, 1971). Plaice are quite commonly caught well above the bottom in midwater trawls and it is generally assumed that the reduction in catch at night is partly or mainly the result of a diurnal vertical migration, whereby at night a proportion of the population is no longer accessible to gear on the seabed. The reversal of the catch ratio during the spawning season may perhaps be attributed to an alteration or even a reversal of this behaviour.

Recent experiments (G.P. Arnold, M. Greer Walker & F.R. Harden Jones, unpublished results) have confirmed a diurnal pattern of vertical migration in plaice on their feeding grounds in the vicinity of Markhams Hole (Fig. 7.5). For

TABLE 7.1

Catches of Plaice by Night in a Midwater Trawl on Consecutive North-going (N) and South-going (S) Tides in the Southern North Sea from November 1974 to February 1977

Months	*No. of paired tows*	*No. of plaice caught on each tide*		*Catch rate on each tide*		*Paired t-test*		
		N	*S*	*N*	*S*	*t*	*d.f.*	*P*
Nov. Dec.	18	35	149	1.9	8.3	4.42	17	<0.001
Feb.	8	84	11	10.5	1.4	4.36	7	<0.01

example, one fish (plaice no. 2, RV CLIONE cruise 6/79), which was tracked for 82 h (Figs. 7.7 and 7.8), made four vertical excursions into midwater. It left the bottom each might at, or shortly after, sunset and swam freely in the water column for periods of 1 – 3 h. Apart from an initial period of eight hours after release, movement was restricted to the periods when the fish was in midwater. It covered a total distance of 26.5 km at an average ground speed of 0.3 km h^{-1}. This was five times less than that of plaice no. 7, whose track has been described above, and which covered an overall distance of 43 km in 26 h by tidal transport. The tides in the Markhams Hole area run northwest and southeast and have peak speeds of 35 and 57 cm s^{-1} at neaps and springs respectively. These speeds are about half those encountered off the East Anglian coast.

Movements of Juvenile Fish

Selective tidal stream transport is probably an innate pattern of behaviour in the plaice, occurring as it does with other rheotactic behaviour at metamorphosis, immediately after the larvae have gone to the seabed. It has been demonstrated for larvae entering the Wadden Sea in March and April through the Marsdiep inlet between Texel and Den Helder. Catches of larvae in an Issacs—Kidd midwater trawl were consistently higher on the flood than on the ebb tide and the catch ratio increased progressively with the age of the fish. Less than 20 per cent of fully metamorphosed Stage 5 larvae were found in midwater on the ebb tide compared with over 65 per cent of the early Stage 4 larvae, which are only just beginning to metamorphose (Creutzberg, Eltink & van Noort, 1978).

After spending their first summer in the Wadden Sea the O-group plaice move out to sea again in the autumn to over-winter offshore. Some return again as I-group fish in the following February and March. These fish also use selective tidal stream transport, moving landwards through the Marsdiep and Vliestroom inlets on the flood tide (de Veen, 1978).

As both O- and I-group fish the plaice in the Wadden Sea ɥsow a tidal feeding migration, collecting in the creeks at low tide but moving up onto the mudflats as they are covered by the rising tide (Kuipers, 1973). Similar tidal migrations occur on sandy beaches (Gibson, 1973).

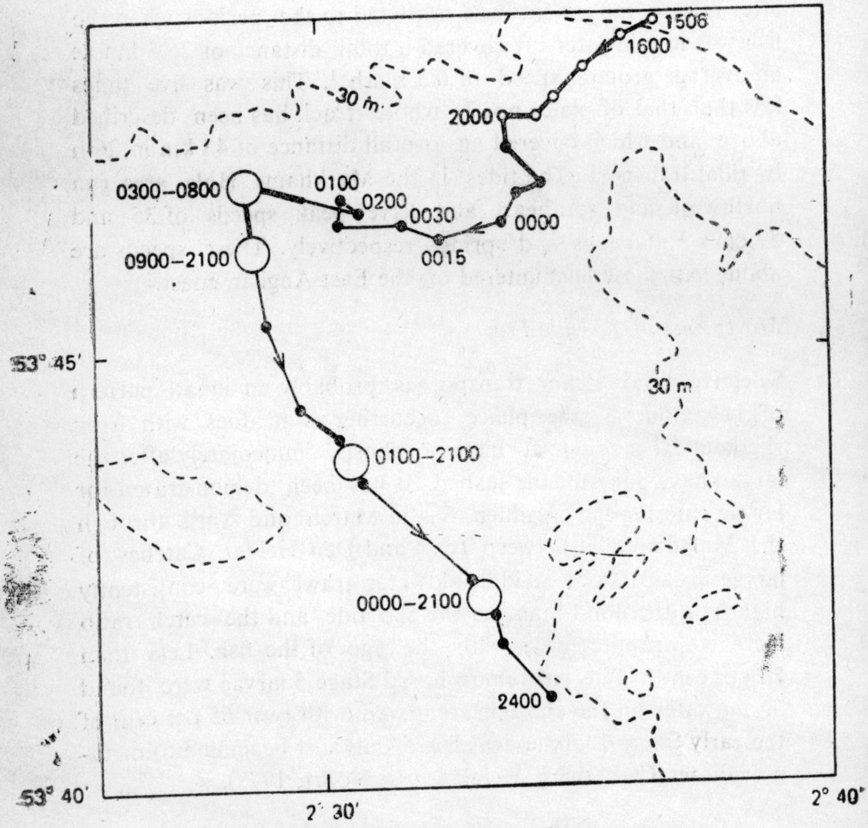

FIG. 7.7: Track chart of plaice no. 2 (RV CLIONE cruise 6/79) released at 1506 h, 27 May 1979. During the day the fish remained within the area indicated by the large open circles. Hourly positions of the fish when it was moving are indicated by the smaller open (day) and closed (night) circles. The portion of the track when the fish was moving on the bottom is indicated by the heavy line. The 30 m depth contour is shown.

Behavioural Problems

The extensive use by the plaice throughout its life history of tidal currents for transport raises a number of behavioural problems. For local movements within a feeding area, direction is probably unimportant and a diurnal vertical migration, occurring independently of the tide, will probably achieve adequate dispersal. But direction—and thus choice of tide—must be of paramount importance when plaice enter or leave a nursery area, or when they migrate between feeding and spawning grounds. The fish must then be able at the outset to 'lock-on' to the appropriate tide.

Newly metamorphosed fish show a diurnal rhythm of activity until they enter the intertidal zone (Gibson, Blaxter & de Groot, 1978). It is only then that a semi-diurnal tidal activity cycle appears and this rapidly reverts spontaneously to the basic diurnal, or circadian, rhythm in the absence of appropriate entraining stimuli (Gibson, 1973, 1976). A fish joining the tidal transport system must switch from a diurnal to a semi-diurnal pattern of vertical migration and it seems that this switching ability must be a basic feature of the plaice's life history. Hydrostatic pressure cycles are at least partially responsible for entraining circatidal cycles of activity in the intertidal zone (Gibson *et. al.*, 1978) but it is not known what the entraining stimuli are in the open sea. Clearly, though, if the fish is to move in the direction appropriate to its migration, the entraining signal must be unambiguous and there must be a distinction between high and low water slack tides. The south-going tide which, in the Southern Bight, transports the ripening fish south to their spawning grounds and the recently metamorphosed fish onto their nursery grounds, is preceded by low water slack. The north-going tide, which returns the spent fish to their feeding grounds, is preceded by high water slack. One possible entraining stimulus for the ripening fish is the conjunction of sunset, full moon and low water slack, which occurs for 2—3 days every month (Harden Jones, 1979). Cessation of feeding might also be a contributory factor. Starvation appears

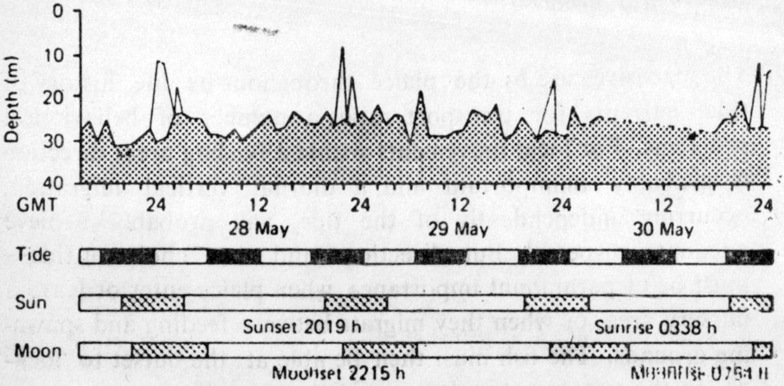

FIG. 7.8: The depth of plaice no. 2 (RV CLIONE cruise 6/79) in relation to the direction of the tide and the time of sunset. The solid bars indicate a tide flowing to the southeast; the open bars a tide flowing to the northwest.

to be an essential precursor of tidal transport in recently metamorphosed larvae. It is not known how these larvae distinguish between ebb and flood tides but, unlike elvers and crabs, it is apparently not in response to changes in odour or salinity.

8

Migration in Fishes

Among the various interesting peculiarities of fishes, migration is perhaps the most remarkable and fascinating. The word "Migration" has been derived from the Latin word (Migrare, meaning to go from one place to another) and is used to represent long journeys undertaken particularly by animals such as fishes and birds.

The term "migration" is frequently employed to express three different kinds of movements. In the early stages of life the eggs and fry may be carried by the currents some distance from the place where they were born. This helpless drift is at first involuntary, but later the young fry are quite able to make their way in and out of currents or remain stationary. Then the young fishes and the adults spend most of their time seeking food; they scatter about within their area; but good feeding grounds or incoming currents rich in plankton food bring many together in a concentration or shoal which persists for some time. These may be called voluntary movements on the part of the fish.

Not all fishes are migratory. In fact, a majority of them restrict their movements to small territorial limits and do not go out of these during any part of their lives. Few species, however, travel over long distances for feeding and spawning either in the same type of water to which they belong or in search of water naturally different than their own, *e.g.*, from freshwater to sea water or *vice versa.*

Rightly considered, the true migration is something very different from the drift and wanderings. Here the fish is urged to leave its ordinary hunting grounds and even to stop feeding, in order to undertake sometimes long and perilous journeys. The call of the blood, a deeper necessity common to all living things, takes possession of the fish, and the thousands and millions in like case forgather year by year almost at the same place and time—a circumstance of great importance both in the past and the present.

Most of the great fisheries are based on the spawning migrations of fishes, and at the present day we can observe how fishing population move from place to place and even from one country to another in pursuit of their calling. History tells us also, how the New Englanders would hardly have surmounted the hardships they encountered in the first years of their endeavours to settle in America but for the rich fisheries of the coast; and again, how the Dutch fishing population spread from one country to another in the seventeenth century, and how England and Holland went to war on a small matter of the fishing industry. In ancient times also the Greek and Phoenician colonists chose their settlements where the fishing was of importance, and the nature of the fishing is sufficiently indicated by the representation of the Tunny on the old medals of Carthage and Cadiz. These under-currents in the life of nations are not apparent in the surface histories, yet Edward Forbes was justified in his quaint comment, "the existence of Alexander may have been determined by the migration of a shell-fish." It is a question whether the leading or the feeding of peoples is of the greater importance.

Heape (1931) has defined migration as "a class of movement which impels migrants to return to the region from which they have migrated."

Migration can be said to typed as following:

(*i*) *Alimental Migration*

For the need of food and water.

(*ii*) *Gameiic Migration*

For reproduction and development.

(*iii*) *Climatic Migration*

For securing suitable climatic condition.

(*iv*) *Osmoregulatory Migration*

For better watery conditions to live as for osmoregulation is concerned.

A fish can make migratory movements by several methods.

A. By Drifting

The fishes are carried passively by water currents. This may result in 'directional movement' if the overall water movement is in one direction.

B. Random Locomatory Movements

Random movements in any direction may lead to distribution or aggregation. If fishes move uniformly from one point to all direction, it is called dispersal, which results in distribution.

C. Oriented Swimming Movements

The fishes swim in a particular direction (i) either towards or away from the source of stimulation, or (ii) at some angle to an imaginary line between them and point of stimulation.

The migration pattern is believed to be related to the currents. The young ones drift with the water current to the nursery grounds. The spawning migration is against with the current. Two terms are used to describe the movements of fishes in relation to water currents. Swimming or migration alongwith the direction of current is called denatant and swimming or migrating against the water current, and the migration of the adult fishes towards the spawning area is called contranatant.

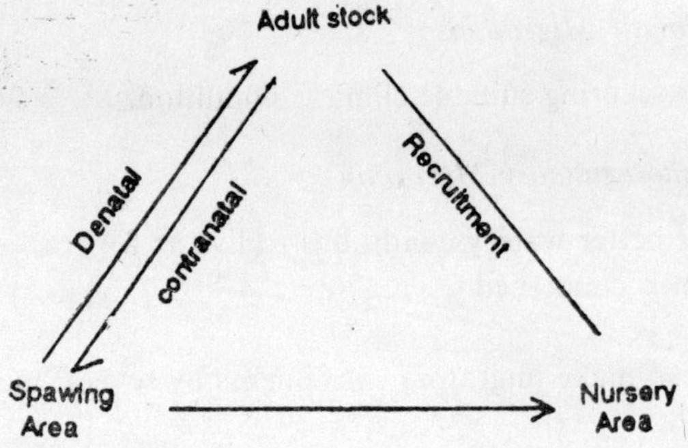

FIG. 8.1: Pattern of fish migration.

Types of Migratory Fishes

(i) *Anadromous.* Those fishes which spend a major part of their lives in the sea but migrate to freshwater during breeding season for spawning. Salmon and Hilsa have been found to travel several thousand miles in the sea and then several hundred miles inland to reach the spawning grounds.

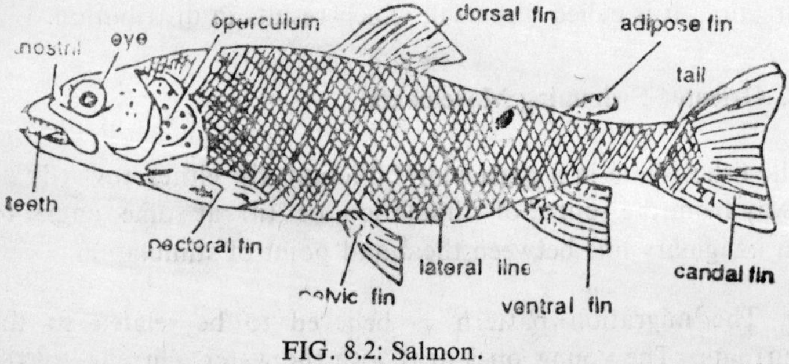

FIG. 8.2: Salmon.

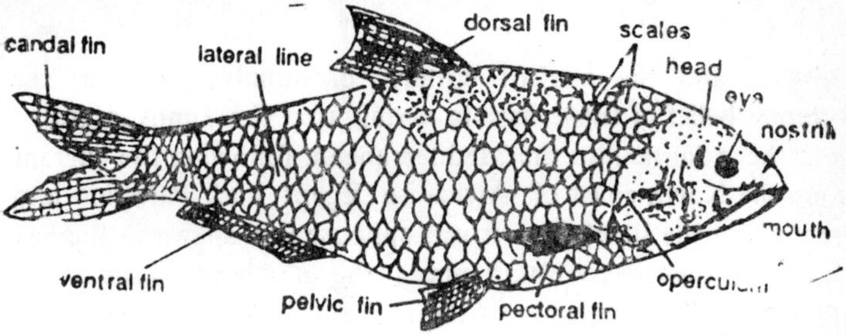

FIG. 8.3: Hilsa.

(ii) *Catadromous.* Those fishes which spend a major part of their lives in freshwater but migrate to the sea for breeding purposes. Anguilla travels several thousand miles starting from the rivers and reaching the spawning grounds in sea.

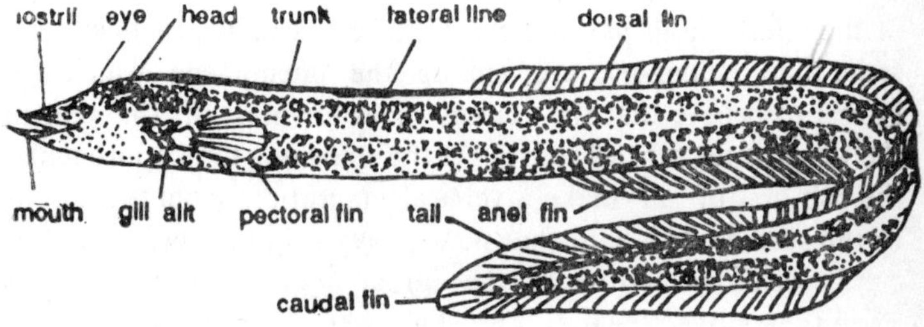

FIG. 8.4: Anguilla.

(iii) *Amphidromous.* Those fishes which migrate from freshwater to the sea, or *vice versa*, not for breeding but regularly at some other stage of the life cycle.

Above three types are together called Diadromous fishes.

(iv) *Potamodromous.* Truely migratory fishes whose migration is confined to freshwater only *e.g.*, Carps and Trout.

(v) *Oceanodromous.* Truely migratory fishes which live and migrate in the sea, *e.g.*, Clupea Scomber.

Migration of Different Fishes

Aristotle and the ancient writers presumably echoed the common belief of those days in saying, that the Tunny migrated from the Atlantic into the Mediterranean, following the current along the North coast of Africa to Egypt and Palestine and thence into the Black Sea. It was supposed to spawn in the Sea of Azov, thence returning along the northern coasts of the Mediterranean out into the Atlantic. Such wide-reaching ideas with regard to the migrations of fishes are the common property of fishermen of every country, for other fishes as well; it is their tribute to the mystery of their calling. But the scientific investigations of the past century have greatly reduced the limits of the migrations, except in one remarkable instance to be noted later.

The presence of the young in all stages and the appearance of the adults in all parts of the Mediterranean at practically the same time (summer), are among the phenomena which led Pavesi to believe that the Tunny lives for most of the year in the deep water off the coasts. As the spawning time approaches, the fish first of all make a vertical migration, from the depths to the surface. It is said that new arrivals in the surface waters can be distinguished from the earlier by their darker colour. Then they move towards the shallower waters to spawn. It is the latter, the horizontal movement, which has given rise to the idea of immigration from the Atlantic and which accounts for the elaborate preparations made by the fishermen for the reception of the Tunny.

These are not spawners; they are the larger fish which, after spawning in the summer on the Spanish or Portuguese coasts, follow the Sardine and Mackerel shoals northwards, and later feed on the autumn Herring of the North Sea. When the trawl with its load of Herring is hauled up, the Tunny follows and snaps at the protruding heads and tails. Possibly it was just as common in earlier years; there is a record by Schonevelde that it was fairly common near Eckernforde in the

Western Baltic as far back as the beginning of the seventeenth century, also in connection with the presence of Mackerel. The Porpoise may often have been blamed for its depredations, but until the trawl was used for the capture of Herrings in large quantities, its occurrence was regarded as a rarity.

A curious thing about the Herring, not known in any other species, is its ability to spawn at all seasons of the year and in fresh as well as salt water, thus under the most diverse conditions as to temperature and salinity. Its variability in this respect suggests that from the beginning it has been a shallow-water or brackish-water form; it can readily adapt itself to any conditions from 0 down to 100 fathoms. There are winter Herrings which spawn in the Fifth of Forth and in the west; spring Herring in Norway, in the Schley, and North of Scotland; summer Herring off the North-east coast; autumn Herring spawning in the southern North Sea, and so on. But the Herring keeps to the temperate zone; in the tropical and subtropical regions the variability has changed it into other species.

Another interesting problem in distribution is furnished by its relative, the Anchovy. This is a southern, subtropical form, with its main spawning places in the Mediterranean and adjacent waters. Yet a northern branch appears regularly year by year, and that in great quantities, in the Zuidersee and neighbouring brackish waters. In southern regions the species spawns in water of comparatively high temperature and salinity; here the water is brackish and cold. There is a wide gap between Spain and Zuidersee; few specimens have been taken in the Channel or Bay of Biscay. Are we to suppose that it migrates this distance yearly, or that it remains somewhere in the Channel, where it eludes capture? In either case the origin of this branch remains a mystery; possibly it represents the original form of Anchovy, as we can more readily understand a migration to the pleasanter conditions of southern waters than the reverse. On the other hand, this particular branch, may have suffered greatly from its enemies, Mackerel, Tunny, and others, and found a safe retreat from their pursuit in the Zuidersee. But no definite solution can be given to such a problem; of greater interest would be the investigation of the structural and developmental differences between forms of the same species growing up under such wide differences.

The principal thing about the species so far mentioned is, that the migrations are from deeper to shallower water in the sea. Some Clupeids, like the Shads (C. finta, alosa, sapidissima, etc.), go further and ascend rivers to spawn in fresh water. Some are permanent inhabitants of the inland seas (Caspian, North Italian Lakes, and Great Lakes of America).

The same phenomena are found among the nearly allied Salmonoids. Whilst some forms, like Argentina, spawn in deep water in the open sea, and others, like Mallotus (the Capelan), come close inshore to spawn, thus constituting a most important source of food, for example, for the Eskimos of Greenland, others like the Smelt enter brackish and fresh water, and the principal species of the group ascend rivers right to their sources, whilst whole families belong only to the fresh water.

The Salmon, it is now believed, does not move far from the mouth of the river to which it belongs; yet, like the Anchovy, it is rarely found in the sea except where it assembles to feed, as in the straits of the Pacific and in the Southern or South-eastern Baltic. The latter case is of near interest; the Salmon are taken on hooks, but some escape and are caught later, with the hooks still attached, in the lakes of Finland. The Salmon is thus able to spend its whole life in fresh or nearly fresh water. In the Rhine again, when it was a clean river, the winter salmon used to remain nearly a whole year in the middle reaches below Basel before proceeding to spawn.

In sounds almost incredible, therefore, that the Salmon does not feed after it has entered fresh water. Yet the evidence from all countries supports this view, and further, the digestive organs undergo a certain amount of degeneration. How then can we explain that the Salmon rises to the fly, and has even been taken with a Trout in its stomach? For one thing, it is certain that the Salmon of different regions, even different rivers, vary greatly in their powers and habits; for another, we have only to take a wider perspective to understand what happens. In the larger Norwegian rivers the Salmon rise to the fly in the lower reaches, but provide no sport in the upper. In the still longer American waters the Salmon provide no sport at all. They feed, therefore, or endeavour to feed only when fresh from the sea.

All the time they are in the fresh water, it may be a whole year, they live on the materials stored up in their body. This has been considered a wonderful provision or adaptation of nature. Regarded from the Salmon's point of view, it seems more like a tragedy; the Salmon whishes to feed but cannot! Certainly, its body must be an accumulator of energy of very good quality and enormous power. The fish ascends rivers against strongly running waters, jumps numerous falls (probably not more than 10 ft. at a single leap however); if a male it has to fight other makes and keep off enemies, then help to prepare a bed in which the female lays its eggs. It is not to be wondered at, when the spawning is over, that the fish is completely exhausted and allows itself to drift tail-first down the stream.

The Atlantic salmons spawn mainly during November and December. They ascend the rivers as the breeding period approaches, travelling several thousand miles in sea and then inland. On entering freshwater thay stop feeding and lose weight. Their bright silvery colour changes to a dull reddish-brown shade and the skin becomes thick and spongy. The body of the male salmon becomes spotted with red, orange and large black spots. After selecting suitable spawning grounds the fish segregate into pairs. Shallow saucer-like depressions are prepared in the river bed by the females where spawning takes place. After spawning, the return journey to the sea begins but few males survive to breed a second time. However, many females are able to reach the sea and start feeding. They soon recover their normal condition and silvery colour. The salmon does not usually spawn more than three times in its life-span of eight or nine years. From the time it reaches freshwater till spawning, the salmon lives on its reserve food accumulated in its beop in the form of fat.

The king salmon of the Pacific coast of North America travels about 2250 miles from the sea in order to reach the rivers to spawn. However, all of them die after spawning. No male or female survives to undertake the return journey.

Studies have shown that young salmon hatched in freshwater cannot return to the sea until their salt-secreting cells have developed. Studies indicate that migratory behaviour in both adults and the juveniles is not due to any special hormone regulation other than that which influences their general metabolism. Changes in their body states and behaviour may expose them to different environmental conditions to which they react in specific ways, such as travelling up freshwater streams by adults and downstream by juveniles.

Several theories have been proposed to account for the unique way salmons are able to return to the very place where they were hatched. Some authorities believe that the fish have an extremely sensitive chemical sense that enable them to perform this feat. Another theory links the accelerated metabolism at spawning time with the need for more oxygen which increases as they ascend the hard waters of a stream. Still another idea stresses the importance of a carbon dioxide gradient in the water as being the determining factor. Recent researches strongly favour the theory that sense of smell in fishes is the determining factor in locating home grounds, for different streams have different odours.

The Hilsa of the Indian Ocean also migrates to rivers for spawning. This species is known to ascend up the Ganges and go up to Varanasi and Allahabad. During the monsoon months Hilsa also migrates to many rivers of southern India. However, the details of its migratory habit are not known.

On the whole, it may be said that the run of Salmon up the river is connected with spawning, yet some important exceptions have to be made. Made specimens will go some distance up a river only to turn back (and this may happen several times), or even remain months in definitely freshwater before proceeding to spawn. There may be a simple explanation of these eccentricities: it must be remembered that the Salmon is not far from being a freshwater fish. According to Paton and Newbigin (1990) the factor determining migration from sea to river is not the "nisus generativus, but the state of nutrition," The spawning comes later.

With the Salmon we may finish this account of the one stream of migratory fishes, those which journey from deeper to shallower and even fresh water in order to fulfil their life purpose. This is by far the most common form of migration; a very large number of marine fishes, even the Sharks, Skates, and Rays, follow in the main the same course. As suggested by Meek, the term anadromous," formerly used only for those entering fresh water, may well be applied to all those fishes which follows the same general rule in their migration, namely, an ascent from deeper to shallower waters. To a very large extent, this at once indicates the nature of the problem to be solved and even suggests the solution.

But there are two groups of apparent exceptions. In the one the fishes exhibit but little tendency to undertake any spawning migrations. They spawn, as it were, where they find themselves. This applies naturally to the stationary fishes, though even among these a certain amount of migration can be noted. Thus, among the Blennies the Butter-fish or Gunnel has a short, anadromous migration of a few miles, whilst the Angler or Frog-fish shows a longer migrations in the reverse direction. Among the Gadoids, again most species prefer the deep water for spawning; but the Cod, if it has any preference, comes from the deep water to shallower banks.

These groups, in which the migratory movements are of little account or in which different tendencies are shown by different species, may be called neutral for the moment. The probability is, indeed, that they are not so much affected by the spawning process. The seeming migrations may be either a return to the neutral home places, following upon the drift of the young away from these, or they may be mixed up with

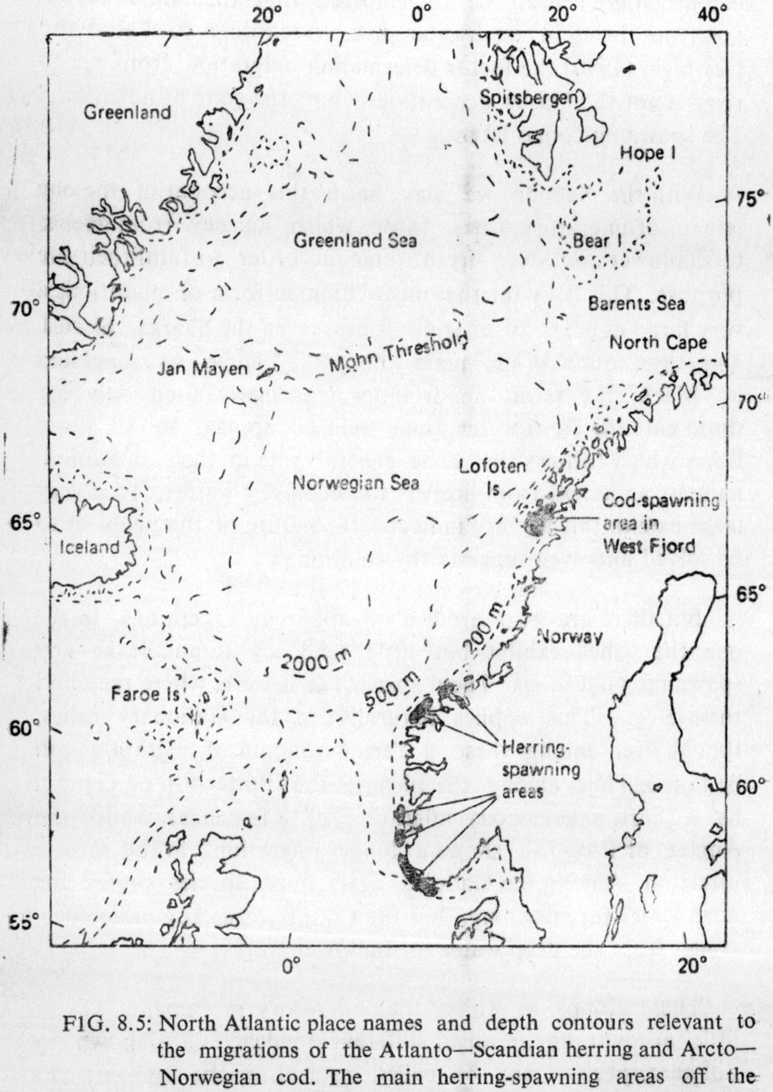

FIG. 8.5: North Atlantic place names and depth contours relevant to the migrations of the Atlanto—Scandian herring and Arcto—Norwegian cod. The main herring-spawning areas off the southwest coast of Norway, and that of the cod within the West Fjord, are indicated.

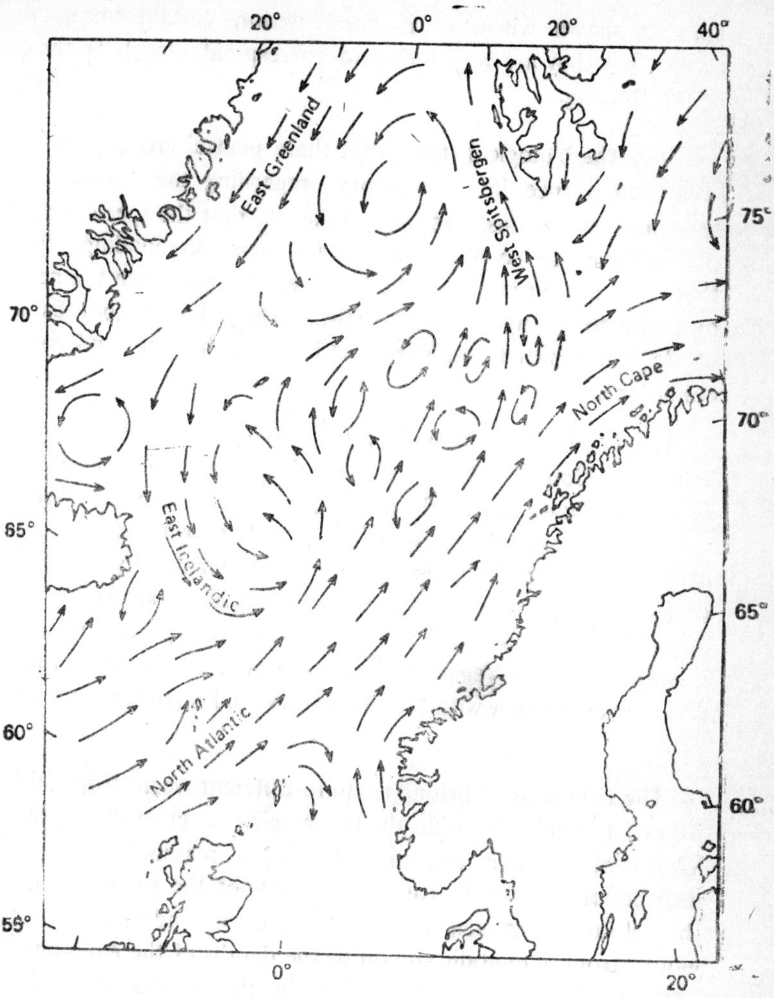

FIG. 8.6: The main surface currents in the North Atlantic relevant to the migrations of the Atlanto—Scandian herring and the Arcto—Norwegian cod.

the movements in search of food. The Cod, for example, is a rover with a circumpolar distribution like the Herring, and it shows a decided preference for the latter as food. It pursues the latter at all seasons of the year and at the proper time it is able to spawn whenever it goes—within certain limits, of course, but they are very wide and correspond closely to those of the Herring.

From the biological standpoint these neutral groups are of great importance in any theory regarding the causes of spawning migrations. We may take it, that the species composing them are so well-adapted to their surroundings that they are not urged to make any great change when the need of spawn comes upon them. Viewing their life history as a whole, they show a better balance between the inner and outer conditions.

The second exception is represented by the Eel group. With the Common Flounder the Eel constitutes the old group of katadromous fishes, those which migrate from the fresh water to spawn in the sea. The Flounder is certainly katadromous like the Plaice seeks water of greater density than its ordinary feeding grounds for the purpose of spawning; but other Flat-fishes like the Soles show the reverse procedure, and it seems better to regard the group as neutral. There is in any case no comparison with the life-history and migrations of the Eel.

The Eel seems to belong to quite different world from other fishes; it would be difficult to imagine a greater contrast. Whilst other fishes are content with a reasonable area of distribution, thus admitting the limits of the piscine nature, the Eel ranges between the most extreme conditions on the globe, from mountain stream to the depths of the Atlantic. It passes through two transformations in the course of its life, when it enters and when it leaves fresh water, other fishes have only one or none at all. The Eel takes some ten to fourteen years (in fresh water, plus three more in salt water) to become mature, other fishes everywhere only one to three.

Other Eels, Conger, Muraena, etc., do not show such a great contrast, so far as we yet know their life-histories. All seem to be katadromous fishes, migrating out from shallow to deeper water, but only the Anguilla genus enters fresh water.

Eels have two distinct phases in their life-history, each morphologically different from the other. Yellow eels represents the feeding and spawning phase while silver eels represent breeding phase. Yellow eels are found both in salt and freshwater and vary in length from a few inches to five or six feet. During autumn a number of yellow eels become silvery and prepare to leave for spawning grounds. They stop feeding, their eyes become large, lips thinner, and the snouts sharper. Their colour becomes silvery along the sides with a blackish back. The reproductive organs develop while the alimentary canal shrinks. The eels migrate down the rivers to reach the sea. They travel about three to four thousand miles in the Atlantic to reach their breeding grounds in the western Atlantic, south of Bermuda. It is believed that the eels spawn at a depth of about 400 metres below the surface and the parents die soon after.

The young larvae that hatch out are called leptocephali. They are flat, leaf like, tiny creatures provided with long needle-shaped teeth for feeding. They begin their homeward journey in the easterly direction. The grow rapidly during the first few months and when reach the coast of western Europe they are about three inches long and two year old. At this point they undergo metamorphosis and stop feeding. The needle-like teeth are lost and the transparent body assumes a cylindrical shape. They are now called elvers, or glass eels. The elvers, when about three years old, start ascending the rivers in large numbers. They are able to cross all obstacles and reach their suitable resting places where they feed and grow for some years to develop into yellow eels. Finally on attaining maturity, they change into silver eels and start moving towards their breeding site in the sea like their parents.

The American species, Anguilla rostrata also has similar breeding habits. In the western Atlantic larvae of both species are found living together. The interesting and mysterious aspects of this strange migratory phenomenon are how the tiny elvers without any guidance are able to find their way across the sea, and why the two species which practically hatch in the same region do not get mixed. The explanation lies in the fact that the American eel grows more rapidly (only in one year as against three years in the case of the European species). Thus, if the larva of the European eel travels westward instead of an easterly direction, it will reach the coast of America long before it is ready to change into an elver. And conversely, if the larva of the American eel migrates in an easterly direction it will undergo metamorphosis in the middle of the Atlantic. In other words, the larval life in each case in geared to the distance it has to travel.

A recent theory, however, suggests that both American and European eels belong to the same species and that the adult European forms die before reaching the spawning grounds. All eggs are laid by American eels but, when hatched, the larvae in the northern part of the spawning grounds are carried away by currents to Europe. Those in the southern part are carried towards the American coast.

By a series of brilliant observations and deductions Johs. Schmidt has at length traced the Eel to its spawning places (1921), in an indirect yet quite convincing manner. The post-larval Eels (Eeptocephali) were found in ever increasing quantities the further he proceeded westwards, but they became gradually smaller. Successive line or zones could thus be marked off, according to the size, until at length he found the smallest larva with the yolk-sac still attached and even the eggs out over the deepest parts of the Atlantic near the West Indies and nearer the American than the European coasts. These were taken in the middle and upper layers, and there is therefore some doubt whether the adults actually spawn at the bottom. The depths on the "spawning grounds" were about 3000 fathoms.

Two more of Schmidt's discoveries are of interest. At the same place where he found the larvae of the European Eel he also took those of the American Eel. The spawning grounds of two forms, living thousands of miles apart, are practically the same, and the young of both species were occasionally taken in the same haul of the net.

The second discovery is even more extraordinary. The two streams of post larvae spread out and move northwards towards the Bermudas; there they separate, one going east to Europe, the other west to America. The latter form spends only one year in the Atlantic and is then ready to enter fresh water; the European form takes three years to cross the Atlantic before it is ready to change its constitution and enter the rivers. How these tiny fishes can settle which direction to take, even allowing for a difference in constitution, is a profound mystery. In illustration of the immense quantities of these young Leptocephali it may be mentioned that Johs. Schmidt took as many as eight hundred specimens is one haul out in mid-Atlantic.

Factor Affecting Migration

Many efforts have been made to connect the migrations of fishes with the external conditions, temperature, salinity, light, etc., and it is probable that each has some influence, but the problem has another side. In the first place, it is essential to distinguish the food concentrations and wanderings from the spawning migrations, a matter not so easy as at first appeared. It would seem, indeed, that a feeding concentration, in many cases if not at all, precedes to actual spawding, and it may be that the storing of foodstuffs leads directly to the internal changes in metabolism which give the stimulus to change of grounds and conditions (Paton, 1900).

Physiologically the feeding and the migratory impulses are very different. In the one the fish is seeking to obtain something, in the other to get rid of something. The latter may be called a morbid, even pathological condition, affacing every part and sufficiently powerful to make the fish ignore terrible injuries and even to over-come the impulse to feed. The digestive organs partially or wholly degenerate, and the fish is then urged to seek relief by unusual or extraordinary means,

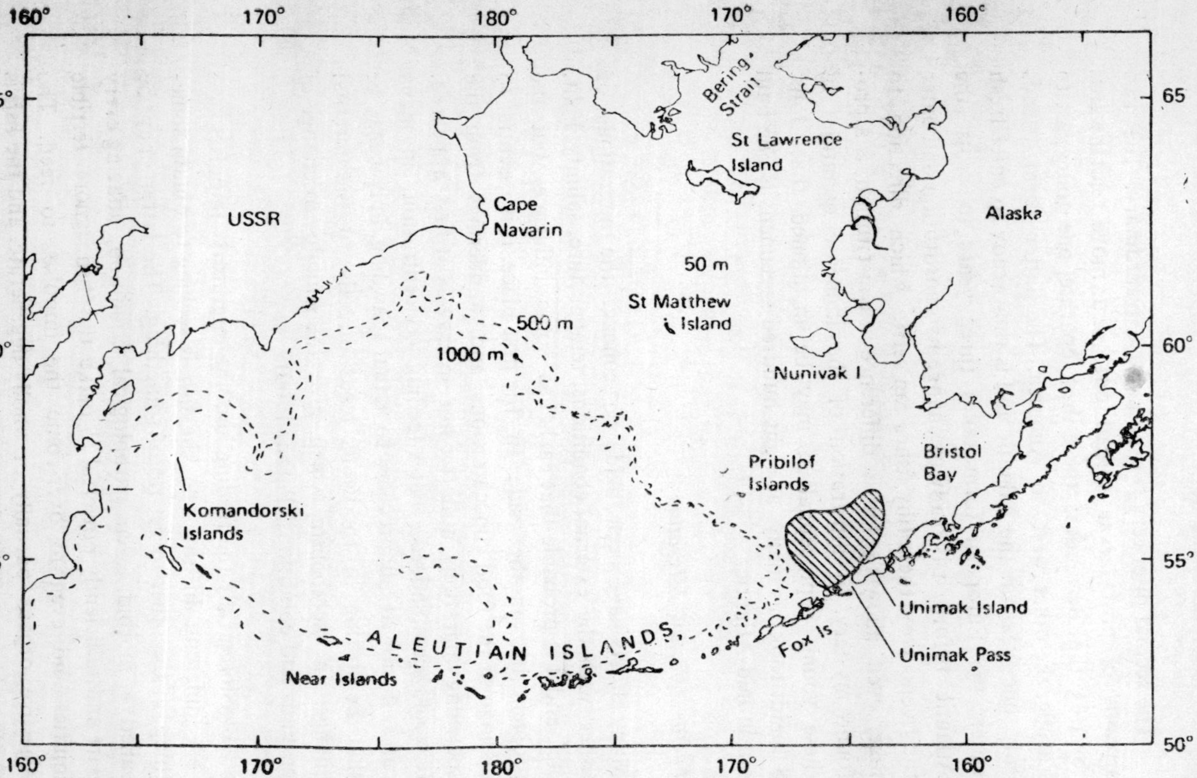

FIG. 8.7: Place names in the Bering Sea relevant to the migrations of the Alaskan pollock. The pollock spawning area to the northwest of Unimak Island is indicated,

that is, to find if possible external conditions in better agreement with its changing or changed constitution.

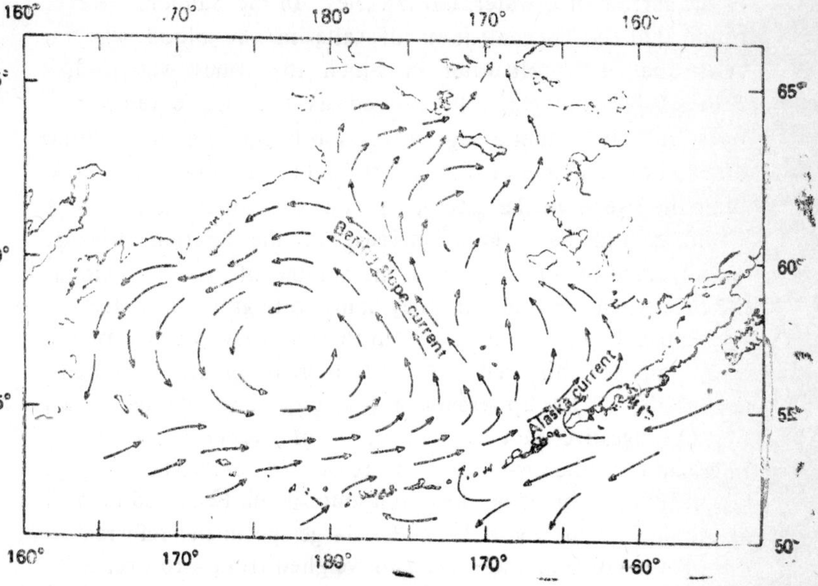

FIG. 8.8: Main surface currents in the Bering. The migrations of older Alaskan pollock are related to the Bering Sea Slope current and the distribution of the larvae to the West Alaska current. The spawning area is indicated on Fig. 8.3.

One cannot say definitely what happens in any particular case, owing mainly to the protean nature of fishes. A Dog-fish, for example, is more in harmony with the surrounding medium than a Salmon, and Stickleback will survive under conditions that kill most fishes. These phenomena depend apparently on their ability to alter the blood-pressure, or concentration of salts, as they pass from one set of conditions to another. And this ability, again, depends upon how far the gill-membrane and the skin permit of the passage of water outwards and inwards, a question that has not yet been sufficiently investigated.

It is possible, therefore, that the internal stimulus to migration comes from some internal change in the blood; probably in the blood-pressure. This is always much less than the pressure of the surrounding water, but it varies. In the Salmon, Greene found that the concentration of salts in the blood was far below that of the salt water in which the fishes was feeding voraciously, and gradually altered as the fish entered fresh water. It may be that a slight reduction in the pressure within the blood is a source of discomfort to the fish and urges it to enter the rivers. In the case of Eel, on the other hand, there would seem to be a great increase of the internal pressure accompanied by an increase in the amount of carbonic acid at the time when the fishes are beginning to migrate from freshto salt water. If they are detained on their journey in any way, the body thickens and swells and even increase greatly in weight, though the fish are apparently not feeding. Petersen has shown that the digestive tract is in process of degeneration at this time and even the air-bladder decreases in size. The average Silver Eel in its breeding dress is about 70 cm. (27 to 28 ins.) in length and weighs about 1½lb. The large specimens referred to were from 110 cm., and more and weighed from 4 to over 8 lb. Most remarkable of all is the great increase in the size of the eyes. These become more than four times their former size and weigh twice as much.

Without being able to say positively that such is the case, one may conclude that these phenomena point to a great increase in the internal concentration of salts and absorption of water. And this increased pressure with seems to be the cause of the Eel leaving the rivers to seek the greater pressures of the salt water. When the impulse to migrate comes upon them, the Eels are restless and ill at ease moving up and down for a time in a manner quite unlike their usual sluggishness, then definitely turning down the rivers. At every stage of their descent they come under increasing pressures without. And it may be recalled here, that the Eels are well-known to be extremely sensitive to the differences of pressure produced by thundery weather.

In the case of marine fishes like the Plaice or God, we know that when the spawning fish are brought up in the trawl. the spawn continues to run even when the fishes are still alive. This indicates that the internal pressure persists for some time at the level of the deep water whence the fishes came. Since the low pressure inshore would help in getting rid of the spawn, the migration of the Plaice from shallow to deep water would be unnecessary for that purpose. Hence, we may conclude, that an increase of pressure within has occurred with the maturation of the sexual organs, and the resulting discomfort has urged the fish to seek deeper water and greater pressures. Similarly in the case of other fishes. Each species is urged to seek the external conditions, whether of lower or greater pressure, which corresponds best to the change in its own constitution.

Migration in Fishes and Birds

Though the migrations of fishes may appear the same superficially, they are widely different in important matters. They are not seasonal migrations in the same sense and the young take no part in them. Fishes have nothing to fear from the winter, not those we are dealing with at any rate; both the Salmon and the Herring spawn during the winter months. Further, the migration brings an advantage or benefit to the birds. On the one hand, they are urged on by the constitutional fever within; on the other, they are enticed forwards by what lies at the end of the journey, this is wanting in the fishes; the urge is there but not the inducement or reward. If an Eel could foresee what lay before it in the depths of the Atlantic, would it undertake the journey?

We may say, then, that the migrations of fishes cannot be regarded as adaptations, as we apply that term to birds. They are on a lower level, more primitive and crude. In all probability they are an inheritance from remote ancestors, which the fishes from lack of sufficient intelligence perhaps have not been able to overcome.

Again, it should be noted that the birds do not show contradictory phenomena; the migrations are always to avoid the severity of the external conditions. We might say this of the summer spawners among fishes; but all sorts and conditions are to be found, and the extreme cases, Salmon and Eel, seem wilfully to be seeking the most dangerous and most deadly conditions to be found on the globe.

How far the fish in its migrations is guided in any way by its own sense organs, we cannot say. The external conditions—temperature, salinity, dissolved gasse, etc.—all make for differences in pressure in diversee ways. Possibly, as Hofer has suggested, the extreme sensibility of the lateral line system to slight variations in the strength of currents, may enable the Salmon to detect the entrance of a river or side-stream. But on the whole, one may doubt whether the senses play much part in the directing of the spawning migrations.

Orientation and Migration

Arnold (this volume) describes some of our recent work with plaice. In the Southern Bight of the North Sea these flatfish appear to use the tidal streams for transport when moving between their feeding and spawning areas, and they come off the bottom to join the appropriate tide. We have always wanted to go a step further and determine the direction in which a plaice was heading when in midwater. With a density of 1.063 g cm^{-3} (Lowndes, 1941) a plaice would have to swim through the water in order to maintain its height above the bottom, but, it would not be expected to maintain a steady course in the absence of visual or tactile clues. Nevertheless, when simple vector analysis was applied to those midwater tracks for which current meter data were available, the results showed that some fish maintained a surprisingly consistent course for periods up to two hours (Greer Walker, Harden Jones and Arnold, 1978, Fig. 8.12).

Our colleagues at the Fisheries Laboratory, Lowestoft, have invented an acoustic transponding compass tag (R.B. Mitson,

T.R. Storeton West and N.D. Pearson, in preparation) so that we can now obtain a virtually continuous record of the position, depth, and heading of a fish in real time. The new tag, which is somewhat larger than the conventional acoustic transponder, is attached to the plaice by two Petersen discs and so aligned along the body axis on the upper right-hand side. The compass reference is provided by a magnet attached to a slotted disc and mounted on a jewelled pivot. Changes in direction are sensed optically using a circular array of 8 infrared emittors and 8 detectors to determine the bearing of the slot—and so the heading of the fish—which can be assigned to one or other of 8 nominally 45° sectors (Fig. 8.9). When the tag is interrogated at 300 kHz, it responds with a reference pulse which can be used to determine the position and the depth of the fish in the usual way (Greer Walker *et. al.*, 1978). The compass heading is indicated by a second pulse, the delay between the two signals (ranging from 24 to 66 ms in seven 6-ms steps) identifying the relevant sector. The two pulses are shown on the B-scan sonar display and, from selected channels of the receiving beam, on an Alden paper recorder. The heading of the fish can also be read directly from a digital decoder.

The track of a plaice fitted with a compass tag and released in the open sea 24 km off Orfordness is shown in Fig. 8.10. The fish was released at 1730 h GMT on 3 June

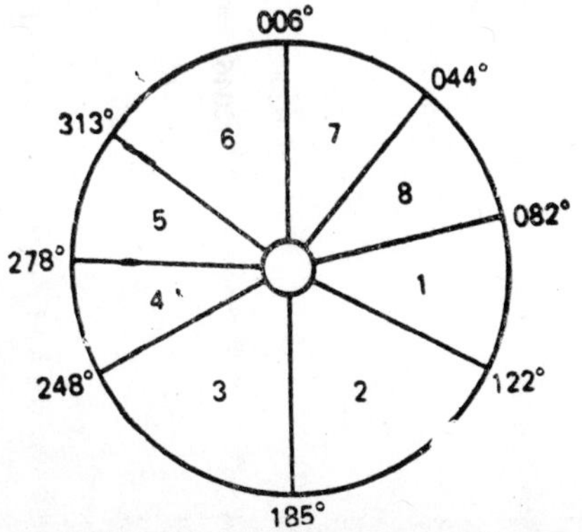

FIG. 8.9: The characteristics of the compass tag fitted to the plaice.

TABLE 8.1

Data Relating to the Track of a Plaice Fitted with an Acoustic Transponding Compass Tag Released off Orfordness

Period in midwater	Fish				Tide	
	Time in hours, GMT		Movement over ground		Direction °T	Speed cm s^{-1}
	off bottom	on bottom	Direction °T	Speed cm s^{-1}		
1	2207	2352	012°	43	N 009°	39
2	0200	0242	175°	95	S 198°	54
3	2046	2338	040°	60	N 015°	62
4	0040	0302	166°	38	S 190°	65
5	2056	2400	038°	66	N—	—

Note: The track of the fish is shown in Fig. 8.11. The tidal velocity was estimated from moored current metres (see Fig. 8.10) which were recovered between midwater periods 4 and 5.

and abandoned at 2400 h on 5 June. It left the bottom during the night and came into midwater for 5 periods as indicated in Fig. 8.11 and Table 8.1 Neither moon nor stars were visible on these nights. When in midwater the plaice was usually from 5 to 20 m above the bottom (total depth 25-30 m) to which it returned several times. These excursions to the bottom rarely exceeded two to three minutes.

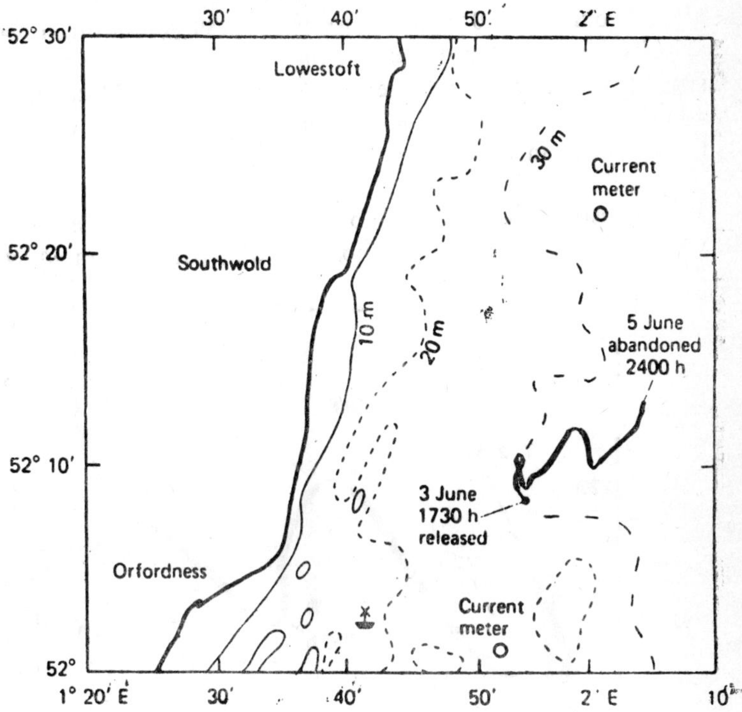

FIG. 8.10: The track of a plaice fitted with an acoustic transponding compass tag and released at 1730 h GMT on 3 June 1979. The 10-, 20-, and 30- mdepth contours and the positions of two moored current-meter stations are indicated.

Fig. 8.12 shows the headings, by sectors, for each of the five midwater periods: they were not randomly distributed. In period 1 the fish was usually heading in sectors 6 and 7 (313°

to 044°) and in periods 2 to 5 in sectors 1 and 2 (082° to 185°). The courses made good over the ground during periods 1 and 3 are markedly different (012° to 040°) although in both instances the direction of the tide was virtually the same (009° and 015°). The difference between the two ground courses would appear to be related to the sector headings maintained by the fish in midwater: in period 1 the fish was heading down-tide and in period 3 the fish was heading across the tide. The ground course of 040° appears to be the resultant of the fish's velocity through the water and that of the tide. The results suggest that a fish in midwater will actively swim and maintain a fairly consistent heading in one direction while being carried by the tide in another.

The distribution patterns shown in Fig. 8.12 are still evident when each midwater period is considered in parts separated by

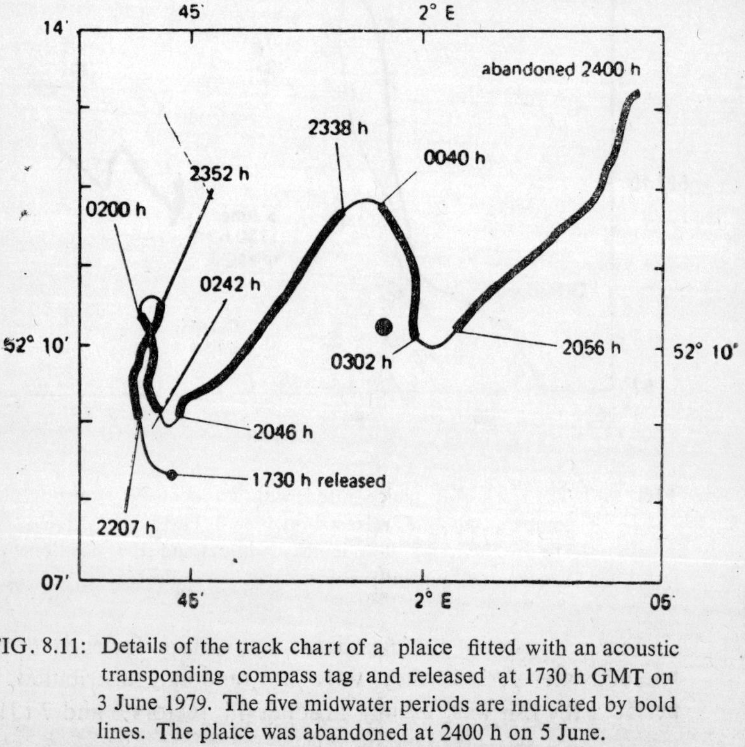

FIG. 8.11: Details of the track chart of a plaice fitted with an acoustic transponding compass tag and released at 1730 h GMT on 3 June 1979. The five midwater periods are indicated by bold lines. The plaice was abandoned at 2400 h on 5 June.

bottom excursions. Two examples are given in Figs. 8.13 and 8.14 which show that the patterns do not vary with average height above the bottom or with the time spent in midwater.

We have looked at the data in greater detail to get some measure of the length of time for which a plaice could maintain its heading when in midwater. With the rather coarse compass reference available the best that could be done was to measure, on the Alden charts, the time that the fish remained in one sector before being recorded in another. The results are summarized in Table 8.2. There were many instances when the fish was recorded as remaining in one sector for less than a minute and at the other end of the scale, there was a single instance when the fish remained heading in one 45° sector (which suggests that the heading was consistent to $\pm 22.5°$) for just over 14 min. When the data were examined on the less stringent basis of consistently keeping in two adjacent sectors, there was one instance when the plaice appeared to maintain a heading to within $\pm 45°$ for 58 min. (Table 8.3).

These results were unexpected. Furthermore, the data may underestimate the ability of a plaice to maintain a steady course as there were instances where the compass signals were regularly alternating between two adjacent sectors which suggested that the fish was keeping to a narrow range of bearings towards the edge of either sector.

If further studies confirm these first results, we are going to have to deal with two new problems: firstly, what is the reference used to select the course when the fish moves into midwater; and secondly, how is the course subsequently maintained.

A lucky observation suggested a possible answer to the first question. Before one midwater period the plaice was seen to be swimming very close to the bottom along the crest of a sand-wave before 'taking off'. When in midwater the fish maintained the same heading: a comparison with an aircraft was irresistible. This observation leads to the suggestion that the heading

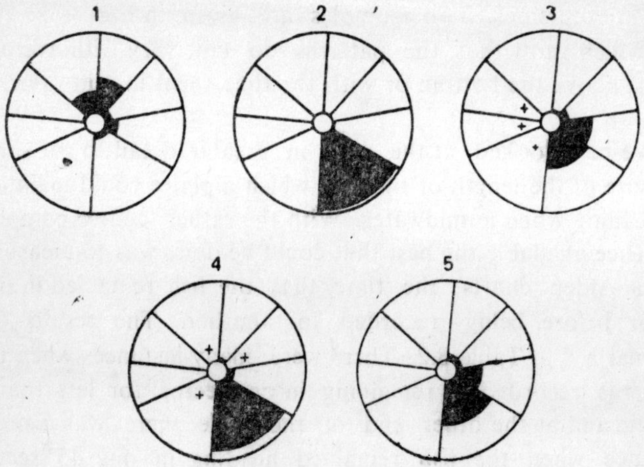

FIG. 8.12: The heading of a midwater plaice as indicated by the compass tag for each of the five midwater periods shown in Fig. 8.11 and summarized in Table 8.1 The time recorded in each sector is expressed as a proportion of the total time for which the fish's heading could be determined for each midwater period. The presentation follows the convention that the part of the radius between the inner and outer circles corresponds to 100 per cent. Total times in all sectors: period 1, 80 min; period 2, 31 min; period 3, 138 min; period 4, 127 min; period 5, 163 min.

maintained in midwater was related to that observed shortly before 'take-off'. With this hypothesis in mind, the sector headings recorded during the hour or so before 'take-off' were broken down into 10-min sections and compared with the patterns characteristic of each of the corresponding five midwater periods. In each case the pattern recorded immediately before 'take-off' was similar to that observed in midwater. Two examples are shown in Figs. 8.15 and 8.16. During the first midwater period the headings of the fish were largely in sectors 6 and 7 (northwest to northeast). Fig. 8.15 shows that in the hour before 'take-off' the plaice turned clockwise from heading southeast to head in a predominantly northwest-to-northerly direction in the final 10 min before moving into midwater. A similar change in heading was shown in the hour before midwater periods 2 and 3. But the fish was already heading in the midwater direction for some time before 'take-off' for periods 4 and 5 and here there were no changes: the data relating to midwater period 4 are shown in Fig. 8.16.

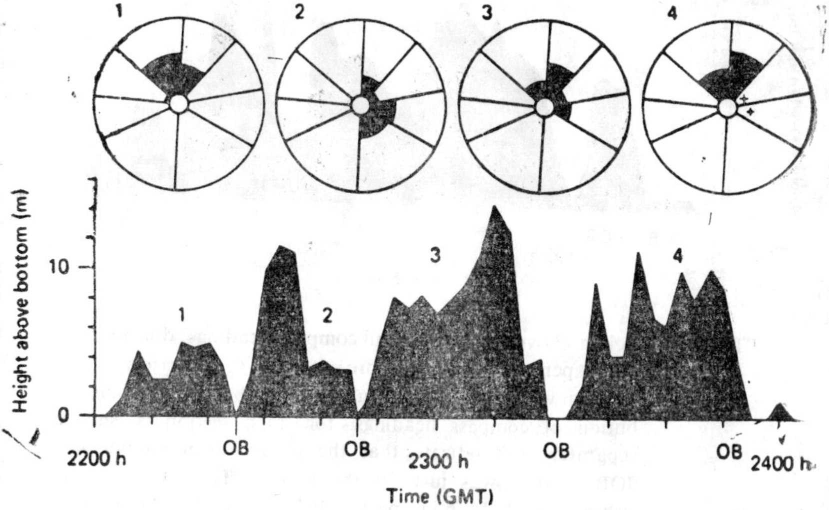

FIG. 8.13: Height above the bottom and compass headings during mid-water period 1 which is broken into 4 separate parts between which the plaice returned to the bottom. The distribution of compass headings for each part is shown separately. Heights above the bottom of one metre or less were included in the category Just Above the Bottom and were not classified as midwater: OB indicates that the plaice was on the bottom. Total time in all sectors: part 1, 14 min; part 2, 16 min; part 3, 28 min; part 4, 22 min.

These observations are consistent with the hypothesis that the heading adopted in midwater is related to the heading of the fish on the bottom before 'take-off' and so, perhaps, to some feature of bottom topography. We think it is possible that sandwaves, whose crests are aligned at 90° to the tidal axis, could provide the bottom reference. The track of a plaice whose movements over the bottom appeared to be related to

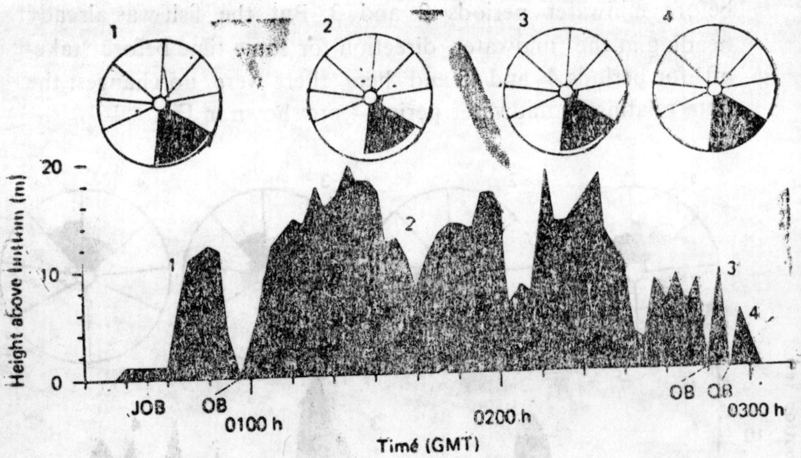

FIG. 8.14: Height above the bottom and compass headings during mid-water period 4 which is broken into 4 separate sections between which the plaice returned to the bottom. The distribution of compass headings for each section is shown separately. OB indicates that the plaice was on the bottom: JOB, that it was just off the bottom, Total time in all sectors: part 1, 14 min; part 2, 104 min; part 3, 3 min; part 4, 6 min.

sandwaves has already been described (Greer Walker *et al.*, 1978).

While we have a reasonable but tentative suggestion for an answer to the first question, we have little to offer in respect of the second. The height of the fish above the bottom in turbid East Anglian waters on nights when neither moon nor stars were visible would seem to preclude the usual rheotropic clues or the use of celestial reference points. The way would appear to be left open for what would have once been dismissed as unlikely contenders: inertial guidance through the labyrinth or some on-line information of the earth's magnetic field through some unidentified sensor. We have given some thought to these matters. While our calculations suggest that the possibility of inertial guidance is not to be dismissed so lightly as it has been in the past, our conclusion is that the need is for more data, from both the sea and the laboratory, and less speculation.

TABLE 8.2

Frequency with which a Midwater Plaice was Recorded as Maintaining its Heading within One Compass Sector (About 45°) for a Number of Consecutive 1-min Periods

No. of Consecutive 1-min Periods	*No. of Observations*
0—1	664
1—2	53
2—3	26
3–4	11
4—5	6
5—6	4
6—7	2
7—8	5
8—9	2
9—10	0
10—11	0
11–12	1
12—13	0
13—14	1

Note: Based on 775 observations.

TABLE 8.3

Frequency with which a Midwater Plaice was Recorded as Maintaining its Heading within Two Compass Sectors (About 90°) for a Number of Consecutive 1-min Periods

No. of Consecutive 1-min Periods	*No. of Observations*	*No. of Consecutive 1-min Periods*	*No. of Observations*
0–1	16	12—13	3
1—2	7	13—14	1
2—3	6	14—15	3
3—4	7	15—16	0
4—5	5	16–17	2
5—6	2	17—18	1
6—7	2	18—19	1
7—8	1	19—20	0
8–9	3	20—21	2
9—10	0	28–29	1
10—11	4	37—38	1
11—12	0	57—58	1

Note: Based on 69 observations.

Dr. Arnold has summarized the lines of evidence which have led us to the conclusion that the plaice of the Southern Bight of the North Sea use tidal transport when on migration. Tidal transport has two potential advantages: it could save

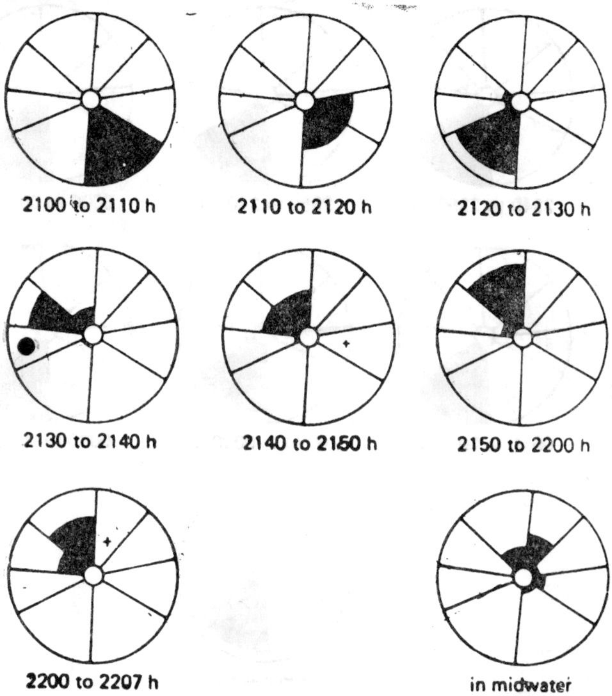

FIG. 8.15: A comparison between the distribution of compass headings recorded when the plaice was on the bottom with that recorded during midwater period 1. The plaice changed its heading from southeast (sectors 1 and 2) to northwest (sectors 5 and 6) before leaving the bottom at 2207 h. When in midwater the compass headings lay between northwest and northeast.

energy; and provided the fish catches and leaves the right tide at the appropriate moment, problems of direction do not arise. So we were surprised to find, in a fish living in shelf waters where tidal streams dominate the current regime, some evidence of an ability to maintain a heading when in midwater. However, these preliminary results are not inconsistent with Baker's (1978, p. 861) least navigation hypothesis': clearly we should give more thought to the design of field and laboratory experiments to see if a similar faculty is evident in other species of fish.

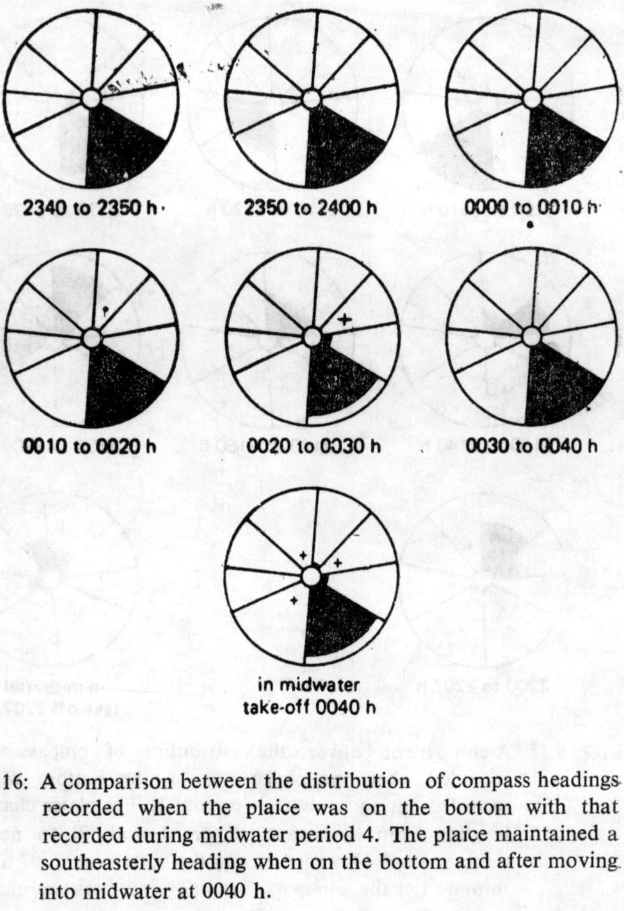

FIG. 8.16: A comparison between the distribution of compass headings recorded when the plaice was on the bottom with that recorded during midwater period 4. The plaice maintained a southeasterly heading when on the bottom and after moving into midwater at 0040 h.

9

Migratory Movement in Snakes

"The way of a serpent upon a rock" is one of four phenomena described in the Book of Proverbs as being beyond understanding. A snake undulates in a smooth and asymmetrical fashion, so that the size and curvature of the loops formed by its body change as the animal glides along. Because the parts of the snake's body that exert motive forces are never as clearly defined as those in species with limbs it is not obvious to a casual observer how such movement propels the animal forward. It would seem that the snake's undulations would be just as likely to move it backward or leave it thrashing in one place. Such considerations make the movement of snakes a challenging subject for investigation.

Slithering and related forms of motion employed by snakes are the consequence of the snakes' evolution. Snakes presumably evolved from animals that lost their appendages when they took up a burrowing existence. Since they lacked limbs, the entire body had to become an organ of locomotion. This function favoured elongation of the trunk and necessitated a repackaging of the viscera. Both of these characteristics were further developed by most snakes when they emerged to radiate into a broad spectrum of terrestrial, arboreal and even aquatic niches. As the trunk lengthened it remained flexible because the number of bony segments in the vertebral column increased. Thus snakes have between 100 and 400 trunk vertebrae, a number that contrasts strikingly with the 16 or so found in a typical mammal. Ophidian trunk vertebrae are also joined by at least one extra set of articulations, and each vertebra bears a pair of ribs that reach down to form a flexible yet strong support for the sides of the elongated visceral cavity.

The increase in the number of the segments gives the snake its flexibility; it also complicates the snake's control problems. Some 30 years ago the late Walter Mosauer of the University of California at Los Angeles showed that a snake's vertebrae are joined in such a way that one vertebra can bend up and down in relation to another through approximately 28 degrees and swing from side to side through about 50 degrees. Although the bulk of tissue probably reduces a living snake's flexibility, the number of degrees through which a snake can bend is still remarkable. Even twisting movements can be approximated by a combination of up-and-down and side-to-side bending, in spite of the fact that the joints between any pair of vertebrae do not allow one segment actually to twist in relation to the other. The posture assumed by a snake at any one moment is established by the muscles that together with the skeleton form the axis of the body. The most important of these muscles are arrayed in 12 or more bundles that connect one vertebra to another and to the ribs on each side of the trunk, and in two to four bands that attach the skin to each side of the snake's main mass.

How does a snake utilize all these elements in order to move? An exact description of slithering requires that the position of each element at any given instant be specified. It is easily seen that investigators who seek such a detailed description must solve a truly staggering number of simultaneous equations. The task is complicated by the fact that as the snake travels its parts move and accelerate at different rates. The solution gains several magnitudes of complexity when one wants to know which of the snake's thousands of muscles are actually exerting the forces that induce and maintain motion.

One method of achieving a simpler solution is to work inward from the outside. The snake is treated as a special kind of "black box" capable of exerting forces on areas within an envelope of space considerably greater than its own volume. One can then deal with the snake's external responses without, for the moment, concerning oneself with the way they are generated internally. One first asks where and how these external forces are exerted, then it can be asked what internal forces and mechanisms are responsible for the externally observed efforts. Ultimately this approach allows investigation of the function of the specific muscle groups, and of the motor unit and their control.

Since the immediate area of contact between the snake and the environment is the skin it is important to consider the texture of the skin's surface. The sides and back of most snakes are covered with longitudinal rows of lozengeshaped and more or less rigid scales. Such scales are generally mounted, so that their trailing edges are free, on a flexible and accordion-pleated intermediate layer of skin that is often lined by a layer of muscle. The skin of a snake has a remarkable capacity for extension, as is evidenced when the animal swallows prey more than twice the diameter of its body. The snake's bottom surface is covered by a series of scutes: wide scales whose free trailing edges overlap like plates of armor. Each scute tends to be associated with a single row of scales on the snake's sides and back, and to a single vertebra and a single set of muscles; the scutes thus reflect the snake's internal organization. The side and back scales often have sensory pits of unknown function and may be ridged; the scutes do not have pits and are normally smooth.

It is generally accepted that the miscellaneous ways by which snakes propel themselves can be subdivided into four distinct patterns: lateral undulation, concertina progression, rectilinear locomotion and sidewinding. These categories are somewhat artificial. Most species of snakes can move in more than one of these patterns; indeed, a snake moves in more than one pattern at the same time. My colleagues and I at the State

University of New York at Buffalo have investigated these patterns of movement. We are currently looking into their control mechanisms.

Lateral undulation is the kind of snake movement most frequently seen. It is the preferred movement not only of snakes but also of limbless lizards. Under certain conditions this kind of movement is even employed by limbed species, but their limbs are first folded against the body. The pattern can be shown to be the most primitive one in vertebrates. Lateral undulation involves the movement of the body along an S-shaped path. In an idealized situation the snake's track can be seen to be a single wavy line scarcely wider than the animal's body; each portion of the snake's trunk has traced this path. The snake begins to undulate laterally by bending the forward part of its body. This movement establishes a wavelike muscular contraction that travels down the snake's trunk.

How does this wave of muscular contraction cause the snake to move? Because human beings walk by exerting force directly on the surface under their feet, it is generally assumed that a snake too propels itself by somehow exerting a force directly on the ground. In actuality undulatory progression operates on a completely different principle. The wavelike contractions flowing down a snake's body do not propel it by acting on the ground under the snake's belly. Instead they cause the snake's body to exert force laterally on irregularities in the snake's path: small elevations and depressions, pebbles, tufts of grass and so on. The snake in effect pushes itself off from such points the way a man sitting in a chair with casters can push himself away from a desk. Since a snake can push only against a limited number of fixed objects, it sets its body in a corresponding pattern of loops, with the outside of each loop forming a contact point. The speed at which a loop progresses down the snake's body toward its tail equals the forward speed of the snake; the moving loops thus seem to be stationary with respect to the ground.

This analysis of lateral undulation has been confirmed by Hans W. Lissmann and Sir James Gray in a series of experiments they conducted to characterize the forces exerted against the ground by certain snakes. Their snakes crawled across a smooth board over which hung a regular array of heavy, pendulum-like pegs. Any lateral force exerted against a peg would cause it to swing, indicating both the magnitude of the applied force and its direction. Calculations of these forces showed them to be consistent with the force and speed of a snake's movement. When the pegs were taken away, the snakes were totally unable to move by lateral undulation; they slipped about on the smooth board because they had no objects to exert lateral force against.

Although a snake propels itself by exerting force laterally, it does apply other forces to the ground. One of them is a vertical force induced by the weight of the animal. Another is the horizontal force due to friction. Only the frictional force is exerted by muscle activity; it is the result of the movement produced when a snake curves its trunk in order to apply propelling forces against surrounding objects.

What effect do these forces have on a snake's movement? Friction is produced by lateral undulation in two ways: by the underside of the body sliding along the ground and by the flanks of the body sliding along objects to the side. The magnitude of such sliding friction equals the force pressing the two surfaces together multiplied by a coefficient determined by their nature. In the first instance this amounts to the weight of the snake times the coefficient of friction between the snake's belly skin and the ground. In the second instance it is a function of the force exerted against the object to the side times the coefficient between the flank skin and the object.

In both instances the frictional forces act in opposition to the force inducing motion; thus in lateral undulation friction has to be overcome by muscular effort. This being so, there is a selective evolutionary pressure to reduce the frictional coefficient of the snake's skin, and various kinds of snake scales have been shown to have the low frictional coefficient of about .35. That friction is unnecessary to the locomotion of snakes can be demonstrated by placing a snake in a virtually frictionless experimental environment: a smooth surface lubricated with powder and studded with vertical pegs that are fixed in position but able to rotate. The snake will traverse such a surface even more rapidly than it can cross one studded with nonrotating pegs.

Continuous forward progression by a snake requires a minimum of three contact sites at all times. Moreover, the contact sites must have a particular spatial arrangement. The snake pushes against two of the sites to generate force and uses the third to balance the forces poduced at the other two so that its body can move in a particular direction. If the snake is to move, the contact points need to be on both sides of its body and the forces on the contact points must be exerted at an acute angle to the rear. Such considerations suggest that as the snake moves from one contact point to another it shifts its position slightly, and this is confirmed in films we have made, The transition from one set of three contact points to the next must be made quickly; otherwise the snake must utilize four or more contact points at a time and will lose efficiency.

The mimimum number of three contact points represents a compromise between two opposing factors. One factor is the snake's flexibility. Since a snake is notably flexible it must expend a certain amount of energy to make its body rigid in order to transmit force from one contact point to another. The amount of energy applied for this purpose can be reduced if the number of contact points is increased; less energy is needed to maintain rigidity over several short distances than is needed over a few long ones. Yet the greater the number of contact points, the shorter the amplitude of the snake's curves, the wider the angle of force application and the greater the waste of force in the snake's direction of motion. The snake balances the requirement for internal force transmission against the requirement for efficient force application by establishing a number of contact points that is consistent with both. The number of points is also affected by the "desired" speed, the relative elongation (diameter divided by length) of the body and the spacing of objects to the side.

Experiments in which snakes of varying sizes are chased through a fixed array of pegs suggest that the speed of a snake is not proportional to its size. In fact, the limitations on the speed of snakes remain obscure. There may well be an optimum peg-spacing for a particular size of snake, where optimum means the facilitation of a maximum velocity rather than a minimum energy expenditure per traverse. No one has yet done experiments with different peg-spacings to see if such spacings would actually affect the snakes' velocities. Nor have there been studies relating speed to work outout or speed to the time or distance a snake is capable of travelling.

Such questions about peg-spacings and efficient movement lead one to ask further questions about control mechanisms. As we have seen, the body of a snake can twist around its long axis to some extent. Thus a snake is capable of lifting a loop over one contact point and making use of another, an ability that enables the snake to select advantageous contact points in its environment. What environmental factors affect such loop placement? Experiments indicate that the key factor may well be the relative resistance of contact points. A snake moving by lateral undulation through grass or sand will increase the size of its loops in order to displace loose surface material until it can exert the maximum force possible under the circumstances. Yet how does the snake monitor information about such resistance, and what is the feedback sequence that shifts the information from one segment of the snake's body to the next as the snake moves past a particular contact point?

The lateral undulatory progression of a snake has the inherent advantage that the capacity for establishing staggered, backward-travelling waves of muscular contraction is phylogenetically much older than snakes. Snakes have only had to develop a feedback sequence for the control of such waves.

This kind of progression is also effective because it permits the snake's body to travel at a relatively constant speed with only limited changes in its velocity. Yet lateral undulation does not work under all circumstances. It is useless for travelling down a tunnel with parallel sides: no backward-directed forces can be exerted against the sides of the tunnel. It is also ineffective on a flat surface that lacks elevations and depressions, and on a rounded surface such as a branch or a fallen tree trunk. Indeed, some snakes find it difficult to use lateral undulation at all. This mode of locomotion is well suited to relatively slender snakes, but it is less effective for short, stout ones. When a snake species has the option of invading an environmental niche that requires a relatively stout body, there is immediate selective pressure for a substitute method of locomotion.

An alternative is rectilinear movement. This mode of locomotion differs from lateral undulation in two respects: it involves the application of force somewhat downward instead of laterally, and it is effective only if friction is established between the snake's skin and the ground. Rectilinear locomotion is made possible by the very loose skin found in many snakes (perhaps as a result of such a specialization as prey constriction). In order to move in this mode the snake fixes several series of scutes and starts to move the skin between them. For example, it pulls together a series of scutes near its head, fixes them against the ground and then moves the rest of its body with respect to this fixed zone and several similar zones behind it. As the body of the snake moves forward the skin is stretched, pulling the forward-most scutes of each series out of contact with the ground, while additional scutes are continuously pulled up to the rear edge of the series. In this way a constant length of belly surface remains fixed to the ground. Normally a snake fixes two or three of these zones to the ground at once, and they can be seen to move continuously to the rear as the snake progresses.

In rectilinear locomotion the two sides of the snake must move symmetrically rarher than in alternation. X-ray motion pictures have confirmed an earlier observation that in this mode of movement the ribs and the vertebrae do not move with respect to one another. The scutes are shifted forward by slender muscles that stretch forward from the sides of each scute to points high on the side of each rib. A second set of muscles runs to the rear at a shallower angle to attach the skin to the bottom ends of the ribs. Contraction of the muscles in the first group pulls the skin forward and up, out of sliding contact with the ground; contraction of the stouter and mechanically better-placed muscles of the second group accelerates the snake's mass and maintains its contant forward movement.

The fact that rectilinear locomotion relies on friction places a certain limitation on this form of movement. The particular form of friction involved is called static friction. Static friction is defined as the force that must be applied parallel to the contact surface between two objects in order to start them sliding past each other. Then static friction becomes sliding friction. It follows that a snake can only exert a force parallel to the ground that is less than the force of static friction; otherwise the animal will slip. A snake in rectilinear movement must acclerate smoothly, because even a temporary imbalance may shift a group of scutes from static to sliding contact and make the effort ineffective.

Even those snakes that have achieved the necessary body form to move by rectilinear progression find this mode of locomotion limiting because of the need to keep some parts in static contact. Rectilinear progression enables the snake to cross flat surfaces and to advance in a straight line when it is stalking prey. Such progression, however, is slow because each scale must establish stationary contact with the ground before it can transmit horizontal forces. It seems that motor control is again of critical importance in keeping the contact zones from sliding.

Most snakes that do not have the muscle and bone structures necessary for rectilinear progression can still use static friction in locomotion by employing concertina progression. In order to move in this way the snake draws itself into an S-shaped curve similar to the posture assumed in lateral undulation, and sets the curved portion of its body in static contact with the ground. Motion begins when the head, the neck and the forward part of the body are extended by forces

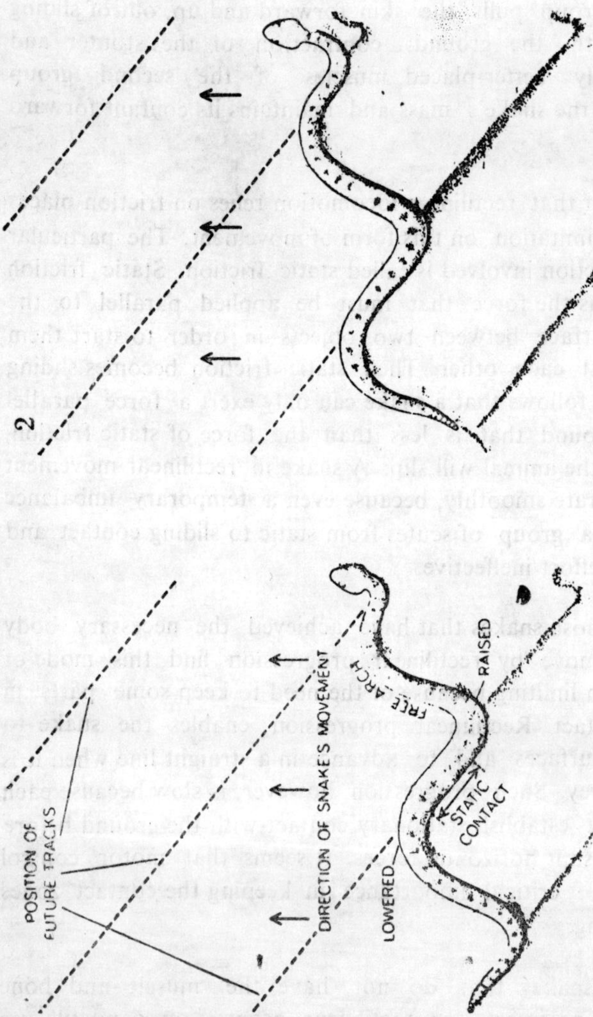

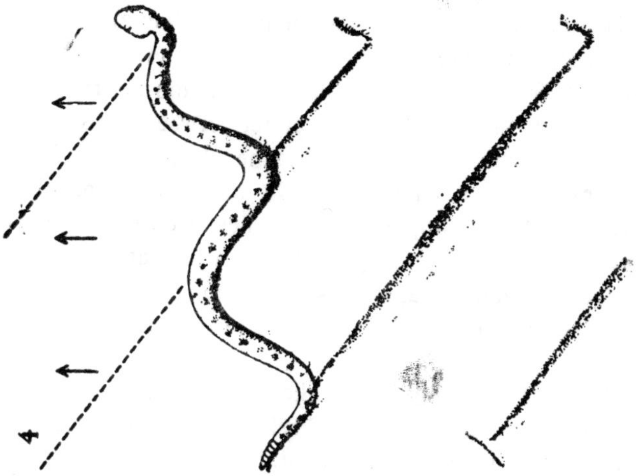

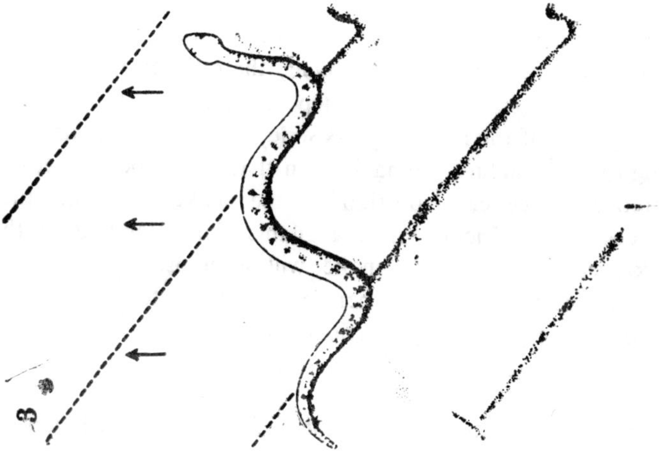

Sidewinding enables desert vipers to move rapidly. In 1 a snake with tail in first track (lower left) has lifted head and forebody from hooked second track (lower right) and arched forward to begin a third track. In 2 head has risen from this new track at right. Meanwhile the snake's body has lifted clear of the first track and begun to move into the second one (broken outline of snake's body indicates earlier position). In 3 snake's body lies in both the second and the third track. In 4 a fourth track has been started.

transmitted to the ground in the zone that remains in stationary contact. These forces produce movement by acting against the force of friction generated by the weight of the snake's body. This reserve of friction is called the static-friction reservoir. It should be noted that the area of contact between the snake's body and the ground does not affect the "capacity" of the reservoir. Friction is a function solely of the force pressing two surfaces together multiplied by a coefficient that represents their roughness. The force available for acceleration in the static-friction reservoir remains constant as long as the snake continues to support its weight on the stationary zone.

After the front end of such a snake has moved forward a short distance it stops. This establishes a new zone of stationary contact in which horizontal forces are exerted against the ground, and the rear end of the snake's body is pulled forward. The concertina progression of a snake is thus rather like the locomotion of an inchworm.

As long as the moving portion of the snake's body can be lifted out of contact with the ground, the ratio of the moving portion to the stationary one depends only on the snake's stability. Yet if the moving parts slide forward or are dragged along they will induce sliding friction that must be overcome. Moreover, since each portion of the snake's weight that includes sliding friction is unavailable for maintaining the static-friction reservoir, the snake will attempt to keep such areas to a minimum. It is therefore possible to predict the approximate fraction of a particular snake's body that can be kept in sliding motion over a particular kind of surface.

Snakes sometimes supplement weight-induced friction by muscular force, thereby enlarging the static-friction reservoir. Such modified concertina progression is often used to traverse a straight-sided channel or to climb tree branches. In the first instance the snake widens the amplitude of its S-shaped loops, actively pushing them into contact with the channel's walls. In the second instance the snake forces its stationary surfaces into contact by constricting around a branch. Such enhancement of concertina progression probably arose among species that were already adapted for it through the accident of having an overdeveloped axial musculature for the constriction of prey.

This observation pays an unexpected literary dividend. In "The adventure of the Speckled Band" Sh erlock Holmes solves a murder mystery by showing that the victim has been killed by a Russell's viper that has climbed up a bell rope. What Holmes did not realize was that Russell's viper is not a constrictor. The snake is therefore incapable of concertina movement and could not have climbed the rope. Either the snake reached its victim some other way or the case remains open.

Concertina progression is almost as common as lateral undulation, and one often sees a long snake combining the two kinds of movement. Various species of rat snake use this combination of movements to climb trees; they use the irregularities in the bark as channels. Many such snakes have a double keel on their belly. The structure enables the snake to push laterally against irregularities in the surface it is climbing. Certain burrowing snakes increase their propulsive force by digging in their tail and then straightening their trunk.

Since concertina progression is effected by a bending of the body, it proceeds by asymmetrical contractions. Observation confirms that these intermittent contraction waves move along the body toward the tail. It thus appears that concertina progression is essentially a low-speed pattern of movement, because the animal must pause while it brings its rear end forward so that it can proceed again. Although concertina and rectilinear progression are suited to surfaces where lateral undulation would be ineffective, the price is lack of speed.

For pursuit—or escape—snakes need a mechancism that enables them to employ static friction without sacrificing speed. What is needed is a means of locomotion that keeps the zones that are in static contact from slipping, allows contraction waves to pass continuously and regularly down the snake's body and prevents minor irregularities in the surface from affecting the pattern. Such a form of locomotion is sidewinding, which was first properly described by Walter Mosauer in 1932. A sidewinding snake achieves firm static contact by moving so that its body lies almost at right angles to the direction of its travel. In this orientation the snake is more likely to encounter rocks and other slight irregularities against which it can brace itself than it would if it were moving straight ahead, for the reason that it is sweeping out an area many times wider than the width of its loops.

The track of a sidewinding snake that has traversed a smooth flat surface appears as a series of straight parallel lines, each inclined some 60 degrees to the snake's direction of motion and each about as long as the snake itself. The rear end of each line is bent into a short, forward-pointing hook. The track of the scutes is well defined; slippage, if any, can therefore only be sideways.

The series of actions producing these tracks can be visualized by assuming that the snake is lying so that its tail points about 60 degrees away from the direction in which the animal is going to move. The snake's head is raised and turned through an obtuse angle, so that it faces in the direction of travel. Only the bend of the snake's neck is in contact with the ground; this produces the hook at the rear end of each track.

As the snake starts to move forward it lifts its head and neck off the ground. In order to reach the next track the forward part of the body has to curve in a smooth loop. The extension of the snake's front end continues until approximately a quarter of the trunk is cantilevered out of contect with the ground. The head then arches downward as the cantilevered trunk remains off the ground; in making the first contact the neck bends at the next track of the sequence, so that another hook is produced. Successive sections of the body and then the tail follow along the new track, which parallels the preceding one. Considerably before the tail has been pulled into the second track the snake's front end starts into the third track. The body of a sidewinding snake thus lies on two or three separate tracks, with the body part between tracks held off the ground.

Although the snake's body swings through loops that are reminiscent of lateral undulations, one can see that the force-transmission pattern is more like that of the concertina sequence. The force for the initial acceleration of the front part of the snake, and for maintaining the velocity of the moving parts, must be transmitted by static friction of the snake's belly. It is this pattern that gives sideminding its peculiar advantages. When a nonsidewinding snake travels over loose sand, its lateral force shifts the sand and its track becomes a trough. Moreover, a snake that exerts a lateral force exceeding the resistance capacity of the force points tends to slip sideways instead of moving forward. Similarly, a snake in rectilinear locomotion will just dig in, slipping within its track, if the force exerted by the contact zones exceeds the limit of the static-friction reservoir. For the sidewinder such slippage proceeds sideways and causes the snake's body to dig in, piling up sand along it length and increasing the forces that produce forward movement. In other words, a slipping sidewinder quickly stabilizes itself by converting sliding friction into static friction.

H. Mendelssohn of the University of Tel Aviv and I are currently analyzing the feedback mechanisms that control sidewinding. The energy required for sidewinding is apparently reflected in the height to which the travelling loops are cantilevered off the ground. The loop height is lowest for those surfaces that are smoothest and these therefore require the least amount of energy. Indeed, when the snakes are allowed to travel over a smooth, low-friction surface such as a polished terrazzo floor, they will slide rather than lift the moving parts of their body. If their tracks were visible, they would tend to be connected. When the snakes traverse a smooth but high-friction surface such as sandpaper, the tracks are quite separate. A rough surface such as a layer of crushed, sharp-edged aggregate causes a sidewinding snake to lift its loops quite high. Frame-by-frame analysis of films suggests that the loops continuously change their height.

Sidewinding has the further advantage that the contraction waves can start at the neck and pass down the snake's lateral musculature. Each motor sequence is thus continuous even though parts of the body stop and start. The snake is accordingly able to cross relatively flat areas, liberated from the need to search for lateral irregularities to propel itself.

Sidewinding seems to be a method of locomotion available to any kind of snake. Various conditions elicit it, and no special structural modification is required. Yet although many snakes sidewind under stress, only a few species do it effectively, and the beautiful control and minimal energy utilization described here were observed only in desert vipers. Such snakes seem to have found this method advantageous for travelling quickly over flat, sandy and rocky surfaces. Sidewinding would seem to be particularly useful when the crossing has to be made at midday, when speed and minimum contact may prevent overheating.

The various sidewinders differ in their track angles and in the number of tracks they occupy at one time. One of the most spectacular of the movement patterns that have been observed so far is seen in juvenile specimens of the South-west African desert viper, a species that has evolved a special escape sequence when it is faced with heat stress. The muscle-contraction waves travel down the snake's body at very high speed, and some specimens jump completely off the ground from track to track.

10

Navigation of the Green Turtle

These animals are known to migrate regularly to and from nesting beach and feeding ground. Now it appears they can also navigate well enough to make an island landfall after 1,400 miles at sea.

One of the stubborn puzzles of animal behaviour is the ability of some animals to travel regularly to remote oceanic islands. The best-known of the blue-water navigators are birds, recently, however, the green turtle (Chelonia mydas) has given evidence of being as keen an island-finder as the sooty tern or the albatross. Because green turtles swim slowly, and do so at the surface of the water or a little below it, they are potentially easier to follow in their journeys than either birds or migratory fishes, seals and whales. The green turtle may therefore prove to be an important experimental subject for students of animal navigation.

The evidence that green turtles can find their way to remote oceanic islands is provided by female green turtles that normally inhabit feeding grounds along the coast of Brazil. It appears that once every two or three years these turtles swim all the way to Ascension Island—a target five miles wide and 1,400 miles away in the South Atlantic—to lay their eggs. By the processes of natural selection this population seems to have evolved the capacity to hold a true course across hundreds of miles of sea, using only animal senses as instruments of navigation. The difficulties facing such a voyage would seem insurmountable if it were not so clear that the turtles are somehow surmounting them.

The green turtle is one of the five kinds of sea turtle found throughout the warmer oceans of the world. Adult green turtles, which may weigh more than 500 pounds, are herbivorous; they feed on the so-called turtle grass that grows abundantly in sheltered tropical shallows. Although the green turtles of the world are separated into reproductively isolated breeding colonies, they show little tendency to evolve into recognizable local races and species. The green turtles of the Pacific, for example, show only minor differences in form and colour from those of the Atlantic. The one area in which what appears to be a well-differentiated species has evolved is the northern coast of Australia, where the form Chelonia depressa is found. Green turtles nest only in places where the average temperature of the surface water during the coldest month of the year is above 68 degrees Fahrenheit. In the Atlantic the northern limit of their nesting range seems to have been Bermuda; early voyagers to the New World destroyed the colony there. The most northerly nesting site known in the Pacific is French Frigate Shoal, an outlier of the Hawaiian Islands.

Until a few years ago what was known about the green turtle consisted mainly of cooking recipes and a sea of folklore from which rose only a few islands of fact. Among these were studies by Edward Banks in the Turtle Islands of Sarawak; by James Hornell in the Seychelles Islands off the east coast of Africa, by P.E.P. Deraniyagala in Ceylon and by F.W. Moorhouse along the Great Barrier Reef of Australia. More recently in Sarawak, Tom Harrisson and John R. Hendrickson have independently uncovered a great deal of information on the nesting ecology and reproductive cycles of the huge breeding colony there. Since 1955 my colleagues at the University of

Florida and I have been working, with the aid of a series of National Science Foundation grants, at the only green turtle nesting beach that remains in the western Caribbean: Tortuguero in Costa Rica. What is now known about Chelonia from all these studies permits the piecing together of a coherent—albeit still somewhat fragmentary—account of its life.

Although green turtles are primarily sea animals, in a few places in the Pacific they sometimes go ashore to bask. They have been seen lying in the sun along with albatrosses and basking monk seals on such small Pacific islands as Pearl and Hermes Reef, Lisianski and Kure Atoll. The females of the Pacific populations sometimes even nest during the day. Neither nesting during the day nor basking appears to be the habit of green turtles in any Atlantic population. Once the Atlantic hatchings leave the beach the males remain at sea for the rest of their lives. The females come ashore only to nest, and they do so after dark. They always return to the same general nesting area and often to the same narrow sector of beach.

The green turtle is one of the few reptiles that are known to reproduce at intervals of more than a year. In Sarawak the entire turtle population nests in a three-year cycle. At Tortuguero about a third of the colony returns to nest every two years and the other two-thirds follows a three-year schedule. In our eight years of tagging turtles at this site no turtle has returned after an absence of only one year. When the Tortuguero turtles finish nesting, they travel back to their home pastures of turtle grass and evidently remain there feeding until their reproductive rhythm directs them to return to the nesting beach. Such a feeding ground may be within a few dozen miles of the nesting beach or many hundreds of miles away.

The time of year at which the nesting season begins and the duration of the season vary from one region to another. In some places nesting is restricted to three or four months of the year; in others it may run through the entire year, with a peak during two or three months or even with two separate peaks. Mating takes place at the nesting ground and apparently nowhere else. Either the males accompany the females during their journey from the home pasture or they make an exactly timed rendezvous with them at the nesting beach. In any case, as soon as the first tracks of nesting turtles appear on the shore the animals can be seen at sea just beyond the surf, courting, fighting and mating. Because these activities are mostly out of sight little is known about them. From a low-flying airplane one sometimes gets a glimpse of two males splashing around a single female. Many females come ashore scratched and lacerated, evidently because of the violent attentions of the males.

Some females apparently mate just before their first nesting trip ashore; others do so later. At Tortuguero mating activity appears to end after the third or fourth week of the nesting season. Even when mating precedes the first landing, it must take place after at least some of the female's eggs have formed shells. It therefore seems unlikely that any of the eggs of the current season are fertilized as a result of current mating. The encounter probably serves to fertilize eggs for the next nesting season, two or three years ahead.

Once at the nesting beach the female goes ashore from three to seven times to deposit clutches of eggs in the sand. The interval between nesting trips is about 12 days, and the number of eggs laid on each occasion is 100 or so. Each egg is about two inches in diameter. The shell is flexible, and when the egg is first laid it has a curious dent in it that no amount of pressing will smooth out. The incubation period is about 57 days.

On hatching and emerging from the nest the young green turtles nearly always set off on a direct course for the sea, even when the water itself is completely hidden by dunes or other obstacles. Experiments with a number of different kinds of marine and freshwater turtles indicate that the mechanism of this orientation is an innate response to some quality of the light over open water. It is clear that no compass-like sense is involved: in a series of tests we flew little turtles from Caribbean nests across Costa Rica and allowed them to emerge from artificial nests on a Pacific beach. Even though the sea now lay in the opposite direction the young turtles reached the water as easily as their siblings on the home beach did.

After the hatchlings enter the water, either their sea-finding drive gives way to other orienting mechanisms or they simply keep swimming until whatever difference they perceive between sea light and land light becomes too slight to provide a guiding stimulus. By this time they will probably have been picked up by a longshore current. Since the adults navigate over long distances, they must have some kind of compass sense. This sense may be latent in the hatchlings: recent laboratory experiments at Duke University by Klaus Fischer seem to show that this is the case. In nature, however, this sense evidently does not come into operation until the light-beacon sense fails. In any case, tests have shown that the sea-finding drive is not lost even when turtles are kept away from water for as long as a year after hatchling. It is also evident that both capacities are present in the mature female. The compass sense must guide her from feeding ground to nesting beach, and the sea-finding drive directs her back to the ocean after she has nested—even though it is out of sight of her nesting place.

A recent experiment demonstrates the kind of ambiguity that can arise in the study of orientation among young green turtles. Several hundred 20-day-old hatchlings from Tortuguero were placed in a circular tank at the Lerner Marine Laboratory on the island of Bimini in the Bahamas. The nearest water was on the bay side of the island, some 40 yards away. The ocean was some 200 yards away in the opposite direction. From turtle's-eye level the rim of the tank blocked any view of either bay or ocean. The skyline was broken by trees and buildings.

The distribution of the young turtles in the tank was recorded over a three-day period, at 9.00 A.M., 4.00 P.M. and 11.00 P.M. each day. During most of that time the wind blew steadily from the bay side of the tank. At night, while the sleeping turtles floated on the surface, the steady breeze piled them up on the ocean side of the tank. Once awake in daylight, however, the turtles showed a marked orientation in the opposite direction. There was active—at times frantic—swimming toward and crowding along the wall nearest the bay. This bias was not simply a tendency to swim upwind; the bay side of the tank was equally favoured during a few windless periods. The response could have been an innate direction preference based on a compass sense, but it was more probably the same light-seeking urge that guides hatchlings from the nest to the water. Two months later, however, when the same test was repeated with the same animals, they showed no seaward orientation at all.

Almost nothing is known about the movements and habits of green turtles during the first year of life. They are mainly carnivorous at this age, but they are able to feed only on small, weak marine invertebrates. Such prey is scarce both at the nesting beaches and at the turtle-grass pastures that feed the grown turtles, and very young turtles are almost never seen in these places. It seems likely that the hatchlings must spend their first months moving from location to location at sea as growth qualifies them to feed on invertebrates of increasing size. All we really know is that the hatchlings disappear.

The only place in the Atlantic and Caribbean regions where we have been able to study green turtles of an age between hatching and maturity is off the west coast of Florida. There a migrating population of young turtles, ranging in weight from 10 to 90 pounds, shows up each April. At this stage most of them have become herbivorous, and they spend the summer months browsing in the turtle-grass flats between Tarpon Springs and the month of the Suwannee River. In November they move away to an unknown destination. These Florida visitors may have come from the Costa Rican nesting ground. We have no proof of this, however, because of the difficulty of devising a marking system that will survive the changes in size and proportion of the growing animals.

As recently as 10 years ago it was not definitely known that green turtles migrated long distances from home range to nesting beach. The first clear evidence of periodic long-range migrations came from the tagging programme conducted at Tortuguero. During the past eight years 3,205 adult turtles have been tagged; 129 of these have later been recovered. Most of the tags have come to us from professional turtle fishermen operating off the coast of Nicaragua, but other recovery sites are distributed over an area more than 1,500 miles across at its widest point. The eight most distant recoveries are as follows: one from the Marquesas Keys off the tip of Florida, one from the northern coast of Cuba, four from the Gulf of Mexico off the Yucatan Peninsula and two from the Gulf of Maracaibo in Venezuela.

These returns furnish grounds for some generalizations that support the reality of migratory travel by Chelonia. One such generalization is that no turtle tagged at Tortuguero has ever been recovered there after the end of the nesting season. Another is that no turtle tagged at Tortuguero has ever been found nesting anywhere else. A third is that there is very little correlation between the time elapsed after tagging and the distance the tagged turtle travelled from Tortuguero. This strongly suggests that the turtles are not random wanderers but migrants following a fixed travel schedule between the nesting beach and their restricted home range. Otherwise the animals would tend to cover the same distance in the same time.

Although such findings make a strong case for periodic migration and homing by the green turtles that nest at Tortuguero, they do not prove a capacity for true navigation, that is, oriented travel involving something more than piloting by landmarks and an ability to keep headed in a fixed direction by using the sun or stars for a compass; such piloting is accomplished by many different kinds of animals. The beach at Tortuguero is part of a maintained shore. The green turtles could simply leave their distant pastures on an initially correct compass heading and, on making landfall, follow the coast until things look, smell or taste in ways that mean the ancestral breeding ground has been reached. In their successive nestings during a single season females often return to the same 200-yard stretch of nesting shore they had used earlier.

Indeed, it appears certain that once the turtles have reached the Tortuguero area, they do search the shore for cues to guide them to a nesting site. At the start of the nesting season distinctive "half-moon" turtle tracks appear, usually toward one end of the 22-mile length of nesting beach. These are semicircular or U-shaped trails left by females that have come out of the water, made a short trip toward the upper beach and then turned back to the surf without nesting. Such behaviour implies that some sort of discriminatory process is involved in the selection of a nesting site. In the course of coming ashore a female nearly always stops in the backwash of the surf and presses her snout deliberately against the sand, sometimes repeating the process as she moves up the wet lower beach. This behaviour appears to be an olfactory assessment of the shore, although it could also be tactile. Little more is known about the senses involved in the selection of nesting sites.

In searching for an instance of turtle migration that clearly involves an ability to make a long, oriented sea voyage in the absence of landmarks, I thought of the green turtle colony that nests on Ascension Island. The capacity for open-sea orientation is the ultimate puzzle in the study of animal navigation. Even human navigators, with their ability to measure the position of the sun and the stars, were unable to calculate positions at sea accurately until the development of precise chronometers in the 18th century. Work with various animals, notably homing pigeons and migratory birds, suggests that there are three inherent aids to navigation: a clock sense, a map sense and a compass sense. On the basis of these senses the navigation feats of animals that migrate overland can be explained, at least in theory. On the featureless open sea, however, the situation is quite different.

One difficulty in the study of open-sea navigation by animals is the dearth of information on routes and schedules of travel. I am not aware of a single instance in which the journey of an oceanic migrant has been traced in detail, with data on headings and speeds. Only when this has been done will it be possible to say that here piloting by visual guideposts is taking place, there celestial navigation can be assumed and elsewhere some cryptic signal must be involved.

Before we could use the green turtle nesting colony on Ascension Island for a study of open-sea navigation it was necessary to establish whether the turtles travelled there from some distance away or were merely a local population. That they were not local residents seemed certain on two grounds:

1. the turtles arrive to nest in substantial numbers each February but disappear by June, and
2. there are no beds of turtle grass anywhere in the vicinity of Ascension.

It was then necessary to find out where the feeding grounds of the Ascension nesting colony were located. A survey of the coasts of Argentina and Brazil in 1957 yielded no evidence that the abundant green turtle population of the Brazilian coast came from nesting grounds on the South American mainland. Turtles were known to nest on the island of Trindade, some 700 miles off the Brazilian coast, but their numbers were small. It therefore seemed likely that the green turtles that nest on Ascension feed along the coast of Brazil. The next step was to test the reality of this apparent instance of open-sea migration by means of a tagging programme.

The green turtles at Ascension, like those at Tortuguero, breed in either two-year or three-year cycles. During February, March and April of 1960 Harold Hirth, then a graduate student at the University of Florida, tagged 206 female turtles at the six nesting beaches on the island. In 1963, when that part of the 1960 population which nests every three years was due to return to Ascension, a tag patrol was set up at the beaches. Three of the turtles Hirth had tagged were found again. In 1964, when the two-year group was due to return for the second time, two more of Hirth's turtles showed up [see lower illustration on next page]. Four of these five females had landed on the same short section of beach where Hirth had tagged them; the fifth went ashore on an adjacent beach recently formed by high seas. During the past five years nine more of Hirth's 1960 tags have been recovered from turtles captured by fishermen along the Brazilian coast.

A skeptical statistician might attribute these findings to random wandering. Short of tracking a turtle all the way, clinching proof that green turtles travel from Brazil to Ascension will be obtained only when a turtle tagged on Ascension is recaptured in Brazilian waters, released there and then captured again on Ascension. This is not likely to happen; by the time a tag reaches the University of Florida from Brazil the turtle that carried it has usually been eaten. Thus the evidence remains circumstantial: turtles tagged on Ascension have been captured off Brazil; others have disappeared for three or four years only to return to the same Ascension beaches on which they were first encountered. This does not prove the reality of the migratory pattern beyond all possible doubt, but I think it does so beyond reasonable doubt.

There are three basic questions to be asked about this particular migratory journey: How was it originally established as a behavioural adaptation? What route does it follow? What guides the turtles? With respect to the first question, the establishment of a nesting colony on a tiny mid-ocean island such as Ascension seems an evolutionary venture most unlikely to succeed. The problem is to visualize the selective process and the survival values in the earliest stages of the evolution of such a migratory pattern. If Ascension had once been a much more extensive area of land, the navigational equipment with which the green turtles now find the island could have been slowly refined by natural selection as the area shrank to a small island. I have found no one, however, who believes the area of Ascension can have become appreciably smaller during the past 50 million years or so. Certainly the water around it today is to deep that such a change seems unlikely.

There is one possibility that may decrease the theoretical difficulty to some degree. Perhaps the island was originally colonized by green turtles that had been accidentally carried from West Africa by the South Equatorial Current, which flows from east to west. Another kind of sea turtle—the West African ridley (Lepidochelys olivacea)—has colonized the coast of the Guianas, north to Brazil, in just this fashion. If egg-bearing green turtles had landed by chance on Ascension and nested there, the process of selection would have had good material to work on. The hatchlings leaving the island would have been carried to the coast of Brazil by the South Equatorial Current and at the same time could have borne with them "imprinted" information that would help them to retrace their journey at maturity. Such a hypothesis of course does no more than get the nesting colony established on Ascension. It makes no attempt to solve the navigation puzzle.

The next question concerns the route followed from Ascension to Brazil and back. Apart from inference, nothing is known about this. The South Equatorial Current presumably carries the hatchlings to Brazil; the shortest path for mature turtles returning to the island would be directly estward from Brazil against the thrust of the same current. Such a route would conform to a classic pattern for aquatic migrations: upstream movement for the strong adult animals and downstream travel for the weak and inexperienced young.

There are two other routes that would allow the entire round trip to be made with the aid of favourable currents. If the Ascension-bound migrants swam northward from Brazil, they would enter the part of the Equatorial Current destined to become the Gulf Stream and could stay in this current until a full circle brought them westward to Ascension from the coast of West Africa. If instead they rode the Brazil Current southward, they could then travel the West Wind Drift to South Africa, catch the northward-flowing Benguela Current and rejoin the South Equatorial Current on its way to Ascension and beyond. The time required for these journeys seems prohibitive. Both would have to be negotiated without food, and the temperature of the West Wind Drift waters can drop to a chilling 40 degrees F. The direct easterly course, or some modification of it, appears to be the most logical route from Brazil to Ascension.

As for the final question, in attempting to judge how the Brazil-to-Ascension migration is guided the first step will be to determine how and where some sort of contact—direct or indirect—is established with the island. Even with a pinpoint target such as Ascension, migrating birds would be able to correct a fairly gross navigational error by visual means. To a turtle, however, an island is out of sight a few miles away. Green Mountain on Ascension stands 5,000 feet high and often has a corona of cloud that rises much higher. The sight of birds converging on the island from miles away at sea might provide an approaching turtle with still another visual guide. Such signals, however, would probably be picked up only in the last 20 miles or so. What cues might there be at longer range?

Very little is known about nonvisual phenomena associated with the presence or direction of an island, nor is anyone sure about what a green turtle can taste, smell or hear. It seems a point in favour of the upstream hypothesis of travel to Ascension, however, that such a route would allow the approaching turtles to detect an olfactory gradient—if indeed such a chemical cue is given off by the island. Perhaps when the hatchling leave the island they take with them an imprinted memory of the taste or smell of Ascension water. Coming back as mature adults they may be able to detect this Ascension effusion in the westbound current far downstream from the island and to follow it until they make a visual landfall. There are two basic weaknesses in this proposal: no one knows how far downstream a green turtle can taste or smell an island, and there is no information to indicate the direction from which the migrants approach Ascension.

Even if we assume that far-reaching cues guide the final homing of the turtles to Ascension, there remain hundreds of miles of open ocean to be crossed—and crossed with precision—before such signs could conceivably be detected. For the greater part of this distance the animals must be navigating, and the most logical assumption appears to be that the navigating is done with information from celestial guideposts. Colin Pennycuick of the University of Cambridge has recently called my attention to an old island-finding technique used by human navigators before the accurate calculation of longitude was possible. Knowing the latitude of his target, the navigator would sail north or south by compass until the noon position of the sun showed that he had reached the desired latitude; he would then simply sail due east or west (depending on the location of his target) until he made landfall.

This human technique seems particularly worth considering in connection with the migration of green turtles to Ascension. Possibly the turtles go north or south along the Brazilian coast until they reach the vicinity of their first landfall as hatchlings. Visual or olfactory impressions from the first contact with the mainland would identify that coastal point, which should be close to the latitude of Ascension. Thereafter compass sense alone might conceivably guide the turtles on a journey due east to a point where cues from the island could be detected.

This proposal too has weaknesses. Any drifting due to wind or current during the hundreds of miles of open-sea swimming could put the migrant hopelessly off course in spite of a constant compass heading. Unless the turtle were able to make corrections by one means or another the landfall at Ascension would never take place. For human navigators today the ability to make such corrections requires finding not only latitude but longitude as well. Latitude can be judged from the height of the sun at noon, but reckoning longitude requires the measurement not only of some celestial body's altitude but also of its azimuth. It is hard to see how this can be done effectively by an animal in the open sea; the azimuth measurement—the horizontal components of the movement of the sun and the stars—could not be made against a featureless horizon. Pennycuick's suggestion nontheless has hypothetical merit in regard to the Ascension migration. Eastbound turtles that were displaced north or south during the journey could conceivably correct for drift on the basis of latitude reckoning alone. Even this, of course, would be an astounding animal adaptation.

11

Migration in Birds

Any movement between two areas is called migration. As a rule, it is a response of an animal population to changes in environmental conditions.

Birds are more uniformly migratory than any other group of animals. Nearly all orders of birds include species which perform migrations; in order vertebrates and in the lower groups of animals the migratory habit occurs in species scattered through a smaller proportion of orders.

Among birds there are two common kinds of migration, daily and seasonal. A daily migration is a movement to and from a familiar place such as a roosting area. A seasonal migration, on the other hand, involves a passage at one season from a place of hatching and a return at another season to the same general area. This section of the book is concerned entirely with seasonal migration.

For centuries the phenomenon of migration has been primarily associated with birds; indeed, judging by the references to the subject in the earliest literature, migration was first observed in birds. Since civilization developed in a temperate region of the world where migratory movements are especially pronounced, it is hardly surprising that bird migration has long received attention. Man could not help noticing the flocks of birds in the spring and fall, the seasonal disappearance of some species and reappearance of others; nor could he help being interested in what he saw and eager to investigate what he could not understand.

Today there is an enormous literature on bird migration, based on extensive studies in Europe and North America. And yet the causes and processes of migration are not fully known. The root of the problem is that bird migration, whether it occurs by day or night, is "an unseen movement." One must investigate its mechanisms indirectly through laboratory research on physiological and environmental influences; the analysis of migrant birds mist-netted or otherwise captured, of birds killed during migration by television towers and other man-made hazards, and of returns from banded birds; mathematical calculations based on kinds and numbers of birds observed through a telescope as they fly across the face of the moon; correlation of meteorological data with known migratory activity; deductions derived from direct field observations, radar surveillance, and tracking by radiotelemetry; and experiments on homing and direction-finding. The following pages present briefly some of the important facts and concepts of migrations and suggest a few studies that students may undertage in conjunction with class work.

Causes of Migration

All modern birds—even those incapable of flight—are descended from volant stock. Early in their history birds had the power of flight and presumably could migrate with facility.

Today many species in the temperate regions of the world are strongly migratory, exhibiting mass movements away from both poles as day-length, temperature, and food supply diminish, then reversing the movements at the season when there is a general augmentation of these environmental factors. Such migrations are more evident in the temperate region of the Northern, or Continental, Hemisphere since more species are involved.

Because the north-south migrations of the Northern Hemisphere include many well-known and conspicuous birds which seem to move in the same general way, there has been a tendency to conclude that migration, like bird flight, has developed along the same line in all species, with deviations for adaptive purposes. A study of the migration phenomenon in all groups of birds soon shows the fallacy of this reasoning. One finds that:

1. Some species migrate in directions other than north-south.
2. Some species migrate irrespective of day-length, as in tropical lands.
3. Some species migrate when the temperature is mild and the food supply ample, and others when the opposite conditions are true.
4. Some species migrate as a result of seasonal alternation in rainfall and drought.
5. In certain species, some populations migrate while others do not.
6. In some populations, some individuals migrate while others do not.
7. Some individuals migrate in some years but not in others.

The only conclusion one can safely reach after considering the above peculiarities is that there is no one line along which migration in all birds developed and that there must be different causes of migration in different groups of birds.

A number of authorities have sought causes of migration in historically factors. Three of the several causes suggested bear mention. (1) Bird migration, at least in the Northern

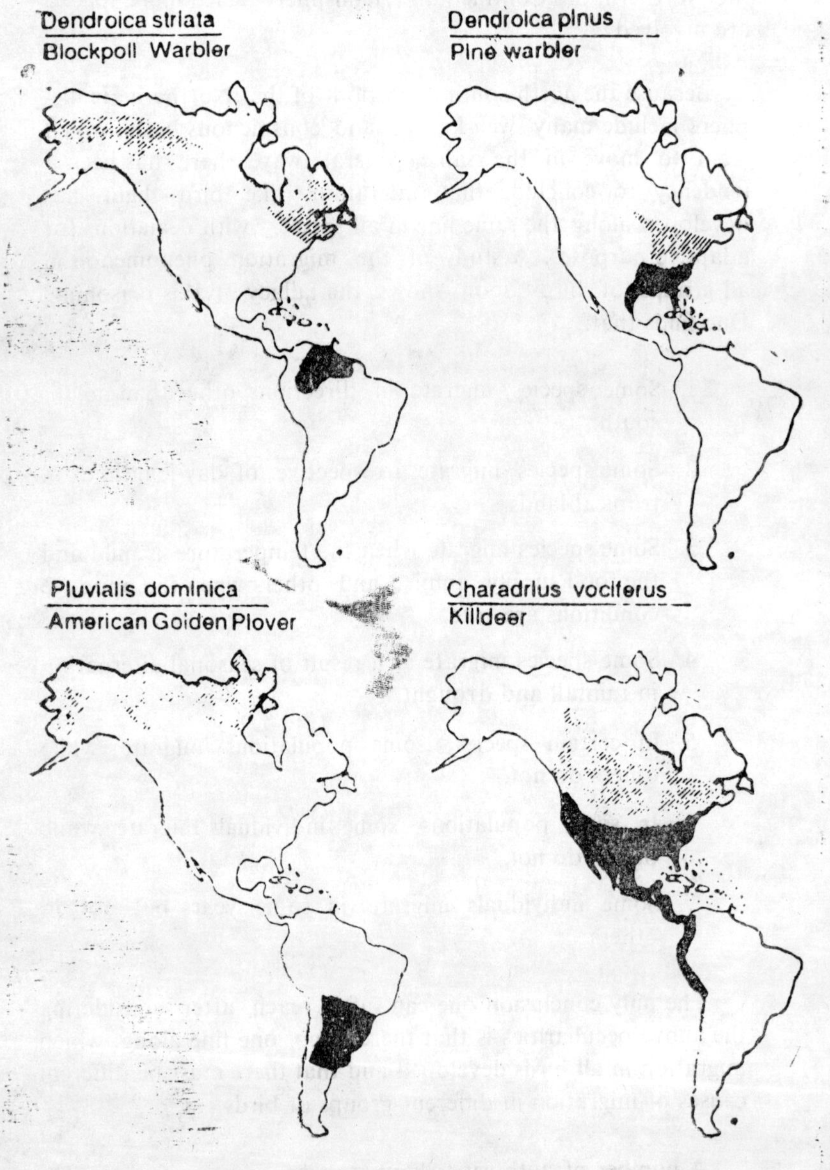

FIG. 11.1: A comparison of long-distance (left) and short-distance migrants within two groups, the warblers (Parulinae, top) and plovers (Charadriidae).

Hemisphere, was initiated by the effects of the Ice Age (Pleistocene). Prior to the coming of the great glaciers to the polar regions, birds lived the year round in the Northern Hemisphere where they originated. Though forced to retreat with the advance of the glaciers, they nevertheless continued to return to nest in the summer because of an innate attachment to their homeland. Objections to this suggestion are several. Many of the birds which are today typical migrants were in existence before the Pleistocene and there is no reason to suppose that they were not already migrating before the glaciers advanced. The retreat from the glaciers does not account for migrations in directions other than north-south, nor does it account for migrations of birds in tropical regions that were never glaciated. (2) Birds migrate in the fall because of the oncoming shortage of food during the winter in their breeding areas. Not that they "know" of the impending lack of food; it is the fact of the food shortage that has caused fall migration to evolve among species which, owing to the shortage, would fail to survive if they stayed for the winter. Birds return in the spring because the northern areas provide more favourable and less crowded conditions for nesting and rearing families. (3) Intraspecific and inter-specific competition for food, territory, nest sites, and so on may have been important factors initiating migration. When, as suggested by Cox, individuals among normally sedentary species could benefit by moving into adjacent areas where the season was favourable and competition reduced, they did so if the hazards of moving did not exceed the gains in their survival and reproduction.

Unquestionably, certain historical factors and various factors prevailing today such as day-length, air temperature, and food supply have influenced migration indirectly by affecting the environment, but no one factor can account for migration or has played the principal role in establishing migration. In view of the current diversity of migratory movements, it is evident that migration has evolved independently in different bird populations through selective pressure. When any resident population experienced an unfavourable situation during a seasonal period in the area it occupied and could gain an advantage by shifting to another area for that period, it gradually developed a migratory pattern, probably with a genetic basis.

Preparation and Stimulus for Migration

Migration is synchronized with the annual seasonal changes, but it will not take place until the bird is internally prepared and outwardly responsive to a stimulus.

Before migrating, a bird must be ready to meet the energy requirements for prolonged flight. It accomplishes this by eating amounts of food in excess of its daily needs and thereby storing energy in the form of subcutaneous fat. At the same time, the bird must become predisposed to migrate by developing a condition commonly called migratory restlessness. In the spring, the physiological process leading toward this state is influenced by the pituitary gland whose activity at this time is stimulated by the total effects of day-length; in the fall, the bird reaches a similar metabolic state during a period when the pituitary is "refractory." When the bird has attained the necessary physiological and behavioural conditions, an outside stimulus is required to trigger migratory behaviour. The stimulus is probably some meteoralogical factor, or combination of factors, such as a change in the temperature of the air, the direction and velocity of the wind, or the onset or passage of a cold front.

Normally an adult migratory bird goes through the special metabolic cycle twice a year, once before the journey to the nesting area and once before the journey away from it. If the bird does not reach the necessary physiological and behavioural condition, it cannot migrate. Furthermore, if the external stimulus for migration is absent, the bird will tend not to migrate even though physiologically and behaviourally capable.

For a summary of knowledge concerning the nature and mechanisms of periodic preparation and stimulus for migration.

Diurnal and Nocturnal Migrations

Migrations proceed by day or night. Many birds—*e.g.*, loons, geese, ducks, gulls, terns, and shore birds—travel by day or night, apparently indifferent to day light or darkness. But this is not the case with many other birds. Herons, hawks, eagles, falcons, crows, hummingbirds, swifts, and swallows migrate only during the day, while nearly all passerine birds (excepting crows, swallows, and a few others) migrate primarily during the night from after sunset until dawn.

At least two explanations have been advanced for the development of nocturnal migration. (1) Movement by night affords birds, which normally live in thick vegeta tional cover and rarely take long flights away from it, the protection of darkness against their diurnal predators. (2) Movement by night affords birds the opportunity of using all the daylight hours of feeding, thereby enabling them to build up sufficient energy resources for sustained long-distance flights.

Effects of Weather on Migration

Normal weather alternates between air and inclement conditions. The movements of the vast majority of migrants show a close correlation with these conditions which are largely governed by barometric pressure patterns, temperature, and wind directions.

To understand how migration takes place with relation to weather, the student must first be acquainted with a few basic facts about weather elements and their sequence over a ground area.

Perpetually sweeping across the continent in an average easterly direction are air masses that vary in velocity, depending on the season and numerous other factors, from 500 to 700 miles a day. In these masses, which may be visualized as roughly circular, are centres of low barometric pressure ("lows"), with generally warmer, more moist air, and centres of high pressure ("highs") with cooler, drier air. Within the lows the air circulates counterclockwise; within the highs, air circulates clockwise. Where the lows and highs adjoin there are boundaries called "fronts." The front of an oncoming high is the cold front, of an oncoming low, the warm front. The area of generally low pressure between the cold and warm fronts is called the warm sector. Any weather map appearing in daily newspapers will show the lines of equal barometric pressure (isobars) as roughly concentric circles around lows and highs, and cold and warm fronts as heavy lines with marks indicating the direction they face. Areas where there has been precipitation are shaded.

The table, "Relation of Weather Elements to Migration," demonstrates in a general way the sequence of spring and fall weather elements when a warm and then a cold front pass over a given area in conterminous United States. The accompanying sketch, "Pattern of Fronts and Wind Directions," shows how the sequence indicated in the table might appear on a weather map. The movements of the elements in both the table and the sketch is to the right (east). Thus the set of weather elements characterzing "Ahead of Warm Front" will be followed by the elements of "Warm Front" and so on. The rapidity with which one set of elements—*e.g.*, the elements of a warm front—arrive, prevail, and finally disappear in an area is dependent on the width of the front and the speed with which the front travels. The width and speed of the front are in turn dependent on the season of the year, the temperature discrepancy between fronts, the wind direction, and continent-wide weather conditions at the time.

Migration in the spring usually takes place with warm weather. Studies of spring migratory movements in eastern United States and Canada show that movements begin at the onset of warm fronts, when barometric pressure is dropping and warm moist air from the Gulf of Mexico and Caribbean Sea is flowing in from a southerly direction. As each low (with mild temperature and southerly winds) passes, movements proceed in full force. On the approach of cold frints, the movements usually slow down—through they may occasional be heavy, as when a cold front advances against a western flank of warm air. When cold fronts arrive, movements stop. Not until the highs have passed will movements begin again.

Migration in the fall usually takes place with cold weather. In the Chicago area migratory movements in September and October start immediately after the passage of cold fronts when there is a flow of continental polar air from a northerly direction. Movements then proceed in full force until the first part of each low passes; thereafter the movements decrease somewhat in intensity and may even cease together as the next cold fronts approach.

The table, "Relation of Weather Elements of Migration" relates spring and fall migration movements to the sequence of weather elements.

Many species are reluctant to initiate migration under an overcast. Air temperature is probably the principal factor in starting migration; wind direction or air flow is a decidedly critical factor in regulating migratory movement. Most movements take place when the wind is favourable—*i.e.* blowing in the direction of flight—and after migrants have been held back for long intervals by such weather conditions as fog, rain, and headwinds. In the spring the weather that ordinarily accompanies cold fronts is especially obstructive to movements, forcing migrants to take to the ground without delay.

The arrival of any migration movement in an area is more often controlled by the weather at the print of departure than during its course. Thus at the height of the migration season, when the weather appears suitable in his area, the student will not always observe migration movement because inclement weather the point of departure of some intervening point has arrested the flight.

The study of weather maps in conjunction with observations on movements is very useful. By means of weather maps one can frequently anticipate a migration movement in a given area. For instance, if during the height of spring migration the map indicates a cold front moving in from the northwest, one may expect its arrival to stop and hold numerous transients in the area. Weather maps also assist in explaining the failure of a migration movement to appear. If there is no pronounced migration movement in an area for a long period, even though the season and weather are favourable, maps may reveal that a pressure area along the migration route has become quasi-stationary, either "damming up" migrants (in case the pressure areas is unfavourable to movement), or (in case it is favourable) allowing migrants to move along from day to day without massing in conspicuous waves.

Exceptional weather conditions with unusually high winds often deflect migrants from their usual routes and at the same time carry many non-migrating birds from their regular ranges. The hurricanes or heavy northeast storms that occasionally move north along the eastern Atlantic seaboard, by the counter-clockwise direction of their winds, force many southbound migrants and sea birds far inland. As a result, ducks, geese, and gulls are reported in great abundance in places where they do not ordinarily occur, and such sea birds as petrels show up far inland. Two very severe northeast storms in December, 1927, and January 1966, bore spectacular numbers of Eurasian Lapwings (Vanellus vanellus) over the North Atlantic from their migration route in western Europe to the vicinity of New-foundland. The student will find instructive the paper by Bagg showing in detail; with a series of weather maps, how the great storms brought the Lapwings to North America.

Regularity of Migratory Travel

Despite the effects of weather on migration, migratory travel over a period of years is regular on the average. This is apparent to a student observing bird population from year to year in any given area of North America north of Mexico.

If one keeps records for several years of the days when common summer-resident species first arrive in the spring, and computes average dates of arrival of each species. he can eventually predict within a few days when these species will appear. Similarly, by keeping records of departure of transient spring species—*i.e.*, dates when species going through the area are last seen—he will known approximately when these species depart each year.

There is much less regularity in the arrival and departure of transient species in the fall, thus the securing of records is difficult. Fall migration is more prolonged. Slight variations in weather conditions have stronger effects. Cool late-summer weather, for instance, may induce species to arrive surprisingly early and warm weather may cause them to linger very long. Keeping track of early and late individuals is complicated by the fact that many species are customarily silent in the fall and have inconspicuous plumage. While it is possible to obtain average dates of arrival and departure, fall dates are apt to be much less useful and meaningful on account of the great discrepancy in annual dates and the problems of finding early and late individuals.

In conterminous United States, north of the southern tier of states and in Canada, there are two so-called migration waves. The first, in the early spring, is made up of "hardy" birds—many fringillids and a species or two from various other bird groups; in the second, a month later, there is a preponderance of insect-eating birds—various species of flycatchers, vireos, warblers, and other groups. The migratory movement has a wave effect in that the total population of birds rises and recedes. Thus soon after the first few individuals, called stragglers, make their appearance, the population steadily approaches maximum density, which may prevail for a day or more. Thereasfer the population dwindles until only those birds which stay for the summer are left. Migration waves also occur in the fall but the migration movements are more prolonged and the crests much less apparent.

In the spring, the migration population of any one shecies, provided it is large in the area studied, will show the wave effect. Almost invariably the species makes its initial appearance with a straggler or two. A few days to a week or more later the population begins to increase and later decreases in the manner described for groups of species. Such a population wave, as it moves northward in the spring, may be many miles in width. The author of this book took a trip southward from Minnesota through Jowa and Missouri in late March, before the Robins (Turdus migratorius) had appeared commonly in Minnesota. He recorded an occasional Robin in northern Iowa; great numbers through central and southern Iowa; and a steady decrease in numbers through northern Missouri until in central Missouri there were only scattered individuals, presumably the birds that were to become summer residents there. He estimated the width of the migration wave to be roughly 225 miles.

In making the first spring and fall studies of birds, the student should pay special attention to the local movements of transient species. He should note the dates each species is seen, make counts or estimates of the number of individuals of each species observed on each date, and keep a record of weather conditions. This information will give him a proper conception of the length of time each species remains in the area, the way a species population rises and recedes giving the waves effect, and some of the relationships of weather conditions to population trends.

Irregular Migrations

The many cases of movements among bird populations, which either do not conform to the usual seasonal migration pattern or are not sufficiently well understood to seem a part of the pattern, are loosely classified as irregular migrations.

In certain permanent-resident species there may be mass movements of particular populations with some periodicity. The Blue Jay (Cyanocitta cristata), especially in the northern part of its range, shows some migratory movement. Each fall, small numbers usually move south past Hawk Mountain in

Pennsylvania: in 1939 there was an exceptionally heavy migration (over 7,000 individuals counted), which may have been due to a shortage of beechnuts and acorns in northern forests. In northwestern Oklahoma, the Bobwhite (Colinus virginianus) shows a distinct seasonal population shift by moving from summer habitats in the uplands to pass the winter in bottom-lands and dunes where there is better cover. Their movements, involving distances up to 26 miles, are apparently heavier during severe winters.

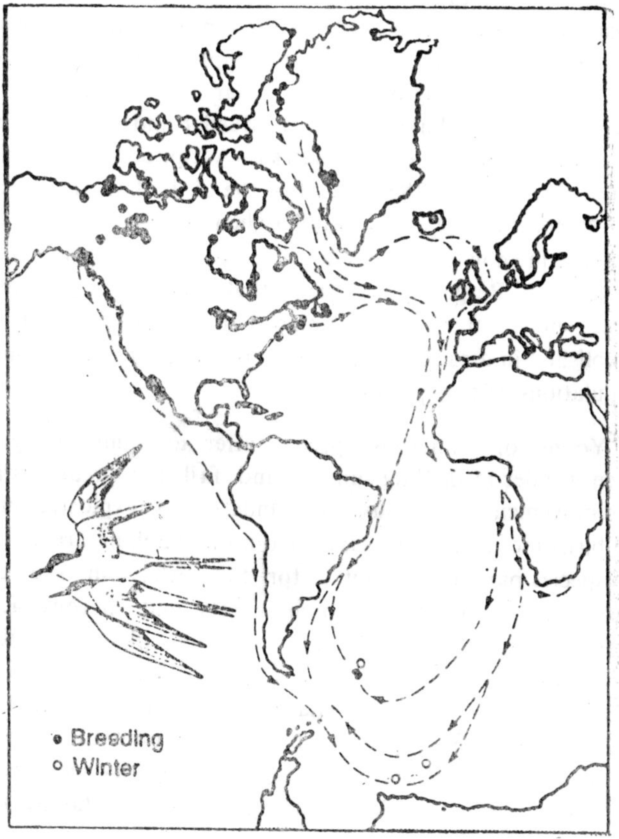

FIG. 11.2: Distribution and migration of Arctic Terns from North America. This species is distinctive for its migratory pathway, which takes it across the Atlantic and southward as far as Antarctica.

Populations of species which are permanent residents may on occasion show a mass movement—invasion or irruption without periodicity. Now and then there is a winter when considerable numbers of Snowy Owls (Nyctea scandiaca) leave the Arctic and invade southern Canada and northern conterminous United States. The cause of this behaviour is sometimes attributed to a sharp reduction in the lemming on which the Snowy Owl preys to a large extent Crossbills (Loxia spp.) and Evening Grosbeaks (Hesperiphona vespertina), normally residents of the Coniferous Forest Biotic Community, occasionally appear in flocks as far south as Florida, presumaably due to a failure of their food supply. Sometimes invasions involve mainly young birds. At Cedar Grove, Wisconsin, on the west shore of Lake Michigan, Mueller and Berger reported a southern invasion of Goshawks (Accipiter gentilis) in the years 1961 through 1963. Most of the birds that they trapped and examined were apparently hatched in 1961 shortly after a decrease in snowshoe hares and grouse, their principal prey. Mueller and Berger hypothesized that these young birds had come south after being displaced by adults already well established in a range that could not support large wintering populations of the species.

Young of numerous species, after attaining full growth, often wander in the late summer and fall for great distances. The movement, called juvenile wandering, is explosive in that the birds move in all directions from the hatching area. Among the species particularly noted for this behaviour are egrets, herons and gulls. Some young egrets and herons actually travel several hundred miles north of their place of hatching in the south. At the conclusion of the breeding season in the big colony of Herring Gulls (Larus argentatus) at Kent Island, New brunswick, an impressive number of immature birds go northward along the coast, although the majority seem to take a southerly direction. Many first-year Herring Gulls, reared in colonies on islands in the Great Lakes, proceed in late December to the coast of Georgia, Florida, and the Gulf of Mexico, and quite a few reach the coast of Mexico; second-year and older Herring Gulls tend to remain on the shores of the Great Lakes within 300 miles of their colonies. Juvenile Sooty Terns (Sterna fuscata) from the large nesting colony on the Dry Tortugas, islands lying directly west of Key West, Florida, move across the Atlantic to the Gulf of Guninea, West Africa, and do not straggle back to the western Atlantic until they approach breeding age; the adults, after nesting, tend to confine their dispersal to the Gulf of Mexico and the Caribbean. For the extensive wandering of young birds, the most plausible explanation is that they cannot compete sufficiently well with older birds for food and must therefore keed moving until they find an adequate supply for themselves.

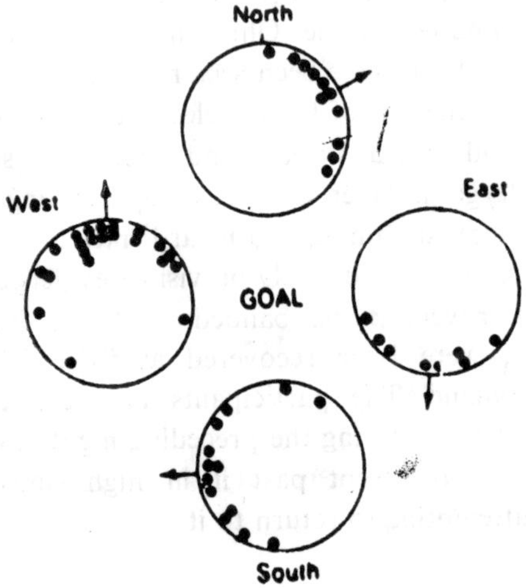

FIG. 11.3: The observed departure bearings of pigeons whose clocks had been advanced by six hours and then were released 30 km to 80 km north, east, south and west of their home lofts. Solid arrows note mean directions.

Reverse Migration

Migratory movements may be reversed, proceeding in a direction opposite the one expected for the season. A good example of reverse migration occurs in the fall on Nantucket Island, Massachusetts, and Block Island, Rhode Island, where many nocturnal migrants (mostly passerines representing well over 50 species) sometimes pass rapidly through during the daytime and leave in a north or northwestward direction for the mainland, into the wind. Baird and Nisbet interpret the movement to be the result of south-bound migrants, carried toward the Atlantic Coast by strong northwest winds, attempting to fly back and redetermine or regain their preferred lanes of passage overland.

Another example with a different interpretation comes in the spring from Point Pelee, a peninsula projecting nine miles southward into western Lake Erie from Ontario, and from Pelee Island that lies about eight and one-half miles southwest of Point Pelee and nearer the Ohio mainland. Time and again many small land birds have been seen returning southward over Lake Erie from the tip of Point Pelee. Lewis reported such a movement in mid-May at Pelee Island. Here for several hours he watched large number of birds representing 35 species (mostly passerine) streaming southward into a headwind. The movements are not based solely on visible evidence; they have actually been proven. Birds, banded at Point Pelee prior to starting south, were later recovered at Pelee Island and on the Ohio mainland. The participants in these reverse flights may be birds which, during the preceding night, overshot their destination or were swept past it in high winds and, consequently, are attempting to return to it.

Rate of Migratory Travel

Most passerine birds fly at ground-speeds averaging 18 to 25 miles per hour (mph). Stronger fliers such as ducks, hawks, falcons, shore birds, and swifts attain much greater speeds. Any bird can accelerate its speed in special circumstances as when frightened or diving earthward. In general, the normal, unhurried, cruising speed of a bird is much shower than suggested by published records, most of which, until recently, were estimated by observers in automobiles or airplanes moving parallel to the bird's line of flight.

Schnell, using Doppler radar equipment similar to that operated by law enforcement agencies in determining speed of automobiles, measured the ground flight-speeds of 17 species of birds in northern Michigan. He recorded on windless days two speeds of the Spotted Sandpiper (Actitis macularia) at 25 mph; four speeds of the Eastern King-bird (Tyrannus tyrannus) at 21 mph and one at 13 mph; three speeds of the Cedar Waxwing (Bombycilla cedrorum) at 21, 23 and 29 mph; and three speeds of the Red-winged Blacdbird (Agelaius phoeniceus), one at 17 mph, and two at 23 mph. Had the wind been blowing, the speeds might well have been slower for birds flying into it and faster for birds flying with it. Strong winds can significantly affect flight-speed as Schnell proved with the 267 speeds of the Herring Gull (Larus argentatus) that he recorded in different wind velocities. Speeds, he found, averaged 25 mph in winds less than 6 mph, but averaged 18 mph (extremes 7 and 39 mph) into winds of 6 to 15 mph and 34 mph (extremes of 21 and 49 mph) with the same winds.

During migration, according to radar surveillance by Bellrose, birds appear to reduce their flight-speed somewhat proportionately to the increase in favourable wind-speed. Apparently they adjust their flight efforts in relation to the degree of wind assistance or resistance. Thus the ground-speeds of migrants tend to remair fairly constant despite variations in wind-speed whereas the ground-speeds of birds in the daily activity flights, as shown by Schnell, are definitely influenced by winds.

Many of the stronger flying birds show great ability for fast migratory travel. Mc-Cube, in an airplane going at an air-speed of 90 miles per hour, was overtaken by two flocks of sandpipers flying at an estimated air-speed of 100 mph. Speirs, once estimated the average ground-speed of the Oldsquaw (Clangula hyemalis) at 61.5 mph and the air-speed at 50.5 mph.

Birds homing to their breeding sites, after being displaced at great distances away, demonstrate impressive ability for sustained speed for many hours. A female Purple Martin (Progne subis), taken from her colony at the University of Michigan Biological Station in northern Lower Michigan and released in Ann Arbor, Michigan, 234 miles to the south, at 10:40 p.m., was back feeding her young at 7:15 a.m., having made the return flight in not more than 8 6 hours at an average speed of 27.2 miles per hour (Southern, 1959). A Manx Shearwater (Puffinus puffinus), removed from its nesting burrow on Shokholm off the west coast of Wales and released in Boston, Massachusetts, reached its burrow after at least 3,200 miles in 12 days and some 13 hours, or an average of 250 miles a day (Mazzeo, 1953). Another sea bird, a Leach's Petrel (Oceanodroma leucorhoa), averaged about 300 miles a day for nine days from its point of release at Prestwick, Scotland, back to its nesting burrow on New Brunswick's Kent Island in the Bay of Fundy. Of the 18 Laysan Albatrosses (Diomedea immutabilis), taken from their nests on Midway Island—one of the Hawaiian Leewards in the north-central Pacific—and released at widely scattered points in the northern Pacific, 14 returned, one from the Philippines, a distance of 4,120 miles in approximately 32 days, and one from Whidby Island off the coast of Washington, a distance of 3,200 miles in 10.1 days at an average speed of 317 miles a day. Presumably, not one of these birds homing to its breeding site had the fat reverses for energy that migrants acquire prior to their long journeys.

There is much additional evidence, obtained by other means, of the bird's ability for sustained speed during long distances of migration. Cochran, Montgomery, and Graber, using radiotelemetry, tracked migrating Hylocichla thrushes nearly all might from Illinois northward into Michigan, Wisconsin, and Minnesota. Although they found considerable variation, most flights were at air-speeds—*i.e.*, speeds with relation to the winds aloft—between 25 and 35 mph. These were usually less than ground-speeds—speeds with relation to the earth—and thus suggested that the birds were aided by favourable winds. One of the most remarkable records of a long, sustained flight is that of a banded Ruddy Turnstone (Arenaria interpres) released by Max C. Thompson at St. George Island, one of the

Pribilofs in the Bering Sea, on August 27, 1965, and shot four days later, on August 31, at French Frigate Shoals in the Hawaiian Leeward Islands. Assuming that this individual covered 2,300 miles between St. George Island and French Frigate Shoals in a steady bee-line flight, its average speed was 575 miles a day.

Recently, radar studies have provided many reliable estimates of the rate at which migratory birds travel. W.R.P. Bourne assessed the air-speed of Lapwings (Vanellus vanellus) in their flights during June over the southern North Sea to England at 35 knots (40 mph). Lee at the Isle of Lewis in the Hebrides, Scotland, showed that the air-speed of Wheatears (Oenanthe oenanthe) from Iceland approximated 20 knots (23 mph); Redwings (Turdus iliacus), from 30 to 35 knots (34.5 to 40 mph); and Grey Lag Geese (Anser anser), from 53 to 55 knots (61 to 63 mph). Bergman and Donner demonstrated that the still-air-speed for the Oldsquaw (Clangula hyemalis) at low altitudes over the Gulf of Finland was 40 knots (46 mph) and for the Common Scoter (Oidemia nigra), 45 knots (52 mph). Once the birds reached a higher altitude inland, their speed increased by about 10 per cent. After recording air-speeds of passerine birds off the coast of Norfolk, England, for a whole year, Tedd and Lack, on analyzing the results, found evidence of a seasonal difference: in the spring, the speed averaged 27 knots (31 mph), 4 knots faster than in the fall. At Cape Cod, Massachusetts, Nisbet and Drury found another seasonal difference in that the directions of migration in the spring were much less diverse than in the fall, thereby suggesting much less time lost in passage.

In the late fall, long-distance migrating ducks commonly pass from breeding area to winter quarters in a short series of mass movements, each of which carries them many hundreds of miles in one continuous flight. They start each flight immediately after the passage of a cold front when temperature has dropped and the sky is clear, but they may overtake bad weather as they proceed. Sometimes, owing to the triggering effect of extremely low temperatures resulting from a strong flow of polar air, the mass movements are spectacular both in numbers of birds involved and distances covered. Bellrose documented one such migration in 1955 that moved with unusual rapidity from the Great Plains of Canada to the marshes of southern Louisiana. The exodus began from Canada or October 31; early on November 1 the flight was in full force through the Dakotas; and on November 2 the vanguards had reached northern Tennessee and Arkansas shortly after sunrise and Louisiana later in the day. Many thousands of ducks made the flight from Canada to southern Louisiana, a distance of 1,200 to 2,000 miles, in two days, or roughly 35 to 50 hours, at an average speed of 40 miles per hour. No doubt some of the birds covered the distance without stopping, accomplishing their migration in one flight.

In undertaking long-distance migrations, many small land birds tend to begin with short flights and complete them with longer flights. Indirect evidence of this procedure was reported by Caldwell, Odum, and Marshall, after comparing the fat reserves of six pieces of tropical-wintering North American passerines killed during fall migration by television towers, one near the Florida Gulf Coast and the other in central Michigan. All the birds killed by the Florida tower showed significantly greater amounts of fat, strongly suggesting that these migrants began with low to moderate fat reserves that allowed only short flights and then increased their reserves until they had acquired a maximum amount for the long, non-stop flights such as across the Gulf of Mexico. European birds migrating south across Africa build up fat reserves of 30 to 40 per cent of their body weight by the time they set out across the Sahara.

By making longer flights as they near their destinations, birds gradually speed up their migrations. Cooke provided evidence for this acceleration when he analyzed migration dates of North American species, mostly passerines, approaching their northern nesting areas. "Sixteen species," he wrote, "maintain a daily average of 40 miles from southern Minnesota to southern Manitoba, and from this point 12 species travel to Lake Athabasca at an average of 72 miles a day, 5 others to

Great Slave Lake at 116 miles a day, and 5 more to Alaska at 150 miles a day."

From all these studies and reports, several generalizations on the rate of migratory travel emerge. Strong winds can affect ground-speed of birds in their daily activity flights but not in migration. Birds make long, sustained flights, usually at increasingly higher speeds at higher altitudes. Spring migration proceeds at a greater rate with less time loss than fall migration. Larger birds such as ducks accomplish their migrations in a short series of a few mass movements, occasionally in one non-stop mass movement. Small and birds, however, tend to begin their migrations in many short flights, gradually building fat reserves for long, non-stop flights, thereby accelerating their migrations.

Mortality in Migration

Migrating is dangerous for all birds. In their long flights over land or water they are likely to meet disaster through vagaries of the weather. When forced to land by cold fronts, frequently they must accept environments where, because of inadequate cover, they are easy victims of predators.

While migrations are adjusted to the normal alteration of fair and inclement weather during spring and fall, sudden and unseasonable changes in the weather occasionally have serious effects. Once, during fall migration in the vicinity of Lake Huron, untold numbers of birds crossing this huge lake were forced into the water and drowned because of a very quick drop in temperature and an exceptionally heavy snowfall. After the storm, one observer reported an estimated 5,000 dead birds washed up on a one-mile stretch of shore. In their flights north in the spring, birds are sometimes caught in severe storms and killed by becoming first exhausted and then being exposed to excessively low temperature coupled with heavy rain or wet snow. After a blinding March snowstorm in Minnesota as many as 750,000 Lapland Longspurs (Calcarius lapponicus) were found dead on the ice of two lakes, each of which covered only a square mile.

Man has created awesome hazards for migrating birds by erecting lighthouse with strong light beams and by illuminating various tall structures such as the Washington Monument in the District of Columbia and the Empire State Building in New York City. During nights in the spring and fall when migration is proceeding at a low elevation because of an unsurmountable cloud layer, passing birds are attracted by the brightness and, approaching it, soon become blinded and fly into its source, killing themselves.

Under certain circumstances, even street lights can be a hazard. Vast numbers of parulid warblers, driven ashore on the Texas coast by a northeast storm while migrating northward across the Gulf of Mexico during a night in early May, met their death by flying into street lights. In a part on Padre Island, James counted more than 900 dead birds under just one light pole plus an estimated 100 on the adjacent pavement. There were nine other light poles in the vicinity with similar tolls.

Airport ceilometers indirectly cause mortality among small, nocturnal migrants. These instruments, which are used to determine the cloud ceiling, direct a narrow, extremely brilliant beam of light straight upward. At night when the clouds are low, they produce a bright spot on the cloud ceiling that can be seen for a considerable distance. On mornings following a heavy, nocturnal migration, numbers of birds varying from three to over a thousand have been found dead near spots where ceilometers are used. Three ornithologists, Howell, Laskey and Tanner, who have investigated many of these accidents, explain the cause as follows: When there is a pronounced migration and a low ceiling, the beam attracts the migrants. After circling through the bright light, many circle back to fly in and about it. While in the beam their bodies reflect light attracting still other migrants. Blinded by the light, the birds die by collision, either with each other, with the ground, or (rarely) with a building.

A direct cause of mortality among small, nocturnal migrants are television towers erected to heights of 900 to 1,000 feet or more and, as is often the case, situated on hills or bluffs where they reach even greater heights above the local terrain. Each tower is supported by guy wires and has a system of steady, flashing red lights, mandatory on all tall structures that are potential hazards to airplanes.

When migrants are flying under a low ceiling, they are attracted to a television tower because of its lighted area and become reluctant to leave. Just as birds, released at night in a lighted room with doors and windows open, continue to fly about in the room rather than escape into the darkness, the migrants fly through the tower framework and circle out of the edge of the lighted area, then return toward the light. Mortality results when the birds strike the dark guy wires while circling. The above observations and explanations come from Graber. For examples of the extent of avian mortality around television towers, see the papers by Tordoff and Mengel and Stoddard and Norris.

Altitudes of Migratory Flight

The recent analysis of migratory flight by means of radar shows not only that birds move at altitudes averaging higher than formerly believed, but also that their altitude varies widely depending on the circumstances.

Nisbet, studying radar heights of nocturnal fall migrants above Cape Cod, Massachusetts, and the outlying ocean, found that the most frequent height was usually between 1,500 and 2,500 feet. About 90 per cent of the birds, probably small passerines, were below 5,000 feet. On some nights they were lower than 2,500 feet, while on others they were up to 6,000 or even 8,000 feet.

The presence or absence of cloud cover may determine the altitude chosen by birds. Bellrose and Graber discovered, during their radar studies of nocturnal migrants in central

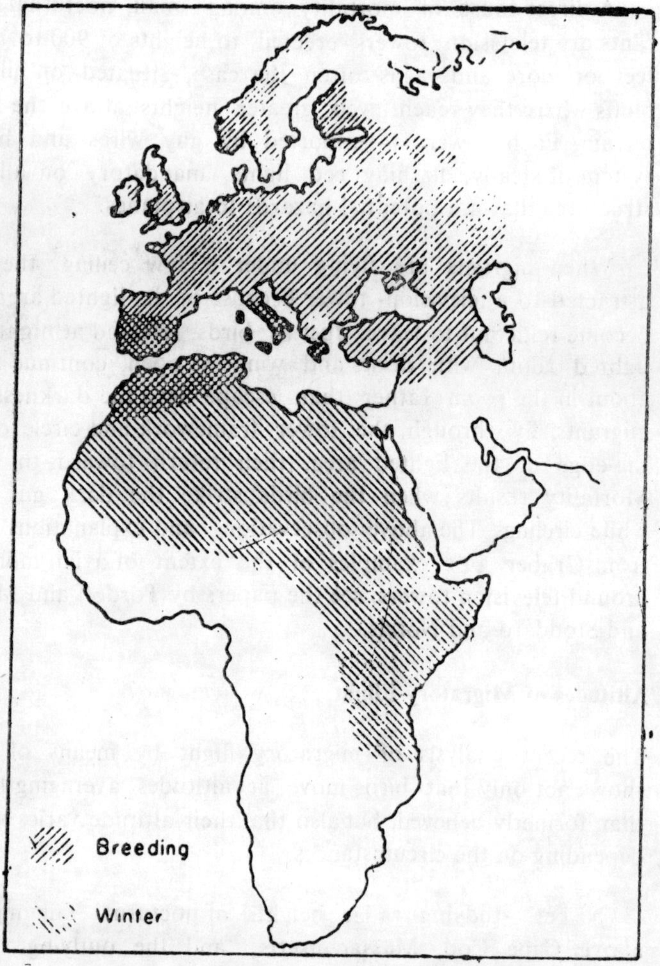

FIG. 11.4: Migration of the blackcap from temperate regions to warm-temperate and tropical Africa. There is some overlap, and a few birds winter further north than shown, for example in southern England.

Illinois, that birds are prone to migrate at higher altitudes when the skies are overcast than when they are clear. If the clouds are not too high, the birds apparently attempt to surmount them; but if the clouds are too high, they usually continue to fly, sometimes in the clouds, although usually immediately under them. When the birds are flying under an overcast and, consequently, at much lower altitude, they can usually be heard from the ground.

Birds fly higher by night than by day. Lack first noted this tendency, by radar, among migrants crossing the southern North Sea between England and the Continent. Later, Eastwood and Rider at the Bushy Hill station in England proved that the tendency is significant. From their considerable data they were able to show that 80 per cent of the birds fly below 5,000 feet at night and 80 per cent below 3,500 during the day. They further demonstrated that migrating birds, in a 24-hour day, have a tendency to fly at the lowest altitudes in the afternoon and the highest just before mid-night. From radar studies and tape recordings to call notes in Illinois, Graber has concluded that migrants reduce their altitudes after mid-night to 1,500 feet or less, although they continue their flight until daylight. After their descent to lower altitude, they increase their calling. This helps to explain why nocturnal migrants can be heard more frequently as dawn approaches than earlier in the night.

There are seasonal variations in altitudes. Bellrose and Graber found that migrating birds in Illinois fly higher during the fall than during the spring, possibly because the winds during the fall are more favourable for southward migration at higher altitudes. Eastwood and Rider, on the other hand, found the reverse to be the case in England. They suggest as one reason for this seasonal difference that flocks of all migrants include many young birds whose fight capabilities are inferior to those of adults and which are, consequently, unable to achieve the higher altitudes of the more mature spring migrants.

Birds migrate higher over land than sea. Common Scoters (Oidemia nigra) and Oldsquaws (Clangula hyemalis), in their spring passage over southern Finland and the Gulf of Finland, were noted by Bergman and Donner to fly at altitudes that averaged 3,400 feet over land and ranged from 300 to 1,00 feet above water. Passerine, migration in passing from sea to land in England, were shown by Eastwood and Rider to make a similarly significant though not as great change in altitude, climbing from a median height of 1,700 feet to a median height of 2,200 feet.

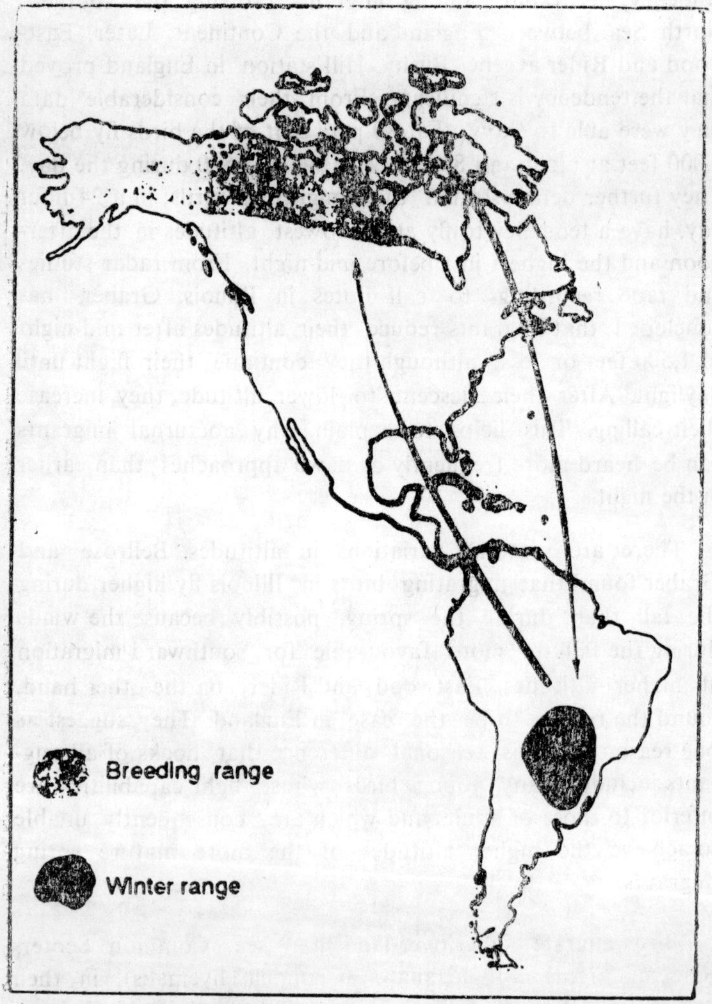

FIG. 11.5: Migration of the American golden plover. The arrows show the approximate trakes of the spring and fall movements.

Birds have long been known to reach very high altitudes during flight. Direct observations from aircraft proved that large birds can fly over the highest mountain ranges—for example, the Himalayas between central Russia and India. The Yellow-billed Chough (Pyrrhocorax graculus) was actually found on Mt. Everest at an elevation of 27,000 feet. The use of radar in determining heights of migration shows that while most birds rarely exceed 8,000 to 10,000 feet, a small proportion of migrants, particularly the stronger fliers, nonetheless attain great heights, in some cases astonishing. In his radar studies of migrants at Cape Cod, Massachusetts, Nisbet recorded a number of birds on several dates in September and October, usually before mid-night, or before and after sunrise, at altitudes between 8,000 and 15,000 feet and a few birds as high as 20,000 feet. These may have been sandpipers and plovers, flying over the ocean, toward the Lesser Antilles and eastern South America. Using an especially powerful and accurate height-finder at Norfolk in southeast England, Lack observed at sunrise on 16 dates in September a thin scattering of birds extending fairly uniformly up to at least 15,000 feet. On seven of these mornings he noted that the greatest height was 19,000 feet and on two mornings it was 21,000 feet. Probably the highest migrants were small shore birds such as the Dunlin (Erolia alpina) which had left Scandinavia the night before.

Bearing in mind that the oxygen content of the air at 18,000 feet is 50 per cent less than the air at sea level, one wonders whether high-flying birds suffer "altitude sickness." When a man sets out to climb a lofty mountain he acclimates himself gradually over a period of days, but birds take off and reach a comparable elevation in a matter of a few hours. Do birds have special adaptations that enable them to avoid altitude sickness? The answer may come someday from experimental studies of bird flight in pressure chambers.

Course of Migration and Migration Routes

Many species breeding in North America north of Mexico have their winter ranges far south and southeast; in southern

Mexico, Central America, and South America. To reach their winter ranges, most of these species proceed at night from their breeding ranges to southern United States in broad fronts without notable regard to topographical features. Radar and—other observations confirm that their movements trend southeast—toward their winter ranges. This direction, coming as it does in the wake of a cold front, is with the wind and, therefore, beneficial to the migrants. In the spring, the direction trends northwest, in reverse, and again with the wind since the movements take place with the onset of a warm front. Many species, however, do not exactly retrace their course in the spring, but fly instead somewhat to the west of their fall passage. Evidence of this elliptical course—going to southern United States one way and coming back another—is borne out by kills at television towers. Certain species are well represented in the spring migration but seldom or not at all in the fall, and *vice versa.*

Owing to the narrowing of the North American continent southward and the intervention of the Gulf of Mexico and Caribbean Sea, all species moving southeastward through the United States toward their wintering ranges in southern Mexico, Central America, and South America, converge on special routes. There are five altogether. Certain species use mainly one route, others two or three; no species is known to use more than three. The routes are as follows:

Route 1: From the coasts of Newfoundland, Nova Scotia, New England, and New Jersey southward over the Atlantic Ocean to the Lesser Antilles and the northeastern coast of South America. A few shore birds use this route.

Route 2: From Florida southward over the Bahamas, Hispaniola, Puerto Rico, and the Lesser Antilles to South America. The few birds which frequent this route are seldom far from land as there are many small islands along the way.

Route 3: From Florida southward over Cuba and Jamaica across 400 miles of the Caribbean Sea to South America.

Route 4: From the shores of the Gulf states across the Gulf of Mexico to the Yucatan Peninsula and southern Mexico. This is the route most frequently used by the many species of birds from eastern United States and Canada.

Route 5: From Texas, New Mexico, Arizona, and California through northern Mexico. The majority of birds from western conterminous United States, western Canada, and Alaska use the western side of this route.

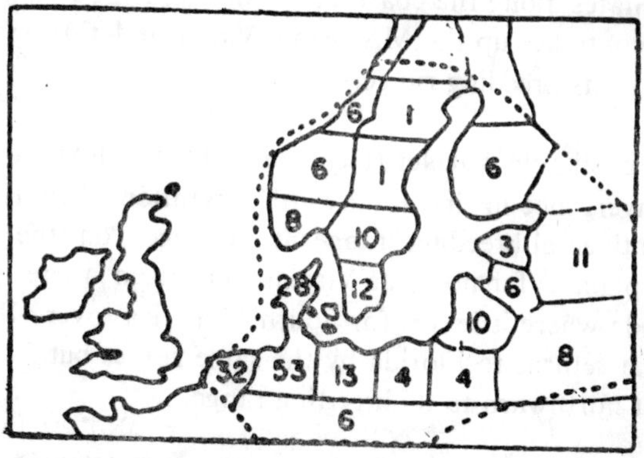

FIG. 11.6: Breeding range of starlings that spend the winter in the British Isles. Numerals indicate the breeding-season recoveries (May—Aug.) of starlings previously ringed during winter (Dec.–Feb.) in the British Isles.

Species vary greatly in their course of migration and use of routes. A few species move south over one route and return by another; a few species move eastward or westward before going south; and a few species have spectacularly long routes that take them as far south as southern South America. Among a few species which breed in western United States—*e.g.*, the Western Kingbird (Tyrannus verticalis) and the Scissor-tailed Flycatcher (Muscivora forficata)— some individuals in the fall move eastward across the Gulf states to winter in Floria whereas most of the population moves directly south into Mexico.

To illustrate the wide variation in migratory movements and the choice of routes, the migrations of six species are described below.

The American Golden Plover (Pluvialis dominica) passes eastward from its breeding range to the Atlantic Coast, where it turns southward over Route 1. Once in South America, it flies directly across the continent to its winter range. It returns, however, by another route, coming up across northwestern South America and the Gulf of Mexico and reaching the United States along the coast of Texas and Louisiana. From there it continues up the Mississippi Valley and through central Canada to its breeding range.

The Blackpoli Warbler (Dendroica striata) shows a remarkable convergence in its southward migration. From its vast transcontinental breeding range it conveges on the Atlantic coastal plain as far north as Virginia and from there proceeds to Florida where it leaves for South America over Routes 2 and 3. It returns to Florida by the same routes but from there fans out northward to its breeding range.

The Mourning Warbler (Oporornis philadelphia) shows a similar convergence. From its breeding range of more modest extent it converges on southern Texas, then goes southward along the eastern portion of Route 5 through eastern Mexico and Central America to its winter range. It returns via the same part of Route 5 and fans out from Texas northward.

The American Redstart (Setophaga ruticilla) shows little convergence. Instead, it passes southward more or less directly over a broad front of nearly 2,500 miles, eventually using Routes 2, 3 and 4. It returns the same way.

The Cobolink (Dolichonyx oryzivorus) migrates an exceptionally long distance, averaging farther than any other passerine species. From its transcontinental breeding range, it converges southward to leave the United States over Routes 3 and 4. Once in South America, in continues directly overland to its winter range in extreme southern Brazil, southeastern Bolivia, Paraguay, and northern Argentina. It returns the same way.

The Connecticut Warbler (Oporornis agilis) migrate in an eccentric manner. From its breeding range it flies directly eastward to New England, then to South America along the Atlantic coastal plain and eventually Route 3. From its winter range in South America it returns over Route 3 to Florida, but from there it passes diagonally to the Mississippi Valley and northward to its breeding range.

Altitudinal Migration

Some bird populations of high mountains in both temperate and tropical regions move down to the lower slopes and valleys when winter sets in at higher elevations. In the descent of a few hundred feet they accomplish what many other populations do in their latitudinal migrations of many hundreds of miles.

Several observers have noticed that mountain birds tend to move to higher slopes after the nesting season. Probably many are young birds which, after getting their growth, always show a great tendency to move about. The reason for this may be the agonistic behaviour of their elders, but more than likely it is another example of juvenile wandering.

Distances in Migratory Travel

Extremes in distances travelled by migrating birds are represented on the one hand by high-mountain species, which merely pass up or down slopes of several hundred to a few thousand feet, and on the other hand by the Arctic Tern (Sterna paradisaea). It makes the longest flight of any bird, migrating thousands of miles from the Arctic, where a part of the population breeds, to its wintering area adjacent to the pack ice around Antarctica. This species nests as far north as the northern tip of Greenland at 83 degrees North Latitude and has been recorded as far south as 74 degrees South Latitude.

Most of the individuals from Greenland, Canada, and northeastern United States undertake the passage by flying across the North Atlantic to the continental shelf of western Europe, and then south over the coastal waters of West Africa and finally across the Antarctica Ocean. They return by the same route.

While a few species breeding in conterminous United States, Canada, and Alaska go as far south as central and southern South America, the majority migrate no farther than northern South America. In fact, many species in northern conterminous United States, Canada, and Alaska move only to the southern states. There appears to be no correlation between the distances travelled by birds and their size, flying abilities, or habits. A number of small species journey to Mexico and Central America, outdistancing a great many species which to all appearances have much greater capacities for travel. Still to be explained is why certain species have developed long migration routes, while other species, sometimes closely related, have routes that are half the length or less. It has been postulated that the long migration of certain species to northern latitudes for nesting is to take advantage of increased daylight and the consequent shortening of the period in which the young are confined to the nest. No satisfactory theory has yet been advanced to account for the arduous passage of such birds as the Bobolink (Dolichonyx oryzivorus) across the tropics from the northern to the southern temperate region, or the Arctic Tern from one polar region to the other.

When a migratory species has an extensive breeding range that includes parts of northern conterminous United States, Canada, and Alaska, the more northern populations of that species move farther south for the winter than the other populations do. For example, among the several subspecies of the Fox Sparrow (Passerella iliaca) breeding along the Pacific Coast from Alaska to Puget Sound, the subspecies nesting farthest north have been found wintering in southern California, passing by other subspecies, which either do not migrate at all or move only to central or northern California. Apparently the more northern breeding populations of a widespread species acquire a stronger migratory habit than the populations breeding in more southern areas that have milder, year-round climate.

In some species there may be sexual differences in the extent of the migration. Howell found that females in the eastern race of the Yellow-bellied Sapsucker (Sphyapicus varius) outnumber the males in the southern part of the winter range by about three and one-half to one.

Flight Lanes and Concentrations

In North America migratory movement is continent-wide. There are probably no areas over which birds do not pass in their latitudinal migrations. Prairies, forests; mountains, lakes, and inland extensions of the oceans fail to stop or divert migration altogether. Even so, as diurnal migrants move northward or southward across Canada and the conterminous United States, many species tend to favour or be influenced by certain topographic features which trend in a north-south direction. Some species fly along ridges; others follow the coasts, large rivers, and chains of lakes. A great many species go through valleys, move along peninsulas, or pass from island to island across large bodies of water. In both Canada and the United States there are many places where the topography is such as to cause narrow flight lanes in which migratory movement is especially conspicuous.

Particularly in the fall, hawks and a few other large birds follow the crests of north-south ridges, riding on the updrafts as they proceed southward. The traffic in these lanes is unusually heavy on clear, windy days following the passage of a cold front, because there is considerable wind deflected upward, thus making these lanes advantageous to travel. One of the best known places is the Kittatinny Ridge in eastern Pennsylvania. At one point, Hawk Mountain, where the ridge becomes suddenly high and slender, the birds are brought together and closer to the ground. As a result of this narrowing of the flight lane, observers at Hawk Mountain have been able to count over 22,000 hawks moving by in a single season.

Large bodies of water constitute barriers to day-migrating birds and thus cause flight lanes to curve around them. The Great Lakes are a good example. In the fall, south-bound hawks approaching their north shores from western Quebec and southern Ontario take the shortest courses around them that geography and air movements will allow. If the birds are migrating in great numbers, as on the second days after cold fronts when there are steady westerly winds and ample sunlight producing thermals, large numbers can be seen in continuous passage at such points as Port Credit on the northwest shore of Lake Ontario, Port Stanley on the north side of Lake Erie, Cedar Grove on the west side of Lake Michigan and Duluth at the westernmost extension of Lake Superior.

Peninsulas projecting into large bodies of water that lie athwart the direction of migration may become funnels for land birds in diurnal passage. Hawks moving northward in the spring through Michigan to Canada converge in large numbers on the northern tip of Lower Michigan at the Straits of Mackinac. Here, if the weather is rainy and windless, they settle on trees and other perches. With the advent of the next clear day and a favouring wind, the hawks begin to spiral higher and higher until one bird peels out and heads northward over the Straits with the others following. Small numbers of north-bound Blue Jays (Cyanocitta cristata), in order to get across Lake Mendota at Madison, Wisconsin, converge first at Picnic Point and then spiral upward until "barely visible to the naked eye" before crossing the 1.7 miles of open water to Fox Bluff on the north shore. Cape May, the southern tip of New Jersey between the Atlantic Ocean and Delaware Bay, is noted for its hordes of land migrants, large and small, which gather from August through November when northerly winds are strong. Some of the migrants held up here are those which regularly follow the Atlantic Coast southward, but many are birds which have been drifted by wind southeastward from their usual flight lines. Frequently, migrants at Cape May may be seen in the day flying north and northwestward into the wind as they skirt Delaware bay before continuing their journey. Similar concentrations may be observed at Cape

Charles, a south-pointing peninsula separating the waters of Chesapeake Bay from the Atlantic.

Night migrants do not follow topographically determined flight lanes to any significant degree. Instead, as radar surveillance shows, birds migrate at night without regard to what lies below.

In the spring, land migrants returning to the United States from across the Gulf of Mexico vary in their manner of arrival in accordance with weather conditions. If the weather is mild and the wind favourable, birds bound for more northern destinations continue inland from the Gulf for considerable distances before coming to land. The coastal area thus appears to be an ornithological "hiatus." But if a cold front with strong northerly winds moves in over the area while migration is in progress, migrants are forced to come down to the first land reached, with the result that the coastal area is flooded with birds that linger here until the weather again becomes favourable. A somewhat similar situation occurs during spring migration at Point Pelee. Here, since it is the nearest land, northbound small land birds, on meeting a cold front from the north while they are over Lake Erie are forced to descend and remain until the cold front abates. At such times Point Pelee swarms with birds.

Flocking During Migration

During migration bird species show wide differences in flocking habit. A number of diurnal migrants, notably many hawks and other predators, are little inclined to move in groups, preferring to travel solitarily. But the majority of migrants diurnal or nocturnal, exhibit the flocking habit.

Certain species migrate in flocks strictly of their own kind. These are usually birds whose flight-speed, feeding habits, or roosting preferences are so individual as to make them incompatible travelling companions. The Common Nighthawk (Chordeiles minor) and Chimney Swift (Chaetura pelagica) are good examples of birds which migrate in their own company. Neither waxwings nor crossbills will migrate with other birds, but Cedar and Bohemian Waxwings (Bombycilla cedrorum and B. garrulus) have been seen in the same flocks and so have Red and While-winged Crossbills (Loxia curvirostra and L. leucoptera). Some of the larger birds—*e.g.*, pelicans, cormorants, storks, swans, geese, and cranes—noted for V-shaped or linear flock formations likewise tend to travel in unmixed groups.

The majority of species travelling in flocks, whether unmixed or mixed, give call notes. This is especially true or nocturnal migrants: some species such as Hylocichla thrushes utter calls heard only in night migration; other species such as the Bobolink and Dickcissel (Spiza americana) give calls that are the same as ones heard in the day-time on their breeding grounds. If the calls are distinctives as in the case of the Bobolink and Dickissel, it is possible to identify the species, but the calls of most species are link and Dickcissel, it is possible to identify the species, but the calls of most species are faint chips and lisps that sound more or less the same to the human ear.

Unmixed or mixed flocks of some species of smaller birds—*e.g.*, certain sandpipers and plovers—fly in compact formations, all individuals in the flcok simultaneously performing almost identical maneuvers. Many more species of smaller birds, including the majority passerines, travel in flocks that are loosely formed, though still cohesive, the individuals proceeding in the same direction. At night, the cohesion of flocks is probably maintained by call notes. At the same time, the call notes serve to space out individuals in each flock so that they will not collide with one another.

No flock of migrants appears to have a persisting leader. The direction taken by a flock, represents a compromise by each individual to the directional preference of the other individuals of the flock. At night, call notes convey directional information from one individual to the other. When a nocturnal flock shifts its direction or lowers its altitude, or when it is disoriented for one reason or another, all its members greatly increase their rate of calling.

Whether or not flocks remain intact for the duration of migration has not been determined. Nor is it known for certain whether flocks remain together during the winter. In all probability, most flocks, if they are comprised of common, widely distributed species, change from day to day during migration and in the winter split up into smaller groups, or combine with other flocks to form larger groups.

Flocking during migration is undoubtedly advantageous to the individuals concerned. Just as flocking among resident birds provides group protection against predation or increases the success in finding and exploiting food sources, flocking in migration greatly facilitates the attainment of destination. Younger birds travelling with more seasoned adults benefit from their experience. As another advantage, Hamilton suggests that groups of birds on an average determine their direction with greater accuracy than single individuals. Thus flocking assists any migrant that goes a long distance, or any migrant required to pinpoint its destination on a small land area in mid-ocean. Hamilton cites as examples:

1. The Board-winged and Swainson's Hawks (Buteo platypterus and B. swainsoni). Both of these North American species travel in large flocks to winter in South America. By contrast, the Red-tailed Hawk (B. jamaicensis) and certain other buteos, which rarely leave the North American continent, seldom move in appreciable groups.
2. The Long-tailed Cuckoo (Urodynamis taitensis).

After breeding in New Zealand, this species gathers in large flocks for the exceedingly long, non-stop passage to winter quarters on tiny islands in the west-central Pacific. By contrast, the Yellow-billed and Black-billed Cuckoos (Coccyzus americanus and C. erythropthalmus) of North America, which have only to reach a broad tropical region of South America for the winter, migrate solitarily.

V-shaped formations help to conserve energy by creating favourable air currents for all individuals in the flock except the leader. When fatigued, the leader drops back and is replaced by another bird in the flock. An alternate advantage hypothesized by Hamilton is that the V-shaped structure serves as a means of communication, enabling the individuals to profit fully from the collective direction-finding of the group. By flying in parallel alignment, each individual moves in the same direction as the leader. If, as in flocks of geese travelling in poor visibility, difficulties in establishing direction arise, the leadership changes frequently in order that collective ''judgement'' may prevail in maintaining the proper course. Sometimes the V-structure gives way temporarily to a crescent form; forward flight remains in the same direction but is slowed until a new leader takes over and the V-shape is resumed.

Unmixed or mixed flocks may contain only immature individuals, only adults, or only adults of one sex; or they may contain individuals of all ages and both sexes. Flocks of geese and cranes may be comprised on one family or several families.

During the fall migration the adults may precede the immature birds, the birds-of-the-year. Hagar reported that adult Hudsonian Godwits (Limosa haemastica) withdraw from their nesting grounds in central and northwestern subarctic Canada in late July; the immatures follow a month later. Among passerine species there is convincing evidence assembled by Murray and Jehl from several thousand migrants, mist-netted in the fall at Island Beach, New Jersey, and analyzed as to age, that adults and immatures travel at approximately the same time. However, in the case of the Least Flycatcher (Empidonax minimus), Hussell, Davis and Montgomerie concluded from an analysis of 182 individuals trapped in the late summer at Long Point, Ontario, that most of the adults migrate in advance of the immatures: the adults during the second half of July and first half of August; the majority of immatures from the second half of August to the end of September.

In the spring migration many of the first flocks of certain species coming north have a preponderance of adult males which reach the breeding grounds and establish territories before the rest of the population arrives. A.A. Allen once carefully studied the spring migration of Red-winged Blackbirds (Agelaius phoeniceus) at Ithaca, New York. Since the males and females have distinct plumages and birds hatched the previous year (*i.e.*, the immature birds) differ sufficiently in colour from the adults, he was able to analyze flocks and work out a migration schedule according to age and sex. While his findings may not hold for all Red-wing populations, it is presented below as a useful guide and basis for comparison.

Migrant adult males	March 13-April 21
Resident adult males	March 25-April 10
Migrant females and immature males	March 29-April 24
Resident adult females	April 10-May 1
Resident immature males	May 6-June 1
Resident immature females	May 10-June 11

In watching spring migration in his own area, the student will find that not all species follow such a schedule. In fact, males and females of quite a few species migrate together. Only rarely, however, do females precede males.

Direction-Finding

Ho , migrating birds determine their direction when migrating or homing by day or night over areas unfamiliar to them is one of the most fascinating aspects of migration. Before considering the subject of direction-finding in migration, it is worthwhile to review some of the problems associated with homing in birds. The basic means by which birds orient themselves, whether migrating or homing, are much the same.

A great many experiments have demonstrated the remarkable ability of wild birds to return to their eggs or young after being removed great distances and released. Years ago Watson tested the homing ability of Sooty and Noddy Terns (Sterna fuscata and Anous stolidus), which nest on the Dry Tortugas, the islands in the Gulf of Mexico west of Florida. These species come to the Tortugas from tropical seas and are seldom seen farther north. Two nesting Sooty Terns and three Noddy Terns were captured, marked, and transported northward in a ship to point off Cape Hatteras, about 1,000 miles by sea from the Tortugas. Here they were released. Just five days later the Solly Terns were back on their nests; one of the Moddy Terns showed up after a few two more days. Some of the more recent experiments demonstrating the sustained speed of homing flight by the Purple Martin, Manx Shearwater, Leech's Petrel, and Laysan Albatross (see above) attest at the same time to their precision in direction-finding.

The inducement of birds to home is not necessarily provided by their eggs or young; it can be the breeding area or home range. Breeding Brown-headed Cowbirds (Molothrus ater), which are brood parasites, will home from maximum distances of 250 to 380 miles. An adult female Bobolink (Dolichonyx oryzivorus), escaping from captivity in September at Berkeley, California, was recaptured the following June at Kenmere, North Dakote, where it was originally tapped as a breeding bird. Six hundred and sixty Golden-crowned and White-crowned Sparrows (Zonotrichia atricapilla and Z. leucophrys), captured while wintering in the San Jose area of California, were immediately carried by plane to Laurel, Mary-land and released. Fifteen were known to have come back the following winter. In the interim they had presumably found their way in the spring to the nesting grounds in northwestern Canada and Alaska, then returned to California in normal migration.

Homing flights are more or less routine with homing pigeons which are Common Pigeons (Columba livia) specially bred for racing. Their precision and speed of return to the home loft is developed by training and experience. Sometimes the experience of only one flight to the home loft is sufficient to determine the proper direction for successive flights. In his work on homing pigeons, Matthews found that certain individuals, trained to maintain a given direction, can adhere to that direction over unfamiliar terrain and that certain other individuals can fly straight toward the home loft from unfamiliar territory regardless of the direction of the home loft. In the course of their training, pigeons must be made familiar with the area around the loft so that they will have a broad or reasonable target.

Although homing has been amply demonstrated in both wild birds and racing pigeons, the perplexing question remains: How do homing birds find their way? In seeking an answer, Griffin and Hock attempted to determine how displaced and released nesting birds find their way by following them in an airplane and watching their behaviour. For their experiment they selected the Gannet (Morus bassanus), a marine bird which rarely occurs inland; being large and white, it was easy to follow. Taking 17 individuals from their nests on an island off the Gaspe Peninsula, Quebec, they carried them to a point in northern Maine, 100 miles from salt water and 215 miles from the home island. They released nine of the birds (the others were used as controls) and, from an airplane, traced their flights from 25 to 230 miles. The investigators were careful to keep the plane, 1,500 feet or more away from the birds so as not to frighten them. The experimental birds flew in all directions with no significant tendency to head directly toward their nests. Their flight paths were generally gradual curves. The first birds to reach the coast within the first few hours were the first to get back to their nests. Although 62.5 per cent of these birds eventually reached their nests, in from one to four days.

The result of Griffin and Hock's work on the Gannets suggests that the ability of birds to home—that is, to return to a known or familiar site such as a nesting area—is dependent on random searching. In a strange territory birds keep circling and exploring by trial and error until they find familiar landmarks within familiar territory. However, the concept of random searching as a means by which all birds find their way cannot be reconciled with the rapid homing exemplified by Sooty Terns and other species already mentioned (see above). Many birds, if not all, obviously have an ability to find their way by orienting themselves—determining their position with respect to their environment –and by following directional cues or navigating.

In taking up the subject of the means by which birds orient themselves and navigate, it is well to consider first the question whether or not birds inherit at least part of their ability to findtheir way. Rowan at Edmonton, Alberta, caught young Common Crows (Corvus brachyrhyncos) in the late summer and kept them in captivity. In November, when winter conditions had begun to set in and all adult Common Grows had left for their winter range in Kansas and Oklahoma, he banded the captive birds and released them. Altogether 54 individuals were set free, and in the next few days he received reports of recoveries. Apparently some of the birds had not travelled very far, but those which had gone an appreciable distance were headed toward their winter range. Schuz at Rossitten on the Baltic Coast of eastern Prussia tried similar banding experiments with White Storks (Ciconia ciconia). In the middle of September, after all the local population had departed, he released 73 young banded birds. Most of them travelled eastward toward the Black Sea, paralleling the normal flight line for the local population, though some of the birds flew in a more southerly direction and three went southwestward to Italy. Perdeck, over a period of four years, caught and banded over, 11,000 fall-migrating Starlings (Sturnus vulgaris) in Holland and released them in Switzerland. Recoveries totaling 354 later showed that the juveniles took their ancestral direction southwest, paralleling the normal route, to a new winter area, whereas the adults soon separated and veered westward toward their ancestral winter range. From these experiments alone, it seems clear that at least some birds must have an innate ability to follow the normal migratory route or one parallel to it. How the birds "know" when they have flown far enough is yet to be determined.

Granting that migrating birds have an innate ability to find their way does not deny that some birds acquire their ability by experience. Young Indigo Buntings (Passerina cyanca), hand-raised in various conditions of isolation, apparently do not attain the accuracy of orientation typical of adult buntings and thus seem to depend, at least partly, one some kind of experience. Young geege and cranes, which migrate in families or groups of families, probably learn the migration routes by following their elders. They no doubt "memorize" features of the landscape when they migrate by day as they often do. But whether or not birds inherit or acquire their ability to find their way, the fact remains that they must depend on cues other than landscape features to orient themselves and navigate when travelling at night, over the sea, or above or in a heavy overcast. What are the cues?

The first break-through came in the early 1950's when it was proved that the sun is a cue to orientation and even a guide in navigation. At Wilhelmshaven, Germany, Kramer placed a small, circular cage high in the centre of a circular pavilion that was completely enclosed except for six windows, each high enough to give only a view of the sky from the cage. Kramer put a hand-reared Starling in the cage at the time in the spring when it would normally migrate, and from below the cage he recorded the direction in which the bird, in its migratory restlessness, showed a tendency to flutter. The bird fluttered persistently in the normal direction of spring, migration if the sun was shining but not if the sky was heavily overcast. When, by an ingenious use of mirrors at each of the windows, Kramer altered the sun's apparent position, the bird deflected its direction in order to maintain the same angle as before. Kramer and his associates soon demonstrated in subsequent experiments that birds have some sort of "internal clock" that enables them to compensate for the sun's daily "movement" across the sky and thus can hold the same direction of flight despite the sun's steadily changing position.

Hoffmann experimented further on the internal clock as a basis for orientation by the sun. He trained caged Starlings to seek a food reward at a particular time of day and always in the same direction. Then he tested them without rewards at the time they had come to expect and at other times. The birds, he found, could allow for the daily movement of the sun. Having proved this, he subjected the trained birds to a regime of light and darkness that shifted their internal clocks six hours ahead or six hours behind the natural day outside. Later he exposed the birds to the natural day. The birds responded by shifting their directions accordingly by 90 degrees, counterclockwise if the clocks were set ahead, clockwise if they were set behind. These and later tests showed that the birds did not orient themselves by the elevation or altitude of the sun in the sky but by its azimuth or position with relation to compass direction. The birds thus determined direction by what is called "sun compass orientation."

Soon after Kramer reported his initial experiments with Starlings, Matthews, found that homing pigeons, which are non-migratory, use the sun as a guide. When he released pigeons in unfamiliar territory under clear skies, they headed in the direction of home. If the sky was overcast, they failed to do so. The concept of the sun as a compass for homing pigeons has since been supported by other experimental investigators using different techniques. For example, Schmidt-Koenig shifted the internal clocks of pigeons by six hours in advance and six hours behind the natural day. On releasing the birds under a clear sky at varying distances and different direction from the home loft, he found that the birds deviated appropriately to the left or right of the correct flight direction taken by the control pigeans whose clocks had not been shifted. This suggests that the sun played a role in their orientation. Experience also plays a role in the orientation. Each time a pigeon is released, the direction it takes is more accurate.

Some birds may show one-directional orientation that is not necessarily related to migration or homing. Matthews, experimenting with non-migratory Mallards (Anas platyrhynchos) at Slimbridge, England, discovered that when birds of any age were displaced during a clear day for any distance in any direction at any season, they demonstrated a strong tendency to fly inone direction, namely, northwest. The flights were always short and seemed to be guided by the sun, but since they had no discernible purpose other than possibly escape, Matthews called the behaviour "nonsense orientation." He believed that it might be innate or perhaps developed at an early age. Other birds may show the same phenomenon. Migratory Mallards wintering in Illinois, on being trapped and released during a clear day or night at points far east and west of their winter-home lakes, consistently flew northward regardless of the direction of their home-lakes. Common Terns (Sterna hirundo), displaced from their breeding colony on Penikese Island Off Cape Cod, Massachusetts, on a clear day, flew southeast, irrespective of where they were released or the direction of Penikese Island.

Adeli Penguins (Pygoscelis adeliae), when displaced from their breeding colonies at Cape Crozier on the coast of Antarctica to the featureless expanse of compacted snow in the interior, oriented themselves by the sun, departed in one direction, and ultimately reached their nest sites. The investigators, J.T. Emlen and Penney, used breeding males, some of whose internal clocks had been artificially shifted in advance. Releasing them individually from three groups, each at widely separated points, they observed that the birds headed straight for the coast on courses which were essentially parallel without any convergence toward Cape Crozier. The birds seemed to have no information on their location at release with respect to their home colonies. They took their courses from the sun and, as shown by the breakdown in orientation among those birds whose clocks had been shifted, maintained their courses by referring to the sun's azimuth. If heavy clouds obscured the sun and eliminated all shadows, their orientation soon deteriorated. Emlen and Penney considered such one-directional courses to be fixed and to be maintained by an inherent time sense; they linkened the procedure to nonsense orientation as described in waterfowl and other birds, but believed that it might be escape orientation with survival value by steering the penguins to off-coast feeding areas whence they would be guided, perhaps in part by familiar landmarks, to their home colonies. Later experiments by the two investigators confirmed their original findings and conclusions.

A few years following Kramer's initial discovery that migratory restlessness in caged Starlings was directed toward the sun. Sauer and his wife, undertook experiments in Germany to determine by what means nocturnal migrants find their way. The Sauers used three species of sylviid warblers, reared in captivity without ever having seen the natural sky. During the period when the species would normally migrate, the Sauers put each birds in a rotatable circular cage and exposed it to the night sky. Under a clear, starry sky, even when their cages were turned periodically, the test birds fluttered persistently in the direction that the species takes at the season of the experiments; under heavily overcast skies the birds fluttered randomly in various directions, obviously disoriented. Then the Sauers placed the cage in a small planetarium with a starry sky, adjusted to the local time and latitude. The birds fluttered in the direction appropriate to that season depicted. There seemed to be no doubt that these birds obtained their information for direction from the starry sky.

The Sauers carried their experiments further. Placing the birds under skies representing the non-migratory seasons, they noticed that some of the birds were completely disoriented. From these experiments and others, Sauer concluded that birds do not rely on the stars themselves for direction but on the azimuth and altitude of the starry sky. This bicoordinate system would give birds the necessary information on their location since azimuth or hour angle would denote longitude and altitude would indicate latitude.

Working with mist-netted Indigo Buntings in the United States, S.T. Emlen repeated many of the Sauers' experiments and made still other well-controlled tests under planetarium skies. The results confirmed some of the Sauers' studies but differed sharply from the others. Emlen found no evidence that the buntings relied on a bicoordinate celestial system for orientation but rather made use of the numerous stars, particularly "the constant, two-dimensional spatial relationships" existing between them. He believed that no single star, with the possible exception of Polaris (the North Star) could give the birds sufficient information for direction. It is the configuration or patterning of the bright stars in the constellations such as Ursa Major (the Big Dipper) that gives the cues. While Emlen could not determine with any certainty which patterns were of special importance or essential to orientation, the evidence seemed to point to the northern astral sky, especially the area within 35 degrees of Polaris.

What prompts birds to take the appropriate seasonal direction? Seeking an answer to this question S.T. Emlen induced physiological readiness for spring and fall migration in two groups of Indigo Buntings and then tested them simultaneously under the spring sky of a planetarium. The group conditioned for spring migration oriented northward; the group in condition for all migration took the opposite direction. This clearly suggested to Emlen that the internal physiological changes in Indigo Buntings, rather than differences in external stimuli (*e.g.*, in the night sky), are responsible for their direction in spring and fall.

Practically all the experiments on the responses of homing and migrating birds to the sun and stars have demonstrated that when skies are heavily overcast the birds show disorientation. The consequent implication is that birds cannot find their way when celestial cues are obscured. Very recent studies, however, indicate that this may be untrue. Keeton has shown that homing pigeons, while using the sun as a compass when available, can navigate accurately under total overcast and even without familiar landmarks. Radar studies reveal that birds migrate successfully when there is a heavy overcast. It the cloud layer is too high to surmount, birds will fly under it or even in it. This being the case, how do birds orient themselves without celestial cues and, if they are flying at night or in a cloud layer, without being able to see landmarks or topographical features familiar to them? The cue may be wind direction.

Nisbet once theorized that the birds might use wind—more specifically atmospheric or wind turbulence—in orientation. In central Illinois, Bellrose, with his associates, has gathered ample experimental evidence to show that Blue-winged Teal (Anas discors), a long-distance migrant that commonly winters in northwestern South America, does indeed use the turbulent structure of the wind for direction. The species migrates by day or night, and radar surveillance and visual sightings by Bellrose confirm its ability to migrate under overcast skies. In the fall and early winter. Bellrose used both hand-raised immature teal that he released at various points up to 100 miles distant from their home pens and wild immature teal, trapped during migration, that he held in pens and released from 10 to 40 days after other Blue-winged Teal had left the region. All his experimental birds were banded. The hand-reared birds had never flown over the area and none of the birds had ever viewed the landscape along the standard migration routes southward. When released under both clear and overcast skies, nearly all the birds tended to start in the same direction—with the wind. Analyzing his data from visual sightings and banding recoveries and drawing upon his knowledge of waterfowl movements from radar studies and airplane tracking, Bellrose concluded that Blue-winged Teal, and probably other species of waterfowl as well, prefer landscape features for orientation, but, if unavailable, they resort to celestial cues. Both being unavailable, they use wind direction, perhaps referring at the outset of flight to landscape cues for information on the sector of the compass from which the wind is blowing.

The present knowledge of direction finding, the highlights of which are briefly reviewed in the foregoing paragraphs, points all too clearly to the danger of generalizing on the means of orientation and navigation in migration. Different groups of birds in their adaptations to different modes of existence may well have developed correspondingly different means of finding their way from one place to another in accordance with the prevailing ecological conditions. It is unlikely that any species orients itself entirely by one cue. A one-directional orientation may suffice for some birds while a bicoordinate system of navigation may be necessary for others. Direction-finding, though appearing to be basically inherent, may actually prove through eventual research to be acquired chiefly by experience. In all probability, direction-finding in strongly migratory birds is a highly complex procedure involving physiological and behavioural responses to not one but several environmental factors—celestial bodies, wind, and perhaps others presently considered as having no relationship to the problem.

Study of Nocturnal Migration

There are at least five field methods of studying noctural migration. (1) By observations of movements across the face of the moon. The principal equipment needed is a telescope with a power of 15 or greater. For methods and procedures, consult Lowery and Nisbet and for the application of accumulated data. (2) By radio-tracking. The equipment needed includes a transmitter and receiver. For general information on the type of equipment required, consult Sclater, Cochran and Lord and Southern. For information of a transmitter adapted to small migrants, as well as an example of methods and procedures in radio-tracking small migrants. (3) By radar surveillance. This requires the use of a special facility operated by skilled technicians. (4) By tape-recording. Besides a tape recorder, this requires a parabolic reflector, microphone, and amplifier. (5) By the study of migrants killed at television towers or migrants killed as a result of airport ceilometers or other man-made interferences with migration.

Of the five listed methods, the most practical for the beginning student is the study of migrants killed at television towers. It requires no special equipment. Practically every city or large community has on its outskirts one or more television towers that almost certainly take a toll of noctunal migrants. Much as these kills are regretted, they nonetheless provide rewarding material for the study of many aspects of migration locally.

An early-morning search of the ground under a television tower and its guy wires following a night during which migration proceeded at low altitude will usually yield birds of many species. If the student chooses to make such a search as often as these accidents occur through the migration saason and on successive seasons, keeping a careful record with dates of the number of individuals of each species killed, he will obtain a true sampling of the nocturnal migratory activity in the area of the tower. He may also discover the presence of species not heretofore reported, or even suspected, in the area because they have always passed through at night unseen.

Given the time necessary, the student may collect specimens for making one or more special studies such as the following:

1. Succession of sexes and age groups in different species: whether males precede females and adults precede immatures, or *vice versa*.
2. Geographic variation in a wide-ranging species migrating through a particular area: whether the colour and measurements of individuals reveal one or more populations or sub-species.
3. Molt with relation to migration: whether certain species are in stages of molt and, if so, which feather tracts are involved. Normally birds do not molt the remiges and their primary coverts during migration, but the rest of the alar tract and all the other tracts may be in the process of molt, depending on the species or the sex and/or stage of maturity of individuals in each species.
4. Weight with relation to the length of night-flight: whether the birds are quite heavy, having been killed soon after take-off, or quite light, having been killed after a long flight.

5. Fat condition with relation to stage of migration and duration of night-flight: whether the birds are quite obese, indicating that their migration was well under way with long night-flights, or quite lean, indicating that their migration was just beginning with short night-flights.

Before collecting any specimens, the student must obtain both a federal and a state For provincial) collecting permit authorizing him to take dead birds "for salvage."

Collect the specimens after a kill as early in the morning as possible since the carcasses not only decompose rapidly but are soon molested by insects or eaten by house cata, crows, gulls, and other creatures hat inevitably find the ground around a television tower a promising source of food. Weigh each speciman without too much delay as the body begins losing weight shortly after death. Mark the weight on a tag and attach it to a leg. Seal all specimens collected in plastic bags to avoid their dehydration, mark with the date, time of day, and place of collection, and put at once in a deep-freezer for study and analysis later.

Study of Diurnal Migration

Whenever opportunity permits, the student should watch the diurnal movements of migrants in his own area. He will soon discover how greatly birds are influenced in their daytime movements by the terrain, vegetation, and waterways. He is almost certain to note that passerine migrants, although continually searching for food, follow much the same courses and in the same direction. If the birds inhabit trees or shrubs, they choose paths in which these grow and avoid, when possible, crossing wide stretches of water and open country. For an idea of how much detailed information one can obtain in a local study and some of the procedures to follow, the student is referred to the classic work by Ball, in which he reported on six fall migrations on a small point of land of the great Gaspe Peninsula, Quebec.

Flight Speed

What are the optimal flight speeds in migrating birds? To answer this question one must look into the theory of bird flight mechanics. For this presentation I will use the model developed by Pennycuick (1969, 1975). Related versions of this model have been presented by Tucker (1973) and Greenwalt (1975). Quite recently, Rayner (1979) tried a new approach to this problem by developing a model on the basis of vortex theory applied to the motion of air in the wake of a flying bird.

A bird in horizontal flight expends power to support its weight (this power component is called induced power), to overcome the drag of the body (parasite power) and of the flapping wings (profile power). Furthermore, power is required for general maintenance, including basal metabolism and extra pumping of blood and ventilation of lungs. These power components can be summed up for various flight velocities (V). Induced power falls approximately in proportion to V^{-1}, while the parasite power increases in proportion to V^3. Basal metabolism and profile power are assumed to be constant irrespective of flight speed. For the profile power this assumption is not theoretically justified, but studies of pigeons flying in a wind tunnel suggest it is a reasonable approximation. As a result, the equation relating total power to flight speed will be of the general form:

$$P=k+aV^{-1}+bV^3$$

where P is power, V is flight velocity and k, a and b are constants that vary due to the bird's mass, form and wing span, to air density and coefficient of drag. This power equation does not apply to extremely low flight speeds or hovering (V=0). For these cases a different method of calculating power must be used.

The power curve will be U-shaped as seen, and there are two specific flight velocities of particular interest. The most obvious is where total power is minimized. This velocity is called V_{mp} (mp is minimum power) and may be calculated by solving $dP/dV=0$. However, for migrating birds the primary interest is not to minimize the absolute rate of energy consumption but rather to minimize energy consumption per unit of distance flown. This will allow migrants to cover the maximum range on a given energy reserve. V_{mr} (mr is maximum range) is found by solving $dP/dV=P/V$, *i.e.*, drawing a tangent from the origin to the power curve. It is easy to see that this gives the minimum possible ratio of P to V.

Studies on flying birds under experimental conditions have yielded empirical approximations for various parameters in the power equation, making it possible to estimate this equation on the basis of the bird's mass and wing span only. In this way I have estimated the power equation for low-altitude flight in a selected number of species of various sizes (Table 11.1), and on the basis of these equations I have calculated V_{mp} and V_{mr}. The estimated power curve is presented in full for the wood pigeon. Note that the power given in this figure refers to power output. The corresponding power input is the rate at which the bird consumes chemical energy (the ratio of power output to input, *i.e.*, the mechanical efficiency of muscles and supporting systems, is about 23%).

The species in Table 11.1 were selected because there are records available of radar observations of flight speeds of migrating flocks and individuals in these birds. From this table it is seen that the mean air speed (flight speed corrected for the influence of wind) of the migrants generally is close to V_{mr}. Hence, in their choice of flight speed migrants seem keen to minimize energy consumption per unit of distance covered. The only exception in the table is the whooper swan, whose air speed is closer to V_{mp} than V_{mr}. One possible explanation is that air speed given in the table is not typical for migrating swans as it is based on radar tracking of only five flocks. These swan flocks were migrating low over the Baltic Sea at the end of April towards their northeasterly breeding quarters. On repeated occasions flocks of swans were observed swimming over deep water in their migratory direction, and one flock of 22 whoopers tracked by radar landed on the water surface and the birds swam purposefully for almost two hours before they set off flying again, and radar tracking was resumed.

This may suggest that swans have difficulty in producing the power required for continuous flight, and that they are forced to fly with minimum power output. Power required for flying in birds with different body masses (M) approximately increases as $M^{7/6}$, while the capacity of power output depends on the mass of flight muscles increasing approximately in proportion to M. This implies a maximum mass in birds above which flight is not possible, and calculations suggest that this size approximates to that of the largest flying birds: about or slightly above 10 kg, for example, for swans, bustards, albatrosses and condors. Hence, it may well be difficult for the whooper swans, presumably loaded with some fat reserves for their migratory journey, to muster up enough power to fly at V_{mp} and it may be impossible for them to fly continuously at V_{mr}.

The birds' adjustment of their flight speed is nicely illustrated in the swift (Bruderer & Weitnauer, 1972). While the mean air speed is 11 m s^{-1} in migrating swifts, it is only 6 m s^{-1} in breeding swifts spending the night aloft. Indeed, these values fall close to the predicted speeds with energy consumption minimal per unit distance (V_{mr}) and per unit time (V_{mp}), respectively (Table 11.1). Furthermore, white-throated sparrows Zonotrichia albicollis released experimentally and tracked by radar, under clear night skies fly off in their normal migratory direction at a mean air speed of 9.2 m s^{-1}, similar to V_{mr}, but when released under overcast conditions, they fly on irregular tracks, with obvious difficulties of orientation, at 6.6 m s^{-1} mean air speed, close to V_{mp} (Emlen & Demong, 1978).

There are some predictions about the optimal adjustment of air speeds in migrating birds that remain to be critically tested

TABLE 11.1

Observed Air Speeds in Migrating Birds, Studied by Radar, in Comparison to Speeds for Minimum Power (V_{m_p}) and Maximum Range (V_{mr}) as Predicted Theoretically from the Power Equation

	Mass *(kg)*	*Wing span* *(m)*	V_{mp} $(m\ s^{-1})$	V_{mr} $(m\ s^{-1})$	*Observed mean air speed* $(m\ s^{-1})$
Chaffinch Fringilla coelebs	0.020	0.30	5.1	9.5	11
Swift Apus apus	0.043	0.45	5.4	9.4	11
Redwing thrush Turdus iliacus	0.065	0.40	6.6	11.5	13
Wood pigeon Columba palumbus	0.50	0.75	9.5	15.7	17
Oystercatcher Haematopus ostralegus	0.55	0.80	9.5	15.7	14
Eider duck Somateria mollissima	2.0	1.1	12.5	20.4	21
Bean goose Anser fabalis	3.5	1.6	12.5	20.3	20
Crane Grus grus	5.5	2.4	11.8	19.4	19
Whooper swan Cygnus cygnus	10	2.4	14.4	23.5	17

Observed speeds refer to horizontal flapping flight. Hence, for the crane, which migrates by soaring over land, the observed speed refers to birds migrating across the Baltic Sea.

(Pennycuick, 1978). One is that V_{mr} increases with increasing flight altitude, another that V_{mr} declines with increasing tailwinds and increases with headwinds. The ground speed (G) of a migrating bird is the vector sum of the bird's air speed and wind speed. Hence, the power equation may be used to calculate V_{mr} under different wind conditions by solving $dP/dV=P/G$. Both radar and optimal measurements strongly indicate that birds indeed do fly relatively faster in headwinds than in tailwinds (Bellrose, 1967; Bruderer, 1971; Tucker & Schmidt-Koenig, 1971; Emlen, 1974; Schnell & Hellack, 1979).

Fuel Economy

Birds store fat for use as fuel during their migratory flights. Depot fat is the most economical type of fuel in terms of a high oxidizing energy per unit mass. The fuel ratio (F) is the friction of take-off mass of a migrating bird that consists of fuel, and it may be as high as 50 per cent in birds performing long non-stop flights. Taking into a account the energy content of the fat load, one can estimate the maximum flight range in still air for birds with different fuel ratios, as seen in Fig. 11.6 (cf. Pennycuick, 1975). Large birds do not store fat to extremely high fuel ratios, because, with too much extra weight, they cannot attain the muscle power equivalent to P_{mr}.

Flight range is greatly affected by winds, and therefore it is especially useful to estimate the total time a bird can spend aloft flying at V_{mr} on a given load of fuel. On the basis of the power equation, the expected relative weight loss due to consumption of fat is about 0.7 per cent body mass per flying hour in small birds like warblers, finches and thrushes, and up to 1-1.5 per cent in large birds like ducks, geese and swans. The rate of weight loss during flight will be less in birds with a wide wing span than in birds with short wings.

Various field estimates of the rate of weight loss in migrating birds are presented in Table 11.2. Excluding the value for Manx shearwater, which uses soaring migration (see below), the estimates fall between 0.6 and 2 per cent loss of body mass

h^{-1}. Some values agree closely with the theoretical predictions, while others are surprisingly high. However, all estimates depend on more-or-less uncertain assumptions and in view of the widely different methods, the variation between the estimates is not unduly high (Nisbet, 1963a). It may be that interspecific differences in effectively storing and using fat are more important than expected from theory, or that a high rate of weight loss in some birds is due to excessive loss of water under certain conditions. Very little is known about water economy and the risk of dehydration in migrating birds.

The time a bird with a certain fuel ratio can fly until all fuel is used up is given by

$$t = k^{-1} \ln (1 - F)$$

where k is the relative weight change due to fuel consumption per flying hour. Hence, with $F = 0.5$ and $k = -0.007$, maximal flying time is about 100 h. These conditions are relevant for *e.g.*, the blackpoll warbler travelling from northeastern North America directly across the Atlantic Ocean to winter quarters in northern South America (Nisbet, 1970). On the basis of radar observations over the Atlantic Ocean, a normal duration of 70 to 100 hours is estimated for birds undertaking this formidable non-stop flight (Larkin, Griffin, Torre-Bueno & Teal, 1979). The above calculation is relevant for a bird flying at V_{mr}; if the bird slows down to V_{mp} it may stay in the air for about 130 hours under the same conditions. Actually, a lot of radar echoes recorded at relatively low altitude over the western Atlantic ocean, which probably were from migrating birds, showed air speeds at or below 5 m s^{-1}, *i.e.*, close to V_{mp} in small birds (Williams, Williams, Ireland & Teal, 1977: Larkin *et al.*, 1979). Perhaps migrants over the ocean were slowing down to save energy, while waiting for more favourable weather and better conditions for orientation?

One of the most impressive feats in animal behaviour is the migration of birds across thousands of kilometres of ecological barriers, like oceans, glaciers and deserts. In the light of today's knowledge of such enomous non-stop journeys, the classical case of migration across the North Sea is no longer so impressive. The distance across the North Sea from Scandinavia to Britain is 500-800 km and the normal flight time is between 10 and 20 hours for small birds. The North Sea is regularly crossed by a multitude of bird species, passerines, shore birds

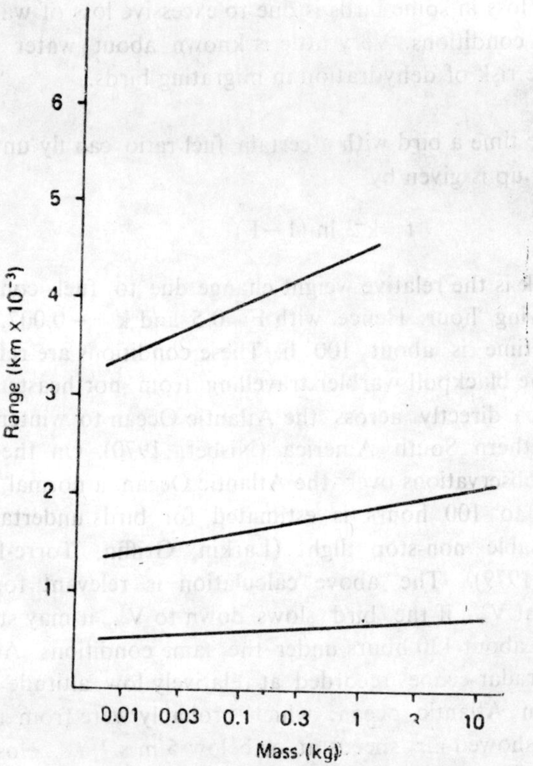

FIG. 11.7: Maximum flight ranges in still air for birds with different fuel ratios. The straight lines of flight ranges in birds of different sizes are rough approximations only, since the birds' dimensions are important for the energy consumption in flight. From upper to lowest lines, F=0.5, 0.25, and 0.1, respectively. By way of example, birds with a relatively wide wing span attain longer flight ranges than birds with short wings. Calculations are based on Pennycuick (1975).

and waterfowl, and the radar studies by David Lack (1959-1963) have demonstrated the very vivid traffic of migrating birds over this region.

Another classical example of a long sea crossing by birds is the migration across the Gulf of Mexico, a distance of about 1000 km. For small birds, this flight normally lasts about 20 h, in poor weather considerably longer (Lowery, 1945; Gauthreaux, 1971). The trans-Gulf migrants, including the ruby-throated hummingbird Archilochus colubris, attain fuel ratios up to 45 per cent (Odum, Connell and Stoddard, 1961), making it possible for the birds to fly a distance, without refuelling, that is at least twice as long as the actual Gulf-crossing.

In this chapter are shown some of the logest non-stop flights by birds in the North Atlantic area. By way of example, snow buntings Plectrophenax nivalis, breeding on East Greenland, cross the Greenland Sea on their way to winter quarters in Russia, while birds of West Greenland populations fly across Davis Strait to North America. Brent geese Branta bernicla, white-frouted geese Anser albifrons, turnstones Arenaria interpres, knots Calidris canutus and wheatears Oenanthe oenanthe breeding in Greenland and parts of the Canadian archipelago fly to and from western Europe, either via Cape Farwell or across the ice-cap of Greenland, sometimes but not always using Iceland as a sto pover site (Salomonsen, 1967). In the autumn, knots travel directly from northeast Greenland to the North Sea countries (Morrison, 1977). In spring, knots in England store a fuel ratio at over 30 per cent for the flight to Iceland. On Iceland the knots refuel, and birds heading for Ellesmere Island attain fuel ratios at 40 per cent. Practically all fat has been consumed when they reach their destination.

The most illuminating ringing results come from Britain, where quite a number of redwings ringed in winter have been recovered during a subsequent winter in the Near East and

TABLE 11.2

Field-Estimates of the Rate of Weight-Loss in Migrating Birds

Species	*Weight-loss* (*% body mass* h^{-1})	*Basis of field estimate*
1. Blackpoll warbler *Dendroica striata*	0.6	Estimated arrival and departure weight for non-stop flight: from New England to Bermuda
2. Robin *Erithacus rubecula*	0.9	from Norway to Scotland
3. Goldcrest *Regulus regulus*	0.7	from Norway to Scotland
4. Wheatear *Oenanthe oenanthe*	0.8	from Greenland to Scotland
	1.3	
5. Knot *Calidris canutus*	1.0	from England to Iceland and from Iceland to Ellesmere Island
6. Manx shearwater *Puffinus puffinus*	0.3	from Wales to Brazil
7. Song sparrow *Melospiza melodia*	0.7	Weight before and after a night of migration, Massachusetts
8. Thrushes *Hylocichla* spp.	1.8	Weight at different times in the night: Migrants killed at TV-tower, Illinois

9. Tennessee warbler Vermivora peregrina	1.8	Migrants killed at TV-tower, Wisconsin
10. Veery Hylocichla fuscescens	1.3	Migrants killed/trapped at lighthouse, Ontario
11. Ovenbird Seiurus aurocapillus	1.0	
12. Chaffinch Fringilla coelebs	2.9 (2.0)	Weights at trapping sites 50 km apart along the migratory pathway, East Baltic

Note: The original estimate for the chaffinch was based on an assumed air speed at 14.5 m s^{-1}.

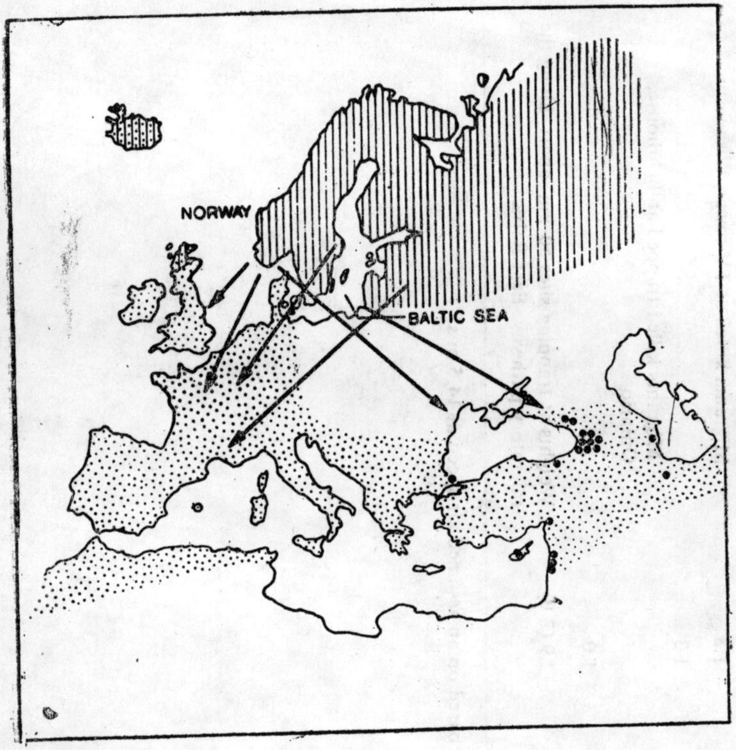

FIG. 11.8: Autumn migration of redwing Turdus iliacus in Europe. Under northerly and easterly winds birds depart from Scandinavia and the Baltic region towards southwest to winter quarters in Britain and west/southwest Europe. With strong winds from the west and northwest behind cold fronts, they depart southeast towards the region around Caspian, Black and eastern Mediterranean Seas. Winter recoveries in this latter region of birds ringed in a previous winter in Britain (Mead, 1976) are indicate on the map (circles). Vertical lines indicate the breeding range, the stippled area, the winter range; lines and stippling, breeding and winter range.

southern Russia (Fig. 11.8). The indicates that winds prevailing when the redwings are prepared to depart from Scandinavia determine which winter quarter will be visited—the birds take advantage of following northeast or northwest winds to travel swiftly either to Britain/west Europe or the Near East/ southeast Europe, respectively. So, an individual redwing may winter in Britain one year and in Soviet Georgia the next.

In general, migrating birds are faithful to their winter as well as to their breeding sites. The exceptional flexibility in the redwing brings the advantage that energy costs for the migratory journey are minimized.

Weather and Timing of Migratory Flights

Migratory intensity varies dramatically between different days and nights, even during the peak period of migration. This variation is due mainly to weather. During recent decades radar has been used extensively to monitor migratory intensity in relation to weather in the nort temperate zone, while next-to nothing is known about this relationship in other climatic regions.

The occurrence of intensive spring and autumn migration in relation to synoptic weather features is shown in this paper Dense spring migration usually takes place with warm southerly winds, to the west of a high pressure and to the east of a low eressure centre. Furthermore, it occurs in the moist, warm sector of a depression, and spring migrants heading northeast also depart during westerly or southwesterly winds behind cold fronts of relatively mature cyclones. In contrast, intensive autumn migration takes place with cold northerly winds to the east of a high pressure centre and west of a cyclone. Migrants on their way towards the southeast depart under brisk northwesterly winds immediately after a cold-front passage while southwesterly migrants often wait for one or two further days, and depart when the anticyclone has moved closer and winds blow from the north or northeast. Spring as well as autumn migration occurs under anticyclonic conditions with calm and fair weather (Richardon, 1978).

In many studies, migratory intensity has been correlated with a great number of different weather variables, and multivariate statistical analyses have been used to investigate which combination is best for predicting migration. Most studies agree that wind and precipitation are key factors: dense migration takes place with following winds and no precipitation. Once these factors are taken into account predictions of migratory intensity do not become appreciably better by considering additional weather variables. This is equivalent to saying that under constant wind and precipitation conditions, only weak correlations exist between migratory intensity and other weather factors.

It is easy to see the selection pressures behind the close association of intensive migration with following winds. Flying time and energy costs for the migratory journey become greatly reduced for birds using tailwinds to achieve a high ground speed. It is advantageous for birds to avoid the stresses of increased heat loss caused by wetting of plumage, and possibly of disorientation, during flight under conditions of precipitation.

Migrating birds' responses to weather may have evolved not only in response to selection pressure for favourable flight conditions but also for arrival at their destinations under favourable living conditions and for departure when conditions at the point of origin worsen. The dominant influence of following winds on migratory intensity, as found in most studies, indicates that flight conditions generally constitute the primary selecting force for the short-term effect of weather on birds migratory activity, while living conditions at departure or destination areas play at most a subsidiary role. However, there are a few interesting exceptions.

In May, intensive northeastward songbird migration takes place in New England. Departures from Massachusetts of birds reaching Maine after a night's migration were observed on radar by Nisbet and Drury (1968). A combination of onshore crosswinds, low but rising humidity, rising temperature, low pressure and absence of rain is most reliable for predicting heavy migration. The synoptic interpretation is that 'migration is concentrated into the zone to the west of the cool dry centre of a high-pressure cell and east of a warm trough of low pressure. This zone usually marks a strong northward flow of warm air; the flow is usually less strong than in the trough to the west, but is more likely to persist and less liable to develop into cyclonic disturbances' (Nisbet and Drury, 1968). As a result, the migrants usually encounter high temperatures and avoid rain in their destination area the next morning. Indeed, the above-mentioned combination of weather variables in the departure area is remarkably good for predicting warm and dry weather at the destination next day. Any reinforced response by the departing birds to tailwinds *per se* would increase the risk of encountering cold and rainy weather after arrival in the northerly areas.

Hence in this case natural selection has made the migrants masters in the task of forecasting weather the next day in areas hundreds of kilometres to the northeast. Selection for favourable living conditions at the destination, warm, dry weather and therefore plenty of available food for the insectivorous birds, overrides the narrow selection for maximal tailwinds en route.

For late-autumn migrants, like pigeons, crows, starlings and finches, a falling temperature is positively correlated with migratory intensity even when the effect of wind is taken into account. Hence, in these migrants living conditions at the take-off site exert a significant selection pressure on the migrants' responses to weather.

During winter and early spring, cold, snowfall and frost release hardweather movements or retrerat, migration.

The use of radar for monitoring bird migration brings two major drawbacks:different species usually cannot be distinguished and migration at extremely low altitudes is often missed because it is below radar coverage. Direct field observations of visible bird migration do not have these drawbacks, and furnish an important complement to radar data. The most serious

Wood Pigeon	Brambling/chaffinch

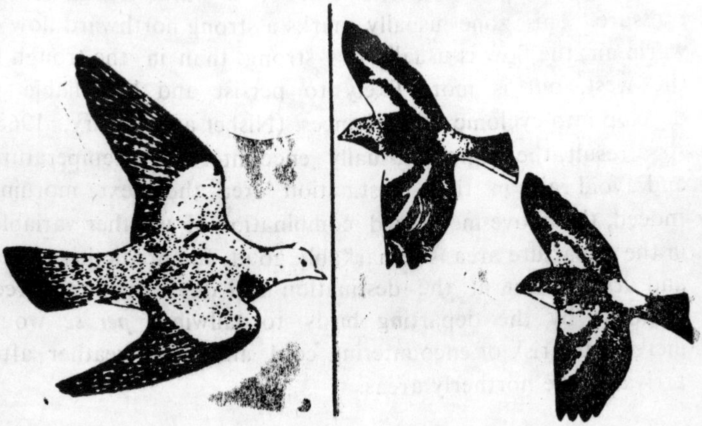

Light and moderate ME winds	Moderate SW winds
Clear sky with scattered cumulus clouds	
No rain, low humidity and good visibility	No rain
High and rising barometric pressure	
Low and sinking temperature	Mild but sinking mimimum temperature
Average no. of migrating wood pigeons	Average no. of migrating chaffinches/bramblings
on days with NE winds: 8900	on days with NE winds: 1500
on days with SW winds: 1600	on days with SW winds: 120500

FIG. 11.9: Weather associated with intensive visible autumn migration of wood pigeons and chaffinches/bramblings (these two species travel in mixed flocks and cannot be counted separately) at Falsterbo. South Sweden. Average numbers of migrants are calculated for the seasonal peak period of migration. Data from Roos (1977, 1978b) and Alerstam (1978).

sources of bias in field studies of visible migration are that high-altitude movements of small birds escape notice, and that migration along leading-lines is overemphasized.

Field observations reveal considerable differences between responses to weather by different species. Weather characteristic of maximal visible autumn migration of wood pigeons and finches (chaffinches Fringilla coelebs and bramblings F. montifringilla) at Falsterbo in South Sweden are compared in Fig. 11.9. Wood pigeons are relatively choosy with regard to following winds, and prefer to migrate after passage of a cold front, under cold northerly winds and clear skies. In contrast, almost a hundred times as many finches are counted by the field observer under headwinds as compared to tailwinds. Rain suppresses finch migration, but there are no significant correlations with cludiness, visibility, humidity, or barometric pressure.

Radar, in combination with observations from aircraft and with vertical telescopes, reveals that great numbers of finches migrate also with following winds, under weather conditions similar to those associated with intensive migration of wood pigeons. Under such conditions the finches fly on a broad front at relatively high altitude, from several hundred up to one or two thousand metres, above the detection range of a field observer. Hence, the large number of finches registered by the field observer under headwinds does not mean that the finches prefer headwinds for their migration, but rather that circumstances force them to take flight under these energetically unfavourable conditions also.

The adaptive values of the different responses to weather can be explained as follows: weather in northwest Europe is characterized by the eastward flow of successive cyclones. Sometimes an intervening anticyclone builds up and there are no cyclones passing for some days. In the east and central part of the high-pressure cell, conditions are optimal for autumn migration, with following winds and fair weather. Migrants of all sorts, diurnal and nocturnal, finches and pigeons, seize this opportunity and depart in vast numbers. Migration takes place on a broad front at relatively high altitudes, where the following winds are strongest. However, especially in late autumn, anticyclonic interludes may seldom occur, so that cyclones pass in rapid unbroken succession. Looking through the daily weather maps for some years. I found that in the southern North Sea area frontal passages occur on the average every third day between mid September and mid November. Regularly there are periods with families of frontal depressions, with fronts passing on five to seven out of ten days. During such periods there are no better occasions for migration than the brief wake of a cyclone just having passed eastwards, before the arrival of the next depression. Finches depart on these occasions, since they would otherwise be delayed for long periods waiting for anticyclones and following winds. In order to minimize the adverse effect of headwinds they fly low over the ground and follow coastlines.

Chaffinches carry out their migratory journey from Scandinavia to Britain or France via the Low Countries by a great number of short hops, each consisting of a few hours' flight in the morning. This seems to be a reasonable strategy in the extremely windy and variable late autumn weather of this region. Scandinavian wood pigeons destined for France and the Iberian peninsula fly at high speed (Table 11.1) and travel a relatively long distance during a morning's flight. They can therefore better afford to wait for anticylones bringing tail-winds. However, this strategy leads to delays in pigeon migration in some autumns, and if tail-winds fail for a long time the pigeons finally give up waiting and depart at low altitude under crosswinds, a compact cloud cover and poor visibility (Roos, 1978b). It is a rule among migrating birds that the threshold in favourable weather for the release of migratory activity decreases as the delay since the previour instance of migration grows longer.

When migrants encounter adverse weather like headwinds, rain or fog in flight they land as soon as possible. The occurrence of sizeable falls of night migrants at the English east coast in the autumn is usually associated with favourable departure weather is Scandinavia and adverse weather over the North Sea off the English coast.

Migrants crossing the Gulf of Mexico in spring normally do not land immediately upon reaching the North American coast but continue 50-150 km across marsh and prairie to the large inland forests before descending. However, when the trans-Gulf migrants meet rain or opposing winds over the Gulf or the North American coast they become grounded, sometimes in tremendous concentrations, on the first available land, barrier islands, marshland and isolated coastal woodlands.

In late summer, arctic waders often fly non-stop across Scandinavia to the shallow North Sea shores. They probably take off from the White Sea region. In years with a lot of northerly and easterly winds the numbers of such waders resting in South Scandinavia are particularly low. It is only in summers with frequent cyclonic passages, brisk southwesterly winds, cool and rainy weather that they occur in large numbers, and many can even be seen on low-altitude coastbound migration into the headwinds. Obviously the waders interrupt their long non-stop flight in years with adverse weather.

Flight Direction and Wind

Do migrating birds drift by the winds or do they compensate for wind drift?

At first thought, one would expect radar to be an ideal instrument for easily solving this problem. If the mean track direction of bird migration is shifted to the right on occasions with winds from the left, and to the left with winds from the right, this seems good evidence that birds are subjected to wind drift. In fact, almost all radar studies reveal that the overall mean track direction of bird migration is shifted in relation to wind in this manner. However, this may also be explained by a phenomenon called pseudodrift different migrants are aloft under different wind conditions.

Pseudodrift may occur because different species prefer different winds for their migratory flights, but also because populations of the same species differ in their choice of winds during migration. Such intraspecific pseudodrift has been confirmed in the eider duck and wood pigeon in South Scandinavia. By way of example, eiders departing eastsoutheast in spring prefer westerly and northwesterly winds, while eiders travelling eastnortheast fly in greatest numbers under southwesterly winds. Hence, the mean track direction of eider migration varies according to which population migrates in largest numbers.

By carefully distinguishing radar echoes from different cohorts of migrants, one may eliminate a large part of the effect of pseudodrift. A few radar studies over land and coastal regions indicate that, after pseudodrift is eliminated in this way, track directions of migrants are constant irrespective of wind—in other words, the birds compensate completely for wind drift. However, other studies under similar conditions indicate that a drift effect remains even after elimination of pseudodrift.

All radar studies agree that true wind drift occurs over the open sea. In South Scandinavia, wood pigeons and cranes compensate completely for wind drift over land, but drift when they migrate across the Baltic Sea. The variation in track directions of birds flying over the sea is, however, less than expected if they drift from a fixed heading. Perhaps migrants over the sea, devoid of landmarks for orientation to compensate wind drift, use the moving sea waves for this purpose, leading to partial drift (Alerstam & Pettersson, 1976)?

Hence, in spite of two decades of radar studies any final solution to the wind drift/compensation problem is not yet in prospect. The issue obviously is much more complex than first suspected: in some instances migrating birds do compensate completely for wind drift over land and coastal areas, but in other instances they are probably drifted to some extent by wind, and over the open sea they fail to compensate completely.

Could it be advantageous for migrants to allow themselves to be wind drifted under certain conditions?

1. Drift will always be disadvantageous if winds blow from a constant direction during the migratory journey. Under these conditions drift leads to a long flight path and flying time to the goal, even if the migrants continually readjust their heading direction towards their goal. The fastest way to reach the goal is to compensate for wind drift and fly along the straight-line track to the destination.

2. However, if winds vary during the migratory journey, the birds can reduce flying time, and costs of transport, if they allow themselves to be drifted by winds from the left during one flight, and wait for favourable winds from the right during subsequent flights. Hence, if winds shift significantly between the separate flights of the migratory journey, the birds should minimize the distance to their destination after each flight, and their optimum strategy is to allow extensive drift far away from their destination and gradually compensate to a higher and higher degree on approaching it, finally compensating completely during the last flight.

3. If the wind pattern shifts in a predictable way along the migratory route, so that wind from the right is succeeded by wind of a similar net force from the left, or *vice versa*, the birds will minimize flying time by travelling on a rather constant heading, allowing themselves to be drifted by wind. Perhaps this is what migrants flying from North to South America over the West Atlantic do. Their heading direction is towards southeast to south-southeast, both off New England, where they drift eastwards when departing under strong westerly to northwesterly winds after passages of cold fronts, and over the Caribbean Sea, where they drift westwards in the zone of easterly trade winds (Williams & Williams, 1978).

4. Migrating birds may also save flying time by allowing themselves to be partially drifted by strong winds at high altitude and correcting for the displacement by overcompensating at low altitude under relatively weaker winds. In fact, overcompensation in the sense that the migrants track directions, when comparing occasions with different winds, are shifted into the winds, has been found in low-altitude visible migration of, *e.g.*, chaffinches and starlings.

In the light of these various possibilities it seems not so odd that a simple answer to the question of whether or not migrants compensate for wind drift is lacking. I think there remains a lot to learn about the migrants' exploitation of winds, involving different strategies in different wind conditions, to achieve a swift and safe migratory journey at small energy costs.

Why do migrating birds fly along coastlines? Bird migration along coastlines sometimes leads to spectacular concentrations of migrants at certain peninsulas, many of which are resorts of bird-watchers' pilgrimages. Migrants' coasting behaviour probably is part of their strategy to get the better of winds. Coastal migration occurs mainly under opposed and cross-winds, while migrants usually fly across the coast and depart over the open sea with following winds. Due to differences in friction, winds generally are stronger over the sea than over land. Migrants minimize the headwind force by following the coast, where they also can use local topography and vegetation to gain additional protection from the wind. Furthermore, over the sea they will be exposed to wind drift, and under certain cross-winds it is beneficial to follow a coastline some distance goalwards rather than to depart directly over the sea.

Final Remarks

The key word when considering the course and timing of bird migration is adaptation. The adaptation of migrating birds' behaviour to hazards and hardships in the environment reaches a level of precision that is highly impressive to the most skilful and experienced human pilot. Much remains to be untravelled about the amazing performances by billions of birds, travelling across continents, deserts, oceans and glaciers in their striving for food, suitable breeding grounds and safety from predators.

12

Celestial Navigation by Birds

A remarkable series of experiments shows that migratory birds navigate by the stars. In a planetarium their flight direction is controlled by the position of the stars in the artificial sky.

In spring and summer the songbirds known as the warblers are familiar residents in the countries of northern Europe. City dwellers know them well, for the small, gray birds find a home to their liking in the shrubs and hedges of gardens and small parks. During the spring breeding season the air is filled with their loud, melodic singing as each male establishes a small territory for himself in noisy battle with a rival. Once the claims are decided, the truculence and the songs subside; the birds proceed to mate, to raise their young, to feed amicably on elderberries and blackberries, to flit about in peace among the bushes. Then in August the birds begin to grow restless; their migratory instinct stirs. Suddently, in one night, the whole resident population is off and away. The next morning the bushes are filled with a new lot of warblers that have flown in from more northern areas; they stay for a few days and then they too fly on to the south. Through the weeks of September and October there is a continuous coming and going of hordes of the migrating warblers. Gradually the number passing through diminishes. The species called the garden warblers disappears first, then the whitethroats, after them the lesser whitethroats and finally the blackcaps.

Where do they go? Ornithologists know exactly where the warblers go, for they have banded these birds for many years and followed them to their winter homes. The warblers travel to various parts of Africa. Some of them migrate as far as from Scandinavia to the southern part of Africa—a distance of thousands of miles. In the spring the birds migrate back to the very same place that they left in the fall.

Most remarkable of all is that each birds finds its own way to its destination! The warblers do not follow a leader or make the journey as a group; they navigate individually. And young birds making their first migration reach their goal as surely as

Lesser Whitethroat migrates between northern Europe and the Nile Valley. In a planetarium it attempts to duplicate its natural journey with respect to the artificial sky.

the experienced travelers. Somehow, purely by instinct, the warblers know exactly how to set their course. Since they fly only at night, we are impelled to ask: Do the warblers navigate by the stars?

The navigation powers of birds have fascinated investigators for more than a century. By now there is a large literature of well-documented testimony to their amazing performances. The late Werner Ruppell of Germany, one of the leading experimenters on bird migration, found that starlings taken from their nests near Berlin and carried away to all points of the compass would find their way back to their nesting places from as far as 1,250 miles away. The Manx shearwater, a sea bird, has astonished investigators with still more spectacular feats; one shearwater, taken from the west coast of England and flown by plane to Boston, was back in its English nest in 12 days, having winged its own way 3,067 miles across the unknown Atlantic. The North American golden plover migrates each fall from its breeding grounds in northern Canada to its winter home in the Hawaiian Islands. This bird, lacking webbed feet, cannot rest on the water as waterfowl do; it must fly on steadily for several weeks to reach its destination over thousands of miles of ocean. If it wandered only slightly off course, it would become lost and exhausted in the vast Pacific, but it finds its way unerringly to Hawaii.

Until recently attempts to explain the incredible navigation feats of birds were almost entirely a matter of speculation [see "The Navigation of Birds," by Donald R. Griffin; Scientific American, December, 1948]. Various theorists proposed that the birds were guided by the earth's magnetic field, by the Coriolis force arising from the earth's rotation, by landmarks, and so on. But more and more ornithologists have been driven to the conclusion that birds must rely mainly on celestial navigation—the sun by day, the constellations by night.

The idea that birds are guided by the sun was suggested as long as half a century ago, but it was not taken seriously until

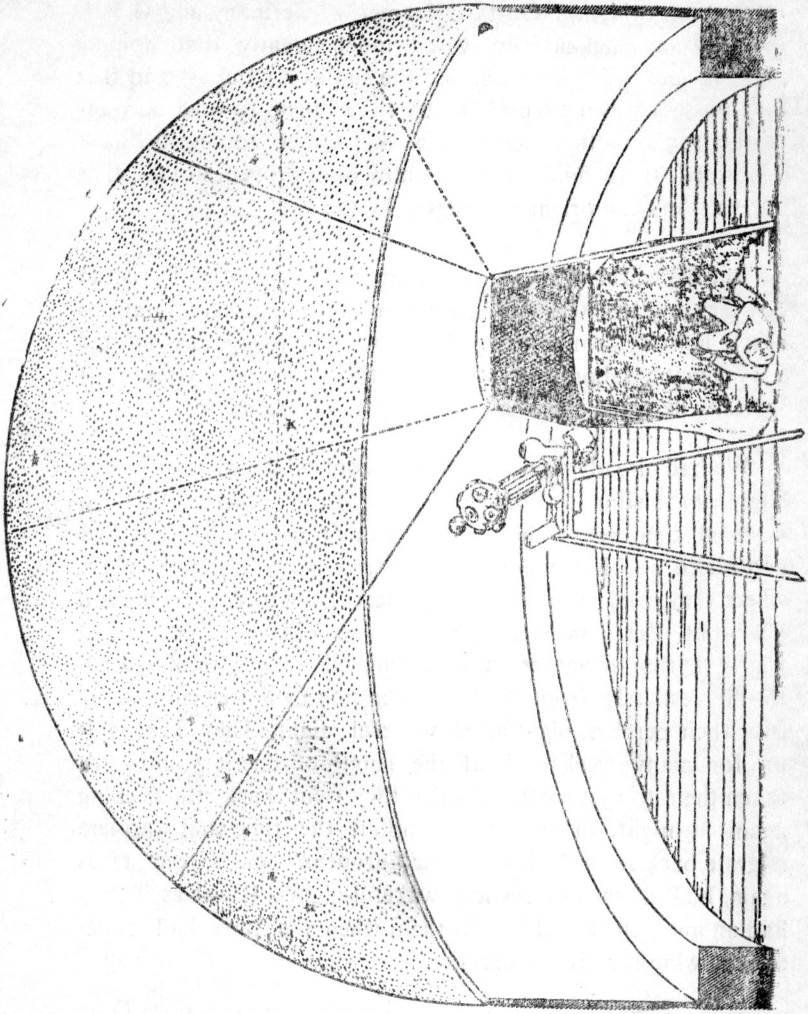

Planetarium Experiment is performed under a 20-foot dome. Dome and bird cage are shown in cross section. A felt cloth from the bottom of the cage to the floor cuts out light from below. Sector of sky visible to the bird is indicated by solid lines. The sector visible from the opposite side of the perch is shown with broken lines. "Flight direction" means direction in which bird faces.

the early 1950s, when experimenters began to turn up some interesting evidence. Gustav Kramer in Germany and G.V.T. Matthews in England discovered independently that homing pigeons and wild birds can use the sun as a compass and that they possess a "sime sense" which allows them to take account of the sun's motion across the sky. Other zoologists have confirmed these findings. It has now been proved, in fact, that our warblers can orient themselves by the sun.

But the warblers fly mainly at night. What sort of system do they use to steer their course in their nocturnal migrations nearly halfway around the globe? Several years ago we started a systematic laboratory study of this question by means of specially designed cages in our aviary at Freiburg.

We had already seen laboratory proof of the stirring of the migratory instinct in these small world-travellers and of a seasonal time sense that governed this urge. We had hatched and raised warblers in completely closed, sound-proof chambers where they lived in the illusion of eternal summer, year in and year out. Yet, although they had no outward cues of the yearly rhythm of nature, in the autumn the birds would begin to flit restlessly from branch to branch or flutter continually over their perches, night after wakeful night. They kept this up for many weeks—about the length of time it would have taken them to fly to Africa. Then they went back to sleeping again at night. In the spring, about the time that warblers migrate back from Africa to their European homes, our birds again had a spell of restless, wakeful nights. It was as if they had an inner clock which told them when the time had come to take wing for distant parts.

To explore the orientation question we now placed warblers in a cage with a glass opening at the top, so that they could see part of the sky (but nothing else of their surroundings). At the season of migration the birds would begin to flutter and, peculiarly enough, each would take up a position pointing in a particular geographic direction, like the needle of a compass. Even when we tried to turn the birds away by rotating their perches, they stubbornly turned back to the preferred direction. The direction in each case was characteristic of the species: the garden warblers, the whitethroats and the blackcaps all pointed toward the southwest, the lesser whitethroats toward the southeast (that is, in the fall; in the spring these directions were reversed). Now these are precisely the directions in which the respective species start their migrations from central Europe to Africa! The lesser whitethroats start southeastward, flying across the Balkans, and then turn south up the Nile Valley; the other species all take off southwestward and fly to Africa by way of Spain and Gibraltar.

Experienced or inexperienced, the birds invariably took up the appropriate direction of flight in the cage. How did they know the direction? Seemingly the only clue available to them was the starry night sky overhead. To explore this theory further we now put them through a series of tests. We found that when the stars were hidden by thick clouds, the birds became completely disoriented. They were likewise confused when only diffuse and strongly polarized light came through their skylight. To adopt and keep to a definite direction they needed a look at the starry sky. Indeed, the birds watched the sky so intently that meteors made them change their direction momentarily.

For still more rigidly controlled experiments we proceeded to test the birds in a cage placed in a planetarium; that is, with a dome showing an artificial replica of the natural starry sky. Again, when the dome was merely illuminated with diffuse light (showing no stars), the warblers were unable to choose a preferred direction. But when the planetarium sky matched the night sky over Germany, the birds took up the proper direction just as if they were seeing the natural sky.

Now our artificial dome permitted us to shift the stars and constellations about. By changing the north-south declination (height) of the stars we could change the apparent geographical latitude, making the birds believe that they were farther south or north than they actually were. Similarly by shifting the sky in the east-west direction we might mislead the birds about their position in longitude. How would they behave under these circumstances?

To illustrate the results I shall describe some experiments with a lesser whitethroat warbler which we will call "Johnny." Recall that the lesser whitethroat normally first travels south-earthward across the Balkans and then turns due south, flying along the Nile to its winter home in the region of the Nile headwaters. In our experiments it turned out that as long as the planetarium sky was adjusted in the approximate latitude of Germany (40 to 50 degrees north), Johnny took up the expected flight position facing southeast. But as we shifted the sky, simulating more southerly latitudes, the bird tended to turn more and more toward the southern direction, until, at the latitude of 15 degrees, it set its course due south!

In other words, Johnny, a bird which had spent all its life in a cage and never travelled under a natural sky, let alone migrated to Africa, still displayed an inborn ability to use the guidance of the stars to follow the usual route of its species, adjusting its direction nicely at each given latitude. Earlier investigators had supposed that these birds used landmarks to find their route: for example, that the coastline at the eastern end of the Mediterranean was the cue which told them to turn south. But our experiments proved that the birds are guided only by the stars.

Now let us see what happened when we shifted the planetarium sky to change the longitude. One night, while Johnny was flapping its wings and heading in the southeast direction, we suddenly presented the bird with a sky shifted back to the configuration five hours and 10 minutes earlier than the local time; in other words, the apparent geographical position of the cage then corresponded to a point about 77 degrees eastward in longitude at this particular time. The bird at once showed that it was deeply disturbed. It looked excitedly at the unfamiliar sky and for almost a full minute stood irresolutely. Then it suddenly turned and took wing in the westward direction.

According to the sky, its position at the moment corresponded to a point near Lake Balkhash in Siberia; Johnny, to correct its displacement, was heading directly toward the usual migration starting point in Germany!

As we reduced its displacement, the bird shifted its heading more and more from due west toward the south. When the displacement was only on hour, corresponding to a position near Vienna, Johnny headed south; when the canopy of stars was restored to the correct configuration at our locality for the season and time of night, the bird took up the normal heading toward the southeast.

Johnny's behaviour, confirmed by experiments with other birds, leaves no doubt that the warblers have a remarkable hereditary mechanism for orienting themselves by the stars—a detailed image of the starry configuration of the sky coupled with a precise time sense which relates the heavenly canopy to the geography of the earth at every time and season. At their very first glimpse of the sky the birds automatically know the right direction. Without benefit of previous experience, with no cue except the stars, the birds are able to locate themselves in time and space and to find their way to their destined homes.

To be sure, the warblers do not have to rely solely on the constellations. In daytime they can guide themselves by the position of the sun. On cloudy nights they get some guidance from mountain ranges, coastlines and river courses gleaming in the pale night shine. Only in almost total darkness, when thick clouds utterly hide the sky, are the birds in trouble: they circle helplessly and sometimes are drawn to lighthouses.

We are going on to study the warblers' orientation system in more detail, systematically removing constellations or stars from our planetarium sky one by one to see if we can reduce the guidance cues to a basic pattern. One very interesting puzzle is the fact that the birds must somehow be able to make adjustments to astronomical evolution, for in the course of time the pattern of constellations in the sky is slowly but constantly changing. Even more difficult to explain is the mystery of how the birds ever came to rely on celestial navigation and to develop their skill in the first place. We know that the warblers are not the only creatures possessing this gift: other birds, insects, crabs and spiders have been found by experiment to be capable of guiding themselves by the sun or stars. But there are many other guidance mechanisms and sign-posts available on earth. What evolutionary process was it that endowed these animals with the highly sophisticated ability to read the stars?

Whatever the answer, we cannot help marveling at the wondrous celestial instinct of the warblers. When fall comes, the little garden warbler, weighing barely three quarters of an ounce, sets off one night on an unbelievable journey. All alone, never in the collective security of a flock, it wings its solitary way southwestward over Germany, France and Spain and then swings south to its distant goal in southern Africa. It flies on unerringly, covering a hundred miles or more in a single night, never once stopping in its course, certain of its goal. In the spring it takes off again and northward retraces its path to its nesting place in a German or Scandinavian thicket—there to give birth to a new generation of little warblers which will grow up, without being taught, with the self-same capacity to follow the same route across continents and oceans by the map of the stars.

Orientational Cues

One of the earliest and most stimulating discoveries in the field of avian orientation and navigation was that celestial cues—the sun and the stars—play a fundamental role in bird orientation. Thus, more than twenty-five years ago, Gustav Kramer (1952) and his students showed that birds possess what has come to be called a sun compass. And, following up Kramer's (1949, 1951) early observations that several species of nocturnal migrants exhibit oriented migratory restlessness (Zugunruhe) in circular cages under clear night skies, Sauer (1957) found that such nocturnal orientation depended on the stars, thus laying the basis for what is now known as the star compass.

The overwhelming preponderance of evidence now indicates that solar cues are used by birds only as a simple compass, not as the basis of a bicoordinate navigation system as proposed by Matthews (1953, 1955) in his imaginative and historically important sun-arc hypothesis (for a summary of the evidence, see Keeton, 1974a). In other words, it appears to be only the sun's azimuth (direction from the observer) that provides the bird with orientational information. But the azimuth can provide such information only if the time is known, so that compensation can be made for the sun's changing position throughout the day. That birds do indeed couple their internal clock (circadian rhythm) with their observation of the sun's azimuth in determining compass directions was clearly demonstrated by Kramer (1953a) and by Hoffmann (1954), working with caged birds. Later, Sehmidt—Koenig (1958, 1960) extended this finding to free-flying homing pigeons. He showed that pigeons whose internal clocks had been experimentally shifted six hours out of phase with true sun time chose initial bearings roughly 90° from those of control pigeons when released at a distant test site (Fig. 12.1a); their clocks had been shifted a quarter of a day, and, as a consequence, they misread the sun compass and chose bearings a quarter of a circle different from those of the controls.

As in the case of the sun, the evidence is overwhelming that the stars provide only compass information for birds, even though they could potentially be used in true bicoordinate navigation (for a thorough discussion, see Emlen, 1975a). Sauer (1957) thought that the star compass, like the sun compass, required time compensation. But in a detailed study of the migratory orientation of the Indigo Bunting (Passerina cyanea), Elmen (1967) found that this species uses star patterns to determine directions, a process that does not require time compensation. It seems likely that this is true of other species as well. Thus the way birds read the star compass differs fundamentally from the way they read the sun compass.

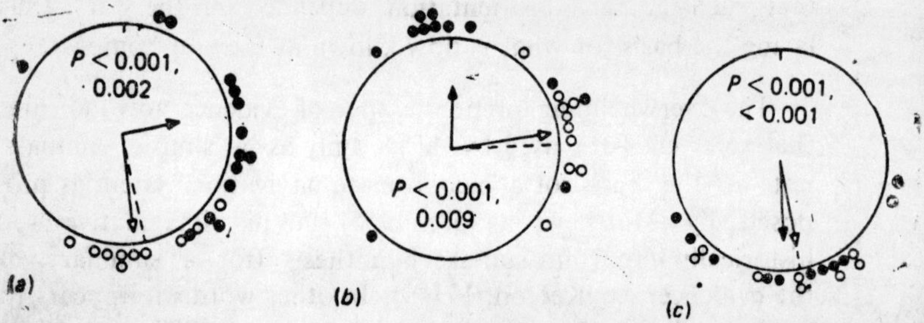

FIG. 12.1: Vanishing bearings of pigeons that have been clock-shifted 6 hours fast.

(a) Experienced pigeons released on a sunny day at a distant site. The mean bearing of the clock-shifted birds is roughly 90° to the left of that of the controls.

(b) Experienced pigeons released on a sunny day at a site less than a mile from home, where the landscape should be completely familiar. Again, the mean bearing of the clock-shifted birds is roughly 90° to the left of that of the controls.

(c) Experienced pigeons released on a totally overcast day at a distant unfamiliar site. Both the clock-shifted and the control birds are homeward oriented, and there is no indication of a difference between them, which suggests that, in the absence of the sun compass, the pigeons use orientational cues that do not require time compensation. In this and laser figures showing bearings, north is indicated by a small line at the top of the circle, and the home direction by a dashed line reaching the perimeter of the circle. The bearing of each individual bird is shown as a small symbol on the outside of the circle; where two treatments are included on a single circle, the bearings of the controls are shown as open symbols and the bearings of the experimental birds as filled symbols. The mean vectors are swown as arrows (with open or filled heads, respectively), whose length is drawn proportional to the tightness of clumping of the bearings (*i.e.* the longer the vector—at maximum reaching the perimeter of the circle—the better oriented the sample of bearings). The uniform probability under the Rayleigh test is given inside the circle; the first value is for the controls and the second for the experimentals.

Although the sun compass during the day and the star compass during the night are certainly dominant orientational cues for birds, there is now abundant evidence that neither is essential for proper orientation. Thus experienced homing pigeons can orient accurately homeward from distant unfamiliar release sites under heavy total overcast. And tracking-radar studies have regularly revealed nocturnal migrants in oriented flight under heavy overcast when the stars are not visible. It is apparent, then, that though birds often use celestial cues when available, they can use alternative cues when necessary. In short, avian orientation systems include redundant, or back-up, cues.

It is also important to point out, as Kramer did long ago, that a compass alone cannot tell a bird where it is or which

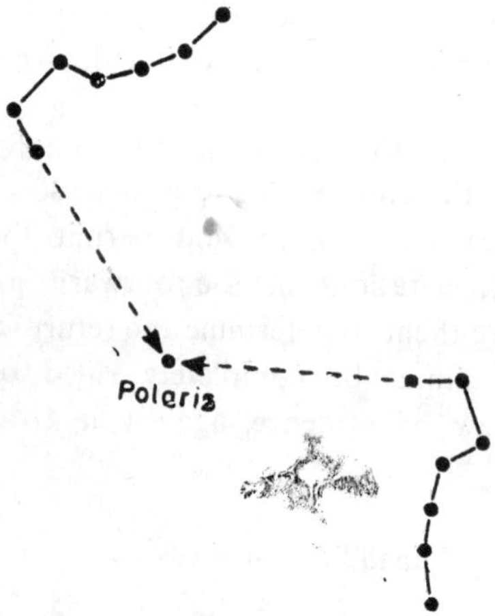

FIG. 12.2: An example of how north can be located by star patterns. If one draws an arrow running through two particular stars in the cup of the big dipper (Ursa Major), it will point toward Polaris. Although the position of the constellations changes during the night (two positions are shown here), the same stars always determine an arrow pointing toward Polaris (*i.e.* toward north, or the pole of the celestial rotation), hence directions can be determined without need of time compensation. Many different star patterns could be used for direction finding in this way.

direction it should fly to reach a particular destination. Other environmental cues must provide goal-orienting birds with an analog of map information. What might these cues be? Certainly the most obvious possibilities would be familiar landmarks, but, curiously enough, birds seem to make only minimal use of these. Radar studies strongly suggest that landmarks play little role in migratory orientation (Emlen, 1975a; Emlen and Demong, 1978). Moreover, pigeons clock-shifted six hours out of phase with true sun time and released half a mile from their home loft, in an area they have flows over every day when exercising, usually choose bearings deflected 90° from the homeward course (Fig 12.1(b)) (Graue, 1963; Keeton, 1974a); the birds act as though they have never before seen the place! Even more remarkable, pigeons wearing frosted-white contact lenses that eliminate image vision beyond two or three meters can orient accurately homeward from distant locations, and can even tell when they have arrived at their goal.

Another possibility that comes to mind in this age of rocketry is that the birds might have some sort of sophisticated inertial guidance system that would permit them to detect all the angular accelerations of the outward journey and then double integrate them to determine the return course. Although this possibility cannot be completely ruled out, it seems very unlikely in view of evidence against it from many kinds of investigations.

Unusual Sensory Capabilities of Birds

Having eliminated the most obvious cues, we turn next to special sensory capabilities (some of them only very recently discovered) that may play a role in the birds amazing navigational feats.

Magnetic Detection

Although it had often been suggested that birds might be able to derive directional information from the earth's magnetic field, the prevailing opinion of this possibility in the scientific community of the mid 1960s was one of intense skepticism. Most investigators thought it very unlikely that any organism could detect a magnetic field as weak as that of the earth (about 0.5 Gauss). But beginning in the mid 1960s, and continuing to the present, a group in Frankfurt, led first by F. Merkel and later by W. Wiltschko, intensively investigated the possible role of magnetic cues in avian orientation.

Having first found that European robins (Erithacus rubecula) exhibit migratorily appropriate orientation in circular test cages when visual cues are unavailable (Merkel, Fromme and Wiltschko, 1964), this group went on to show that the orientation of the robins could be changed in a predictable way by turning the magnetic field (*e.g.*, making magnetic north in the cage coincide with geographic east), using Helmholtz coils positioned around the test cage. Their results were later successfully replicated by Wallraff (1972) with robins, and by Emlen *et al.* (1976) with indigo buntings.

Wiltschko (1972) found, further, that robins apparently pay no need to the polarity of the magnetic field, but rather, in the northern hemisphere, take north as that direction in which the magnetic and gravity vectors form the most acute angle (Fig. 12.3). They are unable to orient in visually cueless test cages when the magnetic field is entirely horizontal (*i.e.*, has no vertical component), as is the case at the equator. In short, the birds' manner of reading the magnetic compass is very different from our own.

The first good evidence of a magnetic effect on pigeon homing came when I found that bar magnets attached to experienced birds' back often cause disorientation on heavily overcast days, whereas they had little effect on sunny days. Walcott & Green (1974), using Helmholtz coils on the pigeons' head and neck, exposed the birds to more homogenous magnetic fields than those produced by my bar magnets; again there was no dramatic effect on sunny days, but on overcast days the birds' orientation was changed (not merely disrupted) in a manner consistent with Wiltschko's formulation of the way the avian magnetic compass works. These results suggest strongly that the disorienting effects of bar magnets I had found earlier were not due merely to some general physiological disturbance.

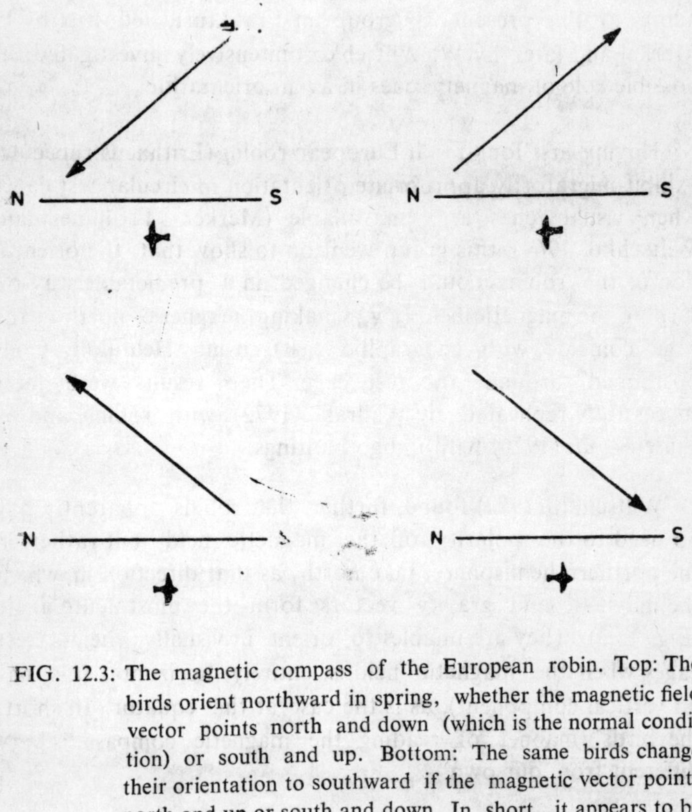

FIG. 12.3: The magnetic compass of the European robin. Top: The birds orient northward in spring, whether the magnetic field vector points north and down (which is the normal condition) or south and up. Bottom: The same birds change their orientation to southward if the magnetic vector points north and up or south and down. In short, it appears to be the alignment of the magnetic vector, not its polarity, that determines the birds' behaviour.

Thus emerged the concept that, in the case of homing pigeons, the magnetic field provides compass information that experienced birds use primarily when the sun compass is not available. While probably generally true, this formulation may be an oversimplification. There is a growing body of evidence that magnetic perturbations alter in a small but consistent way the orientation of pigeons on sunny days. Moreover, there is also some indication that magnetic information obtained during the outward journey to the release site may sometimes influence the initial orientation of pigeons. It therefore seems possible that magnetic cues may play some role in the map aspect of avian navigation.

Lindauer & Martin (1968; see also Martin & Lindauer, 1977) have found that honeybees are sensitive to magnetic changes of less than 10^{-3} Gauss and very probably of less than 10^{-5} Gauss. There is now evidence from several sources that birds are probably equally sensitive. Thus Southern (1972; but see also 1978) has reported for gull chicks, Keeton *et al.*, (Larkin & Keeton, 1977) for homing pigeons, and F. Moore (1977) for free-flying migrants that orientation is influenced by natural magnetic disturbances, due largely to events on the sun, such as solar flares. As B. Moore (1980) has pointed out, the data of Keeton *et. al.*, (1974) suggest that magnetic fluctuations of only one milligauss—roughly a five-hundredth of the earth's natural field —can result in rotations of 10°—40° in pigeons' choice of bearings. Also indicating a very great magnetic sensitivity is evidence reported by Wagner (1976) and by Walcott (1978) that geographic magnetic anomalies may disturb the initial orientation of pigeons.

In view of the abundant evidence that birds are very sensitive to magnetic stimuli, at least when they are orienting, one would hope that we will soon learn the mechanism of their magnetic sense. The recent discover by Walcot, Gould and Kirschvink (1979) of the ferromagnetic mineral magnetite in the heads of pigeons, followed by the finding of Presti and Pettigrew (1980) of what is probably the same material in the necks of pigeons and White-crowned Sparrows (Zonotrichia leucophrys), is very exciting, but it remains to be determined whether this material functions in magnetic detection. Evaluation of this and other posibilities should be facilitated by Bookman's (1977) reported success in training pigeons in a two choice test where different magnetic conditions provide the information on which the choice must be based.

Possible Detection of Gravity Variations

Several years ago, T. Larkin and I noticed that often when animals have shown clear responsiveness to magnetic stimuli they have been simultaneously responding to gravity. Consequently, we sought to determine whether gravity cues might play some role in pigeon homing, especially in situations where the birds are using magnetic information. Being unable to alter gravity in the laboratory, we looked instead for a possible influence of the natural monthly gravitational cycle caused by the changing relative positions of the earth, sun, and moon. We soon found a significant correlation between the pigeon's mean vanishing bearings and the day of the lunar synodic month. Suggestive as these results may be, however, they do not prove a direct effect of gravitational changes on the birds' orientation, because some other environmental variables to which birds might be responsive may also be related to the lunar cycle.

Despite the still uncertain meaning of the synodic lunar rhytem we found, let us consider the possibilities for birds if they could detect minute variations in gravity. Gravity varies not only temporally but also geographically, in both a regular and an irregular manner. The regular variation is in a north—south gradient and is due to the fact that the earth is not a perfect sphere. The irregular uariation is due to the differing densities of the material in the earth's crust at different localities. Gravity cues could, then, potentially be useful in navigation, both because they could indicate the north-south axis and because they could provide another topography in addition to those we normally consider.

If a bird could use the north-south gravitational gradient to determine true (*i.e.*, rotational) north, then magnetic declination (the deviation of magnetic north from true north) might be readable. It happens that declination is one of the very few environmental parameters that vary as a rough analog of longitude, hence its potential usefulness would be very great indeed. Moreover, some of our preliminary results (Larkin and Keeton, in preparation) suggest that declination may actually be the parameter of the magnetic field that most influences pigeon orientation during magnetic disturbances. It is important to emphasize. however, that there is as yet no direct evidence that birds can detect such incredibly tiny differences in gravity (less than ten Gal) as would be necessary to permit use of gravitational cues in long-distance navigation.

Barometric Pressure Detection

One obvious way birds could use a barometric pressure sense is as an altimeter. The sensitivity found in homing pigeons is sufficiently great that they should surely be able to detect a change in altitude of ten metres; indeed it seems likely they can detect much smaller changes. Not only would an altimeter sense be useful when flying in cloud, but we note also that, if birds really possess the ability to detect changes in gravity, as speculated above, an altimeter would be essential because compensation for altitude would be necessary in view of the fall in gravity with increasing elevation.

Human meteorologists find a sensitive barometer very helpful in predicting weather changes; why should not a bird use its built-in barometer in the same way? There is, of course, a long history of reports from bird watchers that the birds in their garden seem to show by behaviour changes that a weather front is approaching, long before the human observer sees any indication of the front. And there is convincing recent evidence that birds about to initiate migratory flight are very good meteorologists indeed (Emlen, 1975 a). Thus, in eastern North America, major autumnal movements tend to occur on the east side of a high-pressure cell following passage of a cold front, and major spring flights tend to concentrate on the west side of a high-pressure area ahead of an advancing low-pressure cell (Richardson, 1978). Many migrants, especially small song birds, seem to be quite accurate at predicting early in the evening, when they are still on the ground, what the wind conditions aloft will be later that night; they go up in greatest numbers when the winds aloft will be favourable.

Finally, the last few years have witnessed a growing interest in the possibility that weather factors such as wind directions, pressure patterns in the atmosphere, and patterns of air turbulence could potentially provide useful orientational information (Griffin, 1969; Emlen, 1975 a), and the detection of these would surely be facilitated by the birds' barometric sense.

Infrasound Detection

Although the literature would lead one to think the lower frequency limit of bird hearing would be about 100 Hz, Yodlowski, Kreithen & Keeton (1977) have found that homing pigeons are sensitive to frequencies lower than 1 Hz. In fact, the audiogram for the pigeon worked out by Kreithen & Quine (1979) shows sensitivity down at least to 0.05 Hz. The pigeons' sensitivity is at the appropriate level for extracting meaningful signals from the background environmental noise.

Detection of infrasound by birds raises an intriguing new orientational possibility. Because attenuation is proportional to the square of frequency, infrasonic frequencies as low as those the birds can detect may travel hundreds or even thousands of miles with little energy loss. Potentially, then, pigeons might be able to monitor distant infrasonic sources, such as mountains whistling because of winds blowing across them, and to use these as a rough system of beacons for determining position.

But if pigeons are to use infrasounds in this way, they should be able to tell from what directions the sounds come, yet binaural comparisons would be impossible for such very long wavelengths (over 3 km at 0.1 Hz). A possible way around this problem comes from the work of Quine (1979) who finds that pigeons' frequency determination in the infrasonic range is sufficiently good so that doppler shifts in apparent frequency induced when a bird flies toward and then away from an infrasound source would be well within the birds' detection range.

Thus pigeons may be able to get directional information from infrasounds while in flight, even if they cannot do so while perching. Field experiments to evaluate this possibility are now in progress at Cornell.

Polarized Light Detection

The discovery by Kreithen & Keeton (1974 b) and by Delius, Perchard & Emmerton (1976) that homing pigeons can detect the polarization of light may mean that these birds, like honey-bees, can continue using a derivative of the sun compass on partially overcast days, when the sun's disk is hidden from view but some blue sky remains. This is possible because of the geometric relationships between the position of the observer, the plane of polarization of sunlight, and the position of the sun, which permits derivation of the sun's position if the polarization can be detected. In addition, there are other more direct ways in which sky-light polarization may be used in orientation; thus, for example, the band of maximum polariza-tion rotates around the celestial north pole during the course of each day, so if the rotation could be observed then true north could be determined directly (Brines, in press). Actual use of polarized light by orienting birds in the field has, how-ever, not yet been studied.

Ultraviolet Light Detection

Yet another addition to our knowledge of avian vision is the recent discovery by Kreithen & Eisner (1978) that homing pigeons can see ultraviolet light (see also Delius & Emmerton, 1978). This raises the question, now under investigation at Cornell, whether the pigeons ordinarily perform their analysis of polar polarization in the ultraviolet wavelengths, as honeybees do.

Olfaction

Olfaction, unlike most of the other sensory capacities discussed above, is not a recently discovered sense. But the possibility that it may play an important role in avian navigation is new. It was in 1972 that Papi. Fiore, Fiaschi & Benvenuti first put forward their olfactory navigation hypothesis. Briefly, they propose that young pigeons at the home loft would learn to associate particular odours with winds from certain directions. Thus odour A might arrive at the loft primarily on winds from the north, odour B on winds from the east, etc. A bird released at a distant site, say north of home, would detect a strong odour of A and therefore determine its position to be north of home. The bird would then use one of its compass systems to locate south, and begin its homeward flight.

Papi and his colleagues have performed a long series of ingenious experiments to test their hypothesis, and have reported consistently positive results. Unfortunately, attempts to repeat some of these experiments at Cornell yielded generally negative results as far as orientation was concerned, though there was sometimes an effect on homing success. Similar attempts at Tubingen in Germany yielded either negative results or ambiguous ones.

In an effort to resolve the differences between the Pisa and Cornell groups, the two groups performed six series of collaborative experiments at Cornell in 1977. The overall result of these experiments was that, with the exception of deflector-loft experiments, to be discussed later, we found no consistent effect of olfactory interference or deprivation on initial orientation, but we did often find an effect on homing success from distant unfamiliar release sites. Unfortunately these results are interpreted one way by the Cornell investigators and another way by the Italian investigators; hence the question of the role of olfaction in avian orientation remains unresolved. Papi and his colleagues feel the poor homing means that olfactory cues play an irreplaceable role in homing from unfamiliar sites. My colleagues and I feel, on the other hand, that since the olfactorily deprived birds usually depart from the release site on a proper course, their poorer homing may have nothing to do with navigation but may merely indicate a diminished motivation that results in the birds' landing when only part way home. We are concerned about the motivational effects of procedures that interfere in any way with the respiratory system, on which flying birds must make very heavy demands. We attempted to evaluate our suggestion by airplane-tracking pigeons in one of the collaborative experiments, but unfortunately, our suggestion proved almost too true—all five of the experimental birds that were tracked soon landed and we were unable to get as much detail about the flight course as we had hoped. However, before they landed, these birds were flying roughly the same course followed by control birds.

In one series of collaborative experiments we were successful in repeating the Italian group's results on initial orientation. This series utilized the deflector-loft technique so imaginatively designed by Baldaccini, Benvenuti, Fiaschi & Papi (1975). In these experiments, pigeons were exposed in their home lofts to winds—and the odours they are presumed to carry—deflected either clockwise or counterclockwise. When tested at release sites, the pigeons chose bearings to the right or left of control pigeons, as predicted by the olfactory hypothesis. These results may indicate that olfactory cues are sometimes used by the Cornell pigeons, but no final decision can be made until experiments that control for the various other orientationally relevant factors altered by the deflectors are completed.

In summary, my appraisal of the role of olfaction in avian navigation is that it is likely that odours constitute one of the many sources of information birds may use in navigating. Olfactory cues appear to be used more by Italian than by Cornell pigeons, for reasons yet to be determined. This is not the first example of major geographic difference in bird orientation. Nocturnal passerine migrants in the southeastern United States nearly always fly downwind whereas the same species in the northeastern part of the country do so only if the wind direction is migratorially appropriate. However, I doubt that olfaction will be found to play as essential a role in avian navigation as Papi and his colleagues have proposed. Indeed, I am not convinced that any single cue so far discovered is

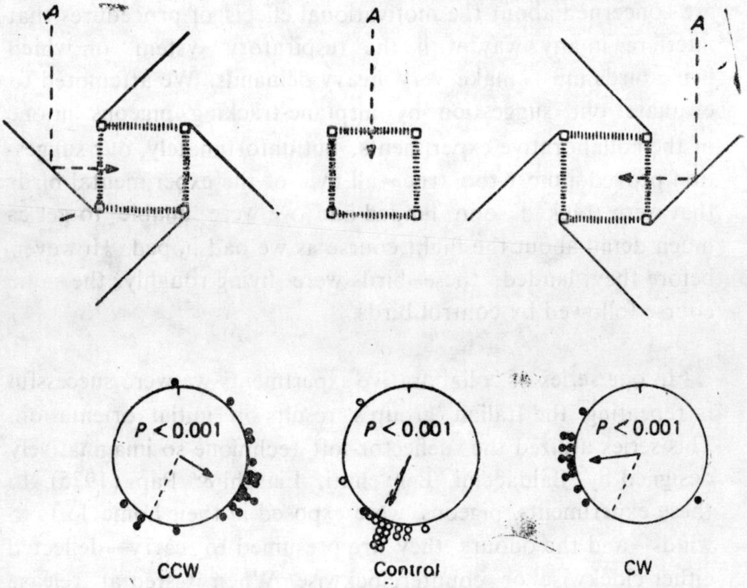

FIG. 12.4: The deflector loft experiments. Top: All three lofts (control loft in centre, lofts with deflectors on each side) have walls that allow free flow of air. Winds from the north, presumably carrying odour A, enter the control loft from the North, but they enter the other lofts from the east and west because of the deflectors. Bottom: Bearings of pigeons released north of home. The control birds, which had earlier experienced normal air flow (*i.e.*, A winds from the north) oriented properly southward, toward home. By contrast, the CW birds, which had experienced A winds from the east, oriented more westerly, and the CCW birds, which had experienced A winds from the west, oriented easterly.

essential; there seems to be so much redundancy in the avian navigation system that experienced birds can orient when only a few of the many possible cues are available.

The Integration of Orientational Cues

From all that has been said above, it should be clear to the reader that birds can integrate the many orientational cues they use in a variety of ways—according to different weighting schemes, if you will—depending on the birds' age, experience, and species, and on weather conditions, the season of the year, and the geographic location. The old hope that a single system would be found to explain all of avian orientation has gone aglimmering. Hence one of the chief thrusts of current orientational research is the attempt to learn how the different cues are integrated and what constraints there are on the amazing flexibility of avian navigation systems.

Studies of Integration: The Ontogeny of Orientational Behaviour

One powerful way of teasing apart the many elements of avian orientation systems is to manipulate the early development of those systems. By so doing, one can often get young birds to omit one or more of the usual cues or to adopt an a typical weighting scheme so that cues that would normally be secondary or tertiary become primary and thus easier to study.

In a series of experiments on the ontogeny of stellar orientation in indigo buntings, Emlen (1972) found that there is a sensitive phase during which young buntings learn to read the star compass. If the buntings have had a view of the starry sky during the weeks preceding the start of their first autumnal migration season, they can orient properly when that season begins. But if they have not seen the night sky until after the first migration season has begun, they never learn to use the star compass, no matter how often they see the sky thereafter.

Emlen (1972, 1975 b) also showed that young buntings respond initially to the apparent rotation of the starry sky during the night. The axis of this rotation is north-south, hence it can provide compass information. But the birds do not long depend on the axis of rotation *per se*; rather they soon learn star patterns that will indicate where the axis is, and thereafter they rely exclusively on those patterns. In other words, the axis of rotation functions in ontogeny only as the reference against which the star compass is initially calibrated. When Emlen exposed hand-reared young buntings to a planetarium sky rotating around an incorrect axis, the buntings learned to use star patterns appropriate to that axis, and consequently they oriented in an inappropriate direction when later tested under a normal sky. When retested a year later, after extensive exposure to the normal sky, they had not corrected their orientation; what they had learned during the sensitive phase in their early life still dominated their behaviour.

In orientation studies with young first-flight pigeons (*i.e.*, very young birds released for their first homing flight), we have found that these birds appear to require the sun compass for orientation; if they are released under heavy total overcast they usually depart randomly, even though experienced pigeons can orient accurately under such conditions. Moreover, first-flight youngsters appear to require magnetic cues also; they usually depart randomly, even on sunny days, when wearing bar magnets attached to their backs. In short, the first-flight birds need both sun and magnetic cues, whereas experienced pigeons need only one or the other. It seems, then, that these young inexperienced pigeons are integrating cues in a manner quite different from that used by experienced birds. Perhaps the effect of experience is merely to enable them to get by with less information, or alternatively, the experience may help them establish a hierarchy of choice, so that they can later deal with situations in which two or more cues give conflicting information.

As an example of an experimental manipulation that results in young birds' omitting a normal cue, we can cite the studies of Wiltschko, Wiltschko, Brown, & Keeton (in preparation) with so-called no-sun pigeons. In these studies young pigeons raised without ever having a chance to see the sun (they were flown for exercise only on overcart days) were found to be able to orient perfectly well when released under total overcast for their first homing flight, even though normal first-flight pigeons vanish randomly under total overcast. Having never viewed the sun, the no-sun pigeons had not incorporated it into their navigation system and hence had no difficulty orienting when it was missing.

Studies of the ontogeny of the sun compass in pigeons have revealed that the coupling of times, directions, and sun azimuths is not inherited but must be learned. Thus young pigeons raised

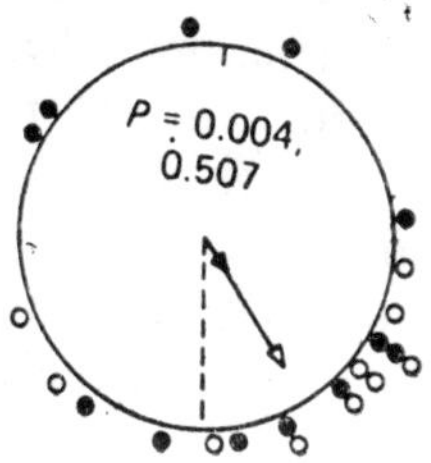

FIG. 12.5: Bearings on a sunny day of first-flight pigeons wearing magnet bars or brass bars. The magnet-laden pigeons (black symbols) vanished randomly, whereas the brass-laden birds (open-symbols) were well oriented.

under a permanently six-hours slow clock-shifted photoperiod orient normally, with no indication of the deflection seen in ordinary clock-shifts (Fig. 12.6 a). The birds appear to have learned that the 'morning' sun is in the south, the 'noon' sun is in the west etc. When these birds are moved to a normal photoperiod and retested after five or six days, they then show bearings deflected 90° from the controls (Fig. 12.6 b) being put in a normal photoperiod has the same effect on them as a six-hours-fast clock-shift has on normal pigeons. These results indicate that the sun compass must be calibrated, which suggests that it may be a derivative compass—that there may be some other more fundamental directional cue that functions as the reference for calibration. One current line of research is the attempt to determine what that reference cue might be.

Studies of Integration: Manipulation of Physiological Condition

Emlen (1969) has pursued the question whether it is the seanal differences in the temporal positions of the stars that determines southward orientation by migratory birds in autumn and northward orientation in spring, or whether the differences in orientation result from corresponding differences in the physiological condition of the birds. By manipulating photoperiods, he contrived to bring one group of male indigo buntings into autumnal condition at the same time as another group was in spring condition. He then tested both groups simultaneously under a spring sky in a planetarium. The birds in autumnal condition oriented southward, whereas the birds in spring condition oriented northward (Fig. 12.7). Since the two groups saw identical star patterns, Emlen concluded that their different directions of orientation were due to their physiological conditions, not to the environmental stimuli. He predicted that the important factor would be found to be hormonal. Later studies by Martin and Meier (1973) appear to support Emlen's prediction by suggesting that by orientation of white-throated sparrows (Zonotrichia albicollis) in circular cages can be reversed by altering the temporal pattern of administration of prolactin and corticosterone.

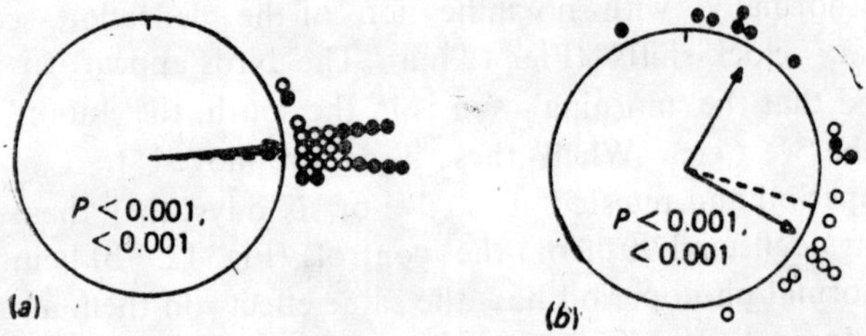

FIG. 12.6: Bearings of pigeons subjected to a 'permanent' six-hour clock-shift. Left: while still living under the shifted photoperiod, the experimental birds orient like the controls, toward home. Right: when retested 5 days after being moved to the normal photoperiod, the experimental birds choose bearings deflected clockwise from those of the controls.

Studies of Integration: Conflicts Between Cues

Another method for getting at the question of the relationship between cues is to pit one cue against another in orientational experiments. Let us examine a few recent examples of investigations using this approach. Wiltschko and Wiltschko (1975) conducted circular-cage experiments in which migratory birds are able to see the starry sky while experiencing a magnetic field that has been turned by Helmholtz coils. In other words, the birds receive conflicting information from the star and magnetic compasses. The Wiltschkos report evidence that their birds (both European robins and warblers) periodically use the magnetic field to recalibrate their star compass. The birds may then orient by the stars for a day or so before again taking a magnetic reading. The process is similar to one a person might use if, after consulting his magnetic compass, he then walked toward a distant tree seen to be in the desired direction: the person might not take another magnetic bearing until he needed to recheck the visual marker or to choose a new one. Wiltschko and Wiltschko (1976) also report that the birds can use the magnetic field to calibrate an entirely artificial 'star' pattern.

It is important to note the difference between these results of the Wiltschkos, in which magnetic cues appear to be used to calibrate the star compass, and the results of Emlen (1970,

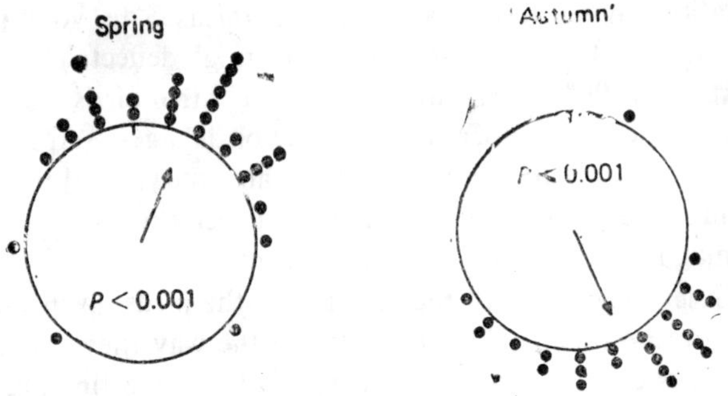

FIG. 12.7: The orientation under a spring planetarium sky of indigo buntings in spring and 'autumnal' physiological conditions. The birds in spring oriented north-northeastward, the usual direction for this species. The birds brought artificially into autumnal condition oriented south-southeastward.

1972), in which the axis of celestial rotation is used. It is possible that the main difference is that Emlen was studying the original calibration of the compass by premigratory young birds, whereas the Wiltschkos, were testing recalibration by actively migrating birds. Nonetheless, it remains unclear why Emlen's buntings, which have often been tested in a planetarium with the celestial axis not aligned along magnetic north-south, have shown no indication of recalibrating their star compass.

Several series of experiments have been conducted at Cornell in which homing pigeons being clock-shifted six hours are permitted exposure to the sun and other natural cues during the shifting process. Adult pigeons allowed to sit in a wire aviary during the overlap period between their shifted day and the real day gave no indication that they derived any orientationally meaningful information from the sun except that the light was on: the timing information potentially available in the sun's position on its are appeared to be ignored.

Extending this approach a step farther, J.R. Alexander and W.T. Keeton (in preparation) tried letting pigeons in the process of being clock-shifted fly for exercise during the overlap period between the shifted and real days. Now there was an effect on orientation. When tested at a distant release site, young birds given this treatment chose initial bearings deflected less than the expected 90° from those of the control birds (the effect differed somewhat depending on the birds' ages). Apparently flight, as opposed to merely sitting in an aviary, had made the birds more responsive to the conflict between their sun compass, their magnetic compass, their visual contact with the loft, and other relevant cues, with the result that the birds given exercise flights during time shifting had altered the way they integrated the various cues. Edrich and Keeton (1978), investigating how much flight is required to produce this effect, found that even pigeons exercised only in fight cages 7.3 m long later chose bearings deflected less than those of normally clock-shifted birds. Further experiments are needed to clarify what cues the clock-shifted birds are using to correct for the erroneous information they get from the sun compass.

Studies of Integration: Radar Tracking of known Individuals

In the past, most radar studies of bird migration have used surveillance radar to monitor migratory movements. While this approach has yielded invaluable information concerning flight paths and concerning the relationships between migratory intensity and accuracy and the weather, it has not been suited to study of the orientation of known individual birds. However, Emlen and Demong (1978) have recently used a large tracking radar at the Wallops Island, Virginia, NASA base to follow individual white-throated sparrows previously captured and assayed for stage of molt, amount of fat, and intensity of Zugunruhe. The sparrows could be released aloft under various weather conditions (including ones when the birds would not normally begin a migratory flight) and after they had been subjected to various manipulations, such as clock-shifting or having a magnet attached. It was thus possible for Emlen and Demong to investigate the decision-making stage of free-flying migration under conditions when different weightings of orientational cues might be expected.

Amongst their many results, Emlen and Demong (1978, and in press) found that information from viewing the setting sun probably plays an important integrative role during the transition between daylight and darkness. Under overcast skies at night, birds that had not seen sunset oriented poorly whereas those that had seen sunset oriented well. This discovery raises a host of new questions for further investigation. For example, might sunset function as a reference for calibration of other cues? And if so, which cues?

Studies of Integration: Release-site Biases

Many investigators have noticed that each release site used for homing pigeons can be characterized by a preferred departure direction that often deviates somewhat from the true home direction; this release-site bias may be only a few degrees to the left or right of home, or at some locations it may be as much as 50-80 degrees. Curiously, the bias is usually not much

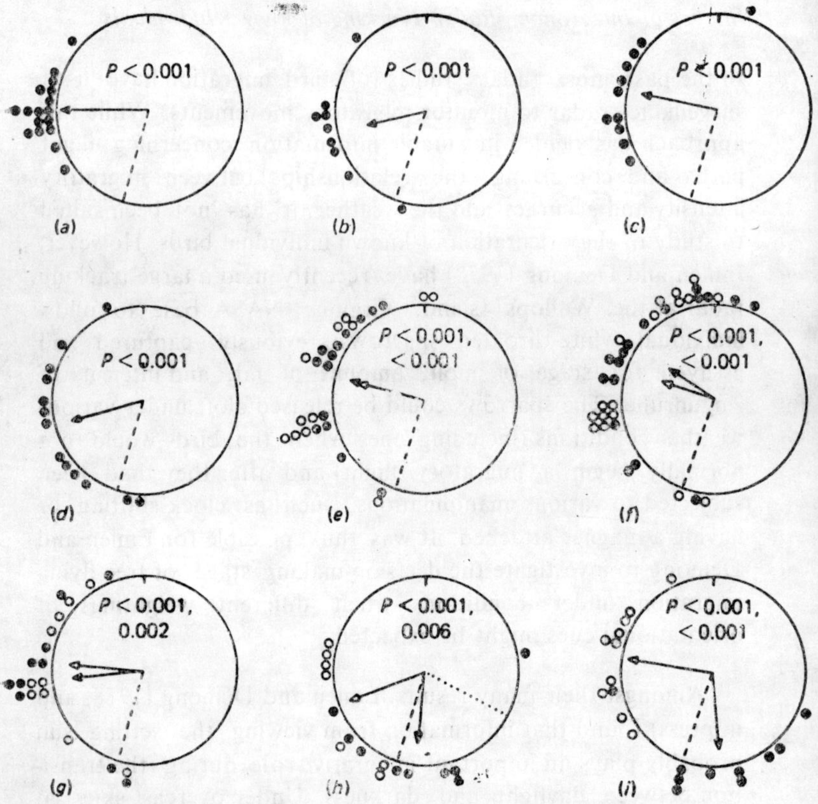

FIG. 12.8: Bearings of Cornell pigeons released at Castor Hill Fire Tower (143 km NNE of the loft), (a) Experienced birds new-to-site customarily depart in a direction 60 to 80 degrees clockwise (westerly) from home. Home line is indicated by dashes. (b) Birds with prior releases from this site usually continue to choose bearings markedly westerly from the home direction. (c) First-flight youngsters also choose westerly bearings. (d) Even under total overcast, experienced birds new-to-site depart in a westerly direction. (e) The bearings of experienced birds new-to-site are nearly identical, whether the birds are wearing magnet bars (black symbols) or brass bars (open bymbols). (f) Experienced birds new-to-site choose westerly bearings, whether they are wearing frosted-white contact lenses (black symbols) or clear lenses (open symbols). (g) The bearings of experienced

birds new-to-site whose beaks and noses had been painted with α-pinene in vasoline (black symbols) did not differ significantly from those of control birds treated with plain vasoline (open symbols). (h) The bearings of pigeons brought from Schenectady, New York (black symbols), were deflected clockwise from their home direction (dotted line) in a manner similiar to the deflection of the bearings of Cornell birds (open symbols) from the direction to Ithaca (dashed line); it therefore appears that some cue basic to pigeon navigation is rotated clockwise at Castor Hill, and that pigeons with different destinations read the cue in the same manner. (i) Although pigeons clock-shifted six hours fast choose bearings (black symbols) that are more nearly homeward oriented than those of normal pigeons (open symbols), their homing success is considerably poorer.

affected by the previous experience of the birds. It remains on overcast days (Fig. 8d) and when the pigeons are wearing magnets; hence the bias is probably not a function of either the sun or magnetic compasses. Tests with frosted contact lenses indicate that the bias is not due to the birds, visual perception of the surrounding landscape. The preferred direction is not altered by application of α-pinene to the birds noses, hence it seems unlikely that the bias has anything to do with olfactory position-determination in the manner proposed by Papi *et al.* (1972). Pigeons from other lofts often choose bearings at these sites that are comparably deflected from their own home directions. Bank swallows (Riparia riparia) show similar biases at these sites, so the biasing factor (or factors), whatever it may be, is apparently not unique to pigeons.

The hope has been that studies of release-site biases (or preferred compass directions, as Wallraff (1978) prefers to call them) would help reveal local factors that might be at least a part of the long-sought navigational map, and would show how those factors are integrated with the orientational cues already known. So far, unfortunately, that hope has not been realized. All efforts to explain the biases have failed. Nonetheless, this approach seems worth continued effort.

13

The Application of Population Dynamics

The Influence of Population Dynamics on the Study of Field Ecology

The theoretical and experimental work detailed in the last two chapters has had a very great influence on the interpretation of observations on natural populations in the field, especially in three important respects. First, the concept usually termed "Gause's theorem" has dominated a whole field of work, particularly connected with the extent to which closely-related species occur together and share common resources. On the whole this idea has generally been vindicated, at least as a pointer to rewarding fields of research.

Secondly, the distinction between density-dependent and density-independent mortality factors has greatly influenced the thoughts of those concerned with biological control and the regulation of numbers of animals.

Thirdly, observed oscillations of the numbers of natural populations have been a challenge to the ecologist and attracted much study and speculation. Early hopes that they would be easily explicable in terms of the equations of competition and predation, although often disappointed, have resulted in probems being tackled which would otherwise have appeared hopelessly complicated. From the few cases of clear-cut oscillations which have been analysed in very abundant species, such as the mammal species which show marked cycles, it is beginning to appear as if oscillations are often the result of intra-specific, rather than inter-specific factors; this may well be untrue of the less common species, however.

Therefore, despite the over-simplifications of the mathematical equations, despite the disappointments which have resulted from a too naive extension of experimental results to field conditions and despite frequent mistakes, it cannot in general be disputed that the theories of population dynamics have stimulated much useful work in a field where guiding principles are sorely needed, have simplified the interpretation of natural phenomena and, for the first time, have introduced the use of measurement in a field where quantitative method was unknown thirty years ago. The extent to which this "leaven" of ideas has produced results in the fields of the interpretation of natural communities, and that of the control of populations of economic importance will be judged. In this chapter we shall consider the factors which limit the size of natural populations and the light-thrown by population dynamics on an understanding of these.

Some Examples of Field-population Studies

In order to provide a basis for the discussion of population-limiting factors some of the few reliable studies of field populations from this point of view will be briefly described as examples. As far as possible the facts in each case will be separated from the authors' interpretations and it should then be possible to judge the kinds of influences at work in these cases, at least.

1. Stored-Products Insects. Solomon provides a useful analysis of population-controlling factors in insects which inhabit grain and other stored products. His article also includes many references to work in this field, arrives at some general principles and can be recommended as an introduction to population control in stored-products populations.

In this particular type of habitat a nearly unique feature is the "competitive" effect of metabolic heat which so raises the temperature of the habitat as to make it uninhabitable when the animals, in the case of Laemophloeus for instance, reach a density of 2.5 animals per gramme. At populations of 5.5 per gramme the temperature rises by 0.7°C per day and, owing to the poor heat-conducting properties of wheat, lethal temperatures are reached in bulk wheat in a few weeks. As a result the populations only reach a high density quite close to the surface of the wheat. Similarly, when individuals of Rhizopertha were released under conditions in which the temperature, moisture and population density could be measured, it was found that this beetle at a depth of five feet in grain is reduced to a density of 0.3 individuals per gramme. The important point made by Solomon is that these densities are well below those at which other density-dependent mortality or fecundity-reducing factors would normally operate to control the population.

Although Callosobruchus fecundity is reduced at low densities due to the infrequency of meeting and of copulation between the sexes, this effect is linearly related to density and there appears to be no confirmation of experiments on Tribolium which indicate a more powerful population-depressing action by this factor.

Solomon provides some interesting estimates of the order of population density at which different factors start to have a significant effect on population increase rate and morality rate. Arbitrary levels of increase in mortality rate or decrease in fecundity are taken and it is shown that, over a wide range of insects, a 20 per cent effect is produced at population densities of 0.25 to 32 insects per gramme. An 80 per cent effect (*i.e.*, increase in mortality or decrease in fecundity) occurs at densities between three and several hundreds per gramme. More precise figures are available for particular species, but even orders of magnitude are useful where so little information is available.

2. *Beetles which Eat St. John's Wort.* Clark's important study of the ecology of two species of beetles which have been introduced to Australia to control St. John's wort raises a number of interesting issues, of which only two can be mentioned here. The beetles have not by themselves proved to be a very satisfactory means of controlling the weed, because they are very near the limits of their physical tolerance range. They suffer density-independent mortality from the effects of frost, high temperatures and soil desiccation, while idiosyncrasies of behaviour prevent them from entering woodland areas where these factors are less severe. The usual history of a local population of Chrysomela gemellata or C. hyperici is that the population gradually expands until it eats out the food supply of Hypericum. The population is then made far more vulnerable to the normally density-independent climatic factors owing to the absence of cover and the more severe microclimate which results. Also the beetles fly far less under Australian conditions than they do in California and thus find their food less easily. There are also more usual density-dependent predators and diseases. The combination of mortality factor results in the complete local elimination of the beetle population, usually through the action of climate or the defoliation of the St. John's wort. The St. John's wort then recovers and becomes a pest long before the beetles have reached a density at which they are able to damage the wort. The violent oscillations in numbers of the beetle would appear to provide an example of oscillations resulting from interaction of a herbivore and its host plant.

3. *The Tawny Owl and its Prey.* Southern gives a preliminary account of the only field study so far published in which populations of both a predator and its prey have been measured. The species concerned are Stryx aluco, the tawny owl, Apodemus sylvaticus, the wood mouse, and Clethrionomys

glareolus, the bank vole. The owl population was followed for twelve years in a 1,000-acre woodland estate and that of the rodents in an area of 250 acres in the midst of the same estate; nearly 20,000 rodents were handled and marked with metal leg rings during this time.

The owl has the convenient habit of swallowing its food whole and then rejecting the inedible parts of the food (fur, bones, etc.) in the form of pellets. These pellets were analysed for their content of food species and also passed under a magnet which removed the metal leg bands.

Owing to "trap addiction" which was discovered during the course of this work, the estimates of the rodent numbers are subject to revision, but Southern was able to obtain a preliminary estimate of their significance by ranking the years in order of availability of rodent food, the best year being scored 10 and the worst year 1. There were three important sources of owl mortality were related to the rodent scores: all three factors, number of pairs breeding, size of clutch and breeding success are correlated with rodent supply, the first two rather more closely than the last, as might be expected. Thoughout the twelve-year period there was a consistent trend in the number of pairs present, namely an increase from 16 to 30 between 1947 and 1958. This trend, involving a halving of the mean territory size, may be due to changes of forestry practice in the wood or to the fact that the population was badly hit by the severe winter at the start of the study and probably consisted mainly of birds which were increasing in age (and hunting skill?) throughout the period of study.

During the first six-year period when there were, on average, less than 25 pairs present in the area breeding success was significantly greater than in the second six years when the average was above 25. This provides further evidence that juvenile mortality could be density-dependent. From the results it is clear that food supply has a density-dependent effect on the owl population despite the very powerful "damping" influence of territorial behaviour which, Southern showed, results in a very high mortality of surplus young adults each year.

The prey populations fluctuated rather violently. From a preliminary examination of the ratios of the two prey species in the pellets it is clear that in 1949, which was a peak year for both species of rodent, the owl fed mainly on Apodemus; this was followed by a marked decline in numbers. In 1950 there was a peak for the Clethrionomys population and the owl turned its attention to this food. Simultaneously the vole population fell to a low level and did not recover for at least a year. In 1951 there were very few of either prey species and the owl breeding season was a failure, only about 25 per cent of the normal number of young being raised. The owl population was low and the rodent numbers recovered during 1951 and 1952.

It is natural to wonder what part the owl had in the reduction of the prey numbers; here we must await further analysis of the data. The general impression given from most small mammal studies is that predation is not of primary importance in the generation of cycles; however, the behaviour of the predator in changing its attack from one rodent to the other until the "crash" phase is virtually complete, may well serve to accentuate the violence of the cycle and possibly to change its phase.

4. Predation by Sparrow Hawks. Tinbergen made an extensive study of the effects, on the populations of four species of passerine birds, of predation by the sparrow-hawk. Since the sparrow-hawk population in Holland is partly migratory it was not possible for him to obtain full data on the predator as Southern has done, but the effect on the prey species is estimated with some accuracy, although it is difficult from Tinbergen's data to be certain that the predator actually controls the prey species because mortality and density are not related throughout the year. The magnitude of the errors was estimated at all stages. The house sparrow suffers most from the attention of the sparrow-hawk, which accounts for over half its mortality. In the chiff-chaff and the great tit this is rather less than half, while in the coal tit the effect of predation on the population is almost negligible. The importance of factors in the birds' behaviour, such as the use of cover and conspicuousness, is demonstrated for these four species and also other birds. The population-limiting factors of the coal tit were not finally decided but Tinbergen believed that parasitism may play a role in controlling this very rapid-breeding species.

Evidence is provided to show that, in 1942, when goshawks were short of their normal food and turned their attention to sparrow-hawks as prey, the resulting reduction in the population of the latter was reflected in an increase in that of the passerines. Tinbergen is insistent that any natural community must suffer many such unpredictable changes as this, which may have a wide influence on the population under study. He therefore deprecates the uncritical application of such theories as Nicholson's to field populations and points to the many population-limiting mechanisms, including territorial behaviour, which are found in birds and which reduce or prevent oscillations in numbers when these might be predicted from theory.

5. Control of the Knapweed Gallfly. Varley carried out an intensive study of the factors which control the population density of the knapweed gallfly (Urophora jaceana) whose larvae are responsible for galls in the flower heads of the knapweed (Centaurea nemoralis). A census of 92 square-metre plots for two years was made and from this the population density of the flies was determined and a kind of balance sheet of mortality was drawn up showing the proportion of flies killed by the various mortality factors at each stage in the life history. A very large proportion of the mortality (some 96 per cent of the larve in the 1935-6 generation) was due to factors which were probably not density-dependent—the eating of larvae in the knapweed heads by caterpillars and by mice and losses due to flooding of the area. However, since the fecundity was found to be between 50 and 70 eggs per female, it would only be necessary for a few insects to suffer a density-dependent mortality for control to be achieved. The only effective density-dependent factor was considered to be parasitism by a Chalcid Eurytoma curta.

Varley made use of Nicholson's "area of discovery" concept and estimated this parameter at 0.25 sq. m. From this he estimated that the parasites could keep the gallfly down to a level of 0.68 per sq m. This level was lower than that actually found for the gallfly. When, however, allowance was made for the reduction in the efficiency of the parasite by factors which operate on both host and on parasite because they kill the parasitized gallflies, the figures were found to fit very satisfactory.

As Varley himself states, the exactness of his figures should not be misinterpreted, for, in a fluctuating environment subject to many external influences, the results predicted by simple theory cannot be expected to be very precise and the precision in fact obtained may have been partly due to good fortune. However, this is most important as a pioneer study in which an attempt is made to measure the magnitude of ecological forces of this type and it is most encouraging. It is also important because it indicated that a population can well be controlled by a factor which is only responsible for a small proportion of the mortality, provided that the factor in question operates in a density-dependent manner. Further, the reduction in efficiency of the parasite by the later-acting indiscriminate mortality factor shows how, by simply killing animals of an undesirable species without regard to its ecology, one may will increase and not reduce the size of the population, an argument, it well be remembered, put forward by Nicholson. Thus, the eating of knapweed heads containing gallflies by mice which killed 64 per cent of the available larvae tended to increase the gallfly population and not to reduce it.

6. Control of German Forest Pests. A great deal of work has been done on the Continent on forest insect pests and the fluctuation in their numbers. The enormous changes exhibited by four species in German spruce forsts can be gauged from the fact that logarithmic graph paper covering five orders of magnitude (from 1 individual per 1,000 sq m to 10 per sq m) is still not able to show all the annual numbers in the case of one of the species. Apart from the original author's useful critique of theories about population fluctuations, which are considered below, Varley's review is valuable for the demonstration that the fluctuations of the different pest species are not significantly correlated and that these four species which at first sight frequently defy "Gause's hypothesis" do in fact have very different life histories, the onset of the different life stages being "staggered" over a period of four months or more. Varley also shows that the periods of high numbers are usually those during which parasitism is at a very high level, highly-specific parasites being involved. This is important because Nicholson concluded from his theoretical study that increasing oscillations will result when a host is controlled by a single parasite. Schwerdtfeger's data do not permit a critical appraisal of the causes of mortality, but there seems little doubt that this situation, in which crops of a single forest tree carry a greatly simplified fauna, is especially likely to favour violent fluctuations in this way: this topic is discussed by Varley.

7. *Biological Control of a Coconut Mining Beetle.* An example reported by Taylor which indicates the unexpected repercussions which may follow an apparently irrelevant change in an ecological community and which illustrates the kind of factor which can decide whether a population is to remain at an economically unimportant level or to achieve serious outbreaks is that of the control of the leaf mining beetle of coconut palms in the Eiji islands. This is but one of many examples of the successful applications of biological control methods in these islands.

The beetle, Promecotheca reichei, is indigenous to Fiji, and so are anumber of hymenopterous parasites, including Oligosita utilis and Elasmus hispidarium. Until 1920, promecotheca was never important enough as an economic pest to cause alarm because small local outbreaks were promptly suppressed by the parasite species. Under these circumstances, the generations of Promecotheca overlap one another at the same time, which results in the parasites, which have rather short life histories, being always able to find hosts. However, in 1921 a very widespread predatory mite, Pediculoides ventricosus, was first reported in the islands; it had probably arrived some years previously but remained unnoticed. Soon after this, in the late 1920s, local but very severe attacks, resulting in complete defoliation of the palms, were made by Promecotheca, especially on the small islands of the Lau group.

The characteristic history of these attacks was found to be that Promecotheca was attacked by the mite Pediculodies, which thus appears at first sight to be beneficial to man. However, the mite was an extremely successful predator during the dry season but suffered heavy mortality by the flooding of the galleries in the wet season. Thus, having reduced the Promecotheca population to an extremely low level, the mite itself suffered a severe "crash," often, apparently, being completely eliminated. During the long period of virtual absence of Promecotheca the natural parasites died out through lack of hosts.

Under these circumstances the stage was now all set for a completely unhindered revival of the Promecotheca populations. A few individuals which survived as resting stages or were transported between the many small islands of the group by human agency, started to increase and, in the absence of parasites or predators, caused great economic damage.

The story of the successful control of this situation can be only briefly mentioned. Evidently, if the indigenous parasites had had longer resting stages, resulting in a more prolonged life history, it would have been possible for them to remain on a given island until the Promecotheca populations returned or built up again and thus to overtake the pest species. An insect which would be a specific parasite for Promecotheca and would fit this specification was sought and eventually found in Java in the form of another Hymenopteron, Pleurotropis parvulus, and, when introduced, this species was successful in preventing the fluctuations and restoring promecotheca to a very low level of no economic importance.

The history of this encounter provides many lessons in ecology, not least of which are the unexpected repercussions which may follow from an accidental introduction and one way in which population control mechanisms may work and may break down.

8. Cycles in Canadian Fur-bearing Mammals. In a useful review of population cycles in Canadian fur-bearing mammals, Butler relates the occurrences of population peaks in eight species during thirty-seven years. The most outstanding point made by this comparison is that the snowshoe rabbit's rather regular peaks are always followed by those of the predatory mammals and that in all cases but one the "good" years for lynx, arctic fox, fisher, coyote and wolf begin about the middle of the snowshoe rabbit peak and continue some years longer. In the one case there is a delay of one year after the end of the snowshoe rabbit peak. There is a strong tendency for the predators to succeed the rabbits in the order given above. The muskrat and the mink cycles are independent of those just considered but are closely related to each other.

It has been shown that the peak years in different parts of Canada do not coincide, but follow a sequence with the central provinces leading, followed by the western and then usually the eastern provinces. Butler puts forward a theory to account for this involving the increased efficiency of reproduction at higher densities and the emigration of animals from regions of higher density. The evidence for the first factor is not very strong, but he is certainly justified in criticizing those who would base conclusions on the study of single small areas. One point which certainly seems to follow from the existence of this progressive delay in the population cycles and from the extreme regularity of the cycles is that single climatic or other external factors are ruled out as direct causes of these cycles.

9. The Great Tit in Holland. An excellent example of the operation of an internal population-limiting mechanism is provided by Kluijver's study of the great tit. This was based on a population which roosted in winter almost entirely in artificial nest boxes and could therefore be counted quite easily. Kluijver found very distinct annual and local variations in the production of nestings, due to changes in fecundity and in clutch size. Thus, young per brood were obtained and were correlated with population density, both fecundity and clutch size being lower in denser populations. Kluijver believes that disturbances due to fighting are responsible for the reduced fecundity. The strong tendency for higher breeding populations to be followed by lower ones and *vice versa* is attributed to the combined effects of the reduced tendency to emigrate and the higher reproductive rate in the lower population. Thus, annual fluctuations occur which are sometimes synchronous and sometimes not in different parts of the country. Kluijver invokes a theory similar to that of Chitty to account for the synchronizing effect of major climatic disturbances.

10. Sewage-bed Flies. A number of enlightening studies have been made of sewage beds and their fauna from an ecological standpoint. Of these, Hawkes' study of factors controlling pupulation size in the fly Anisopus fenestralis is particularly relevant to the question of competition in natural populations. The populations of fly larvae, pupae and eggs were determined by sieving samples of the sewage bed material; it was found that they were of the order of 150 larvae and 100 pupae in a 27 sq in. sample, 6 in. deep. The actual numbers showed great seasonal fluctuations due to the successive generations of the flies. The magnitude of the peaks was determined by the amount of food available and intense intra-specific competition occurred at the higher population levels. This resulted in a steady reduction of the fauna as the season advanced and the available food was reduced. At higher temperatures the frequency of generations and, therefore, of peaks was increased owing to accelerated development. It was shown that when the sewage beds were treated with insecticide the seasonal fall in the population did not occur because the food supply remained unconsumed and it was necessary to continue the application of insecticide at high rates to prevent the flies becoming a nuisance. At the same time, the efficiency of the filters was impaired owing to the accumulation of organic matter. There was evidence of inter-specific competition also and other grazing organisms, which are more susceptible than the fly to injury by chemicals in the sewage effluent due to industrial processes, were sometimes responsible for depletion of the food supply. On the basis of this study, which could provide a fascinating source of information on the population dynamics of the species concerned if studied from this point of view, Hawkes made practical recommendations for eliminating the fly nuisance without heavy expenditure on insecticides and without impairing the efficiency of the sewage beds.

11. Inter-specific Competition among Wood Ants. One of the very few studies of inter-specific competition in the field is that of Brian on a natural ant population. This work, which the author regards as a preliminary survey, combines observations of the normal distribution, population size and habits of the ants with an experimental study designed to change relevant environmental influences and observe their effects on the balance of competition. It provides a useful indication of the kinds of factors which determine the occurrence of competition in four species of ants, namely Formica fusca, Myrmica scabrinodis, M. rubra and Leptothorax acervorum. Only a few aspects of this work can be mentioned here. The experimental alteration of the habitat, by laying slates on the ground and in other ways, resulted in the replacement of one species by another by direct aggression or by infiltration during temporary absence. The main factors involved in this process are a clearly-defined "replacement order" due to superior fighting powers, differences in climatic tolerance and differences in building abilities which permitted the Myrmica species to improve otherwise poor sites.

Formica can always compete successfully against the other species in acquiring the best sites, but is unable to occupy the poorer sites owing to inferior building ability. Leptothorax is also a poor architect and is vanquished in battle by Formica but it has the unique ability to burrow into harder wood, where, being rather small, it constructs systems of galleries in which it is safe from its larger competitors. The Myrmica species both build mounds and can thus occupy damper and colder sites than Formica, but of these two M. rubra is inferior to M. scabrinodis in direct competition but extends into moister places.

This work clearly demonstrates how it is possible for four species to live in the same locality, their populations interspersed and their food the same and it is unique in combining a description of the results of competition with an analysis of the actual processes of competition at work. It has recently been supplemented by an analytical field study on two species of Lasius.

12. Rose Thrips in South Australia. A study by Davidson and Andrewartha of populations of rose thrips in a garden in Adelaide is of special interest because it appears at first sight to contradict Howard and Fiske's and Nicholson's and Smith's conception of the importance of density-dependent factors in population control and has sometimes been quoted in that context.

The thrips populations, which were collected from roses throughout six years, were at a low level of the order of tens per rose for the greater part of the year, but suddenly increased dramatically to numbers approaching a thousand per rose in early December. The peak period altogether spanned hardly two months. The rise in numbers was fitted by the authors to a logistic curve and the peak densities in the different years were mathematically correlated with weather factors in the previous autumn and spring. The duration of the peaks was shown to be determined by the time of onset of the dry weather which resulted in a sudden fall to low population levels once more.

During the dry period the impoverished thrips population was restricted to very few localities of less severe elimate, one of these being the garden in which the samples were taken.

Davidson and Andrewartha were concerned to point out that Nicholson's theories did not apply to the thrips populations, for here was a population whose size is manifestly determined by climate which according to Nicholson, is a density-independent factor and therefore cannot control population size.

These arguments have since been challenged by a number of authors. It is hardly surprising that the release of severe climatic restraints should be followed by the multiplication of the thrips and their extension into a much wider environment, nor that the size and duration of the subsequent population "peaks" should be related to weather conditions; the real question at issue is the nature of the factors which determine the year-to-year abundance of the population, especially in the unfavourable season. As has been repeatedly demonstrated, the control factors are not necessarily those which kill most animals; in the case of the knapweed gallfly and Varley's later experiments density-dependent factors whose incidence varied with population size normally killed only a small proportion of animals. By re-analysing Davidson and Andrewartha's data Smith has demonstrated that factors of this nature were, in fact, present throughout the unfavourable season of the year because mortality at that time was inversely correlated with population size whereas there was no significant correlation between population growth and weather. It follows that these data far from supporting the authors' thesis, in fact provide evidence which contradicts it.

The Importance of Five Theoretical Questions in Field Ecology

The above accounts of work on the factors determining population size in nature provide a background for a consideration of the relevance of the principles derived from theory and experiment to conditions in the field. It is important to realize, however, that these accounts are greatly abbreviated and that very few ecological studies have yet been made on these lines; it is, therefore, most probable that tentative conclusions reached here will be modified in the near future and it is certainly essential that any serious student of ecology should read the original work and form his own conclusions in this vitally important field of study. In order to simplify our discussion it will be planned on the basis of the answers to the following set of questions:

1. What is the relative importance of density-dependent and density-independent factors?
2. Are competitive (and co-operative) relationships significant and what form do they take?
3. Are predator-prey relationships significant?
4. Are parasite-host relationships significant?
5. Are inter-specific or intra-specific factors the more important?

The following paragraphs are designed to answer these questions with regard, first to control of population size and then to causes of fluctuations in each case.

1. The Relative Importance of Density-Dependent and Density-Independent Factors. The thesis has already been argued that theoretically only density-dependent factors can "control" a population in the sense of maintaining it at a level where it continues to exist but does not attain an undesirably high economic density. This argument was first widely applied to the practical field of biological control by Smith, and he and his colleagues have since provide many examples in the field of "biological control" to vindicate his arguments. By means of exclusion sleeves black-scale insects on citrus trees were protected from the attacks of the artificially-introduced Metaphycus helvolus and protected branches compared with unprotected. Thus the mortality due to Metaphycus, which is both a parasite and a predator and came originally from Japan, was shown to be decisive.

A great many examples of this sort, are available from the literature of biological control and a useful comparison of different views on the subject from the practical point of view is given by Solomon.

Alternative views are given, for example, by Bodenheimer, Clark and Andrewartha and Birch. All the examples described in this chapter give evidence of the efficacy of density-dependent factors and almost all provide numerical information on the magnitude of the effects concerned. A feature, which may at first sight appear rather surprising, is that a density-dependent factor which causes but a small proportion of the total population mortality can be effective in control, while a density-independent factor, causing a much larger proportion of mortality, is ineffective. This is probable, for instance, in the case of the knapweed gallfly described by Varley.

The relevant feature here, of course, is that the excess reproductive capacity is automatically compensated by the density-dependent factor because its intensity of operation is adjusted to the host population size; thus, if the density-independent factors fell short of balancing the population increase by only a few per cent there would be no check on the gallfly. For instance, the numbers of animals in successive generations derived from 100 parents in which a 10 per cent increase is not compensated would, at the end of successive

Generation	0	1	2	3	4	5 . . .
	10	20	30	40 . . .		
No. of Individuals	100	110	121	133	146	161 . . .
	259	1,746	4,529	10,165 . . .		

generations, become—This, of course, is simply the exponential series once more. Davidson and Andrewartha's work, which at first sight appeared to indicate control by a density-independent factor, is irrelevant for the reasons given. Serious analysis of mortality usually reveals the presence of factors operating in opposition to density; the negative contention that density-dependent factors don't exist because not sought is unsatisfactory. Analysis of mortality is difficult in erratic populations and theoretical interpretation seems to correlate closely with the kind of population from which one is (perhaps subconsciously) generalizing.

The evidence that fluctuations in numbers are often caused by density-independent factors is. of course, overwhelming, but in this context it is useful to recall the distinction between "fluctuations"—regular or irregular variations—"oscillations" —showing self-restoring characteristics—and "cycles"—field phenomena thought, but not proved, to be oscillatory. That cycles can be the result of climatic effects of equal regularity has often been argued, but it is a theory whose vogue is steadily decreasing as more and more phenomena get out of step with the climatic changes with which they are supposed to be correlated. The theory that sunspot cycles acted in this way, at one time applied to many groups, has now been shown to be inadequate.

2. The Significance of Inter-Species Competition and Co-operation. Of the examples given above, those of Brian's ant study and Clark's St. John's wort beetle investigations are the most relevant here. It is evident that inter-species competition can be a limiting factor in determining population size and that local elimination of species can occur as a result of such competition. Many other studies, provide strong circumstantial evidence of the importance of competition in modifying the habits of birds and other animals because, as was shown in these discussions, the ecological niches of species are modified according to the presence or absence of closely-related competitors. Presumably such modifications occurred through the effect of competition on natural selection. However, the question of how exactly species in the field influence each other in competition remains one of the most obscure and little-studied in population dynamics and it is perhaps fair to say that some of the attention lavished on laboratory experiments might be more profitably employed on field experiments. There is a tendency for the animal ecologist to think only of competition acting through a rise in mortality or a fall in birth rate. In plants, as Harper has stressed, growth and reproductive rare are more flexible and some animals, particularly ant colonies must resemble plants in this respect.

It will be remembered that we showed that competition is the one type of population interaction which does not, according to theory, result in oscillations in the simple models used. Either one species eliminates the other or else, when the overlap in their ecological niches is insufficiently great for this to happen, both continue to survive indefinitely. This is conveniently summarized in "Gause's theorem." Elton's attempt to test the validity of this by a comparison of fauna lists from single communities with those from wider areas has been widely discussed. The observation work of Hartley and a field study of thrips by Broadhead provide much circumstantial evidence, however, which illuminates the situation after competition has occurred, rather than the process itself.

Although Allee's book contains a number of examples of cooperation, as of passerine bird flocks hunting together, and though there are many examples of species co-operating on an individual basis in the form of commensalism and symbiosis, it does not seem that inter-species co-operation at the population level is sufficiently frequent to warrant further discussion.

3. *The Significance of Predator-prey Relationships.* The relevant examples here are the studies by Southern on the tawny owl-rodent system in woodland, by Tinbergen on the sparrow-hawk-passerine population relationships and the work on Canadian mammal populations. An interesting study of the efficiency of human control measures, as an example of predation, against the tsetse fly should be mentioned also, in which it is shown that numbers caught decrease as the prey population is rarefied in accordance with simple mathematical assumptions.

These examples show clearly that the size of predator and prey populations can be closely linked in the field and Southern's results show that the predator may have a reverse effect on the prey. The shortage of examples is probably no reflection on the rarity of this phenomenon, but rather on the great labour involved in collecting the relevant data for two or more species. Even in Southern's account, in which owls' population size is partly limited by territorial behaviour there is probably a tendency for the production of oscillations due to over-eating of the food supply, although this is partly offset by the way in which the owl turned its attention from one rodent species to the other with changing abundance. In the case of the Canadian mammals it can hardly be disputed that the fluctuations of the numbers of the predators depend on those of their prey. However, in this case, there seems to be no evidence of a reciprocal effect as the oscillations in the snowshoe rabbit populations are apparently internally generated (see below), and to this extent the example does not resemble Gause's experiments.

An important point made by Varley is that the direct effect of predators (or parasites) on the prey (or host) population is not strictly comparable with the reciprocal effect of the prey on the predator through reduction of its population and reduction of the predator's food supply. One difference is that this effect is felt only after a considerable delay, with the result that the ratios between the two populations cannot be the same when the predator population is increasing as when it is falling; that is, in the type of curve first introduced by Lotka, in which numbers of predator are plotted against numbers of prey, the ratio of the two populations follows a closed loop, indicating a hysteresis effect. This kind of effect is absent, however in the case of Protozoa which reproduce mainly by fission, and as pointed out by Nicholson, oscillations about a balance level do not occur under such conditions. Varley points out that, so far, in field studies of insect populations the effects of prey on predator have not been measured, nor have the time delays involved. Only by measuring these quantities will it be possible to analyse the oscillations which owe their origin to this cause. Again, the transfer function method would appear to be appropriate to such an analysis. Varley also lists the criteria which must be fulfilled by a parasite if it is to control its host—

(i) it must reproduce faster at high host densities than at low, *i.e.*, its reproduction rate must be related to host density;

(ii) it must find enough food under all conditions for the adult to mature its ovaries;

(iii) it must find many hosts at a sub-economic density;

(iv) it must search for hosts both when and where they are available;

(v) it must have a low mortality rate under the conditions in question.

In general, then, the control of population size by interaction between predator and prey undoubtedly occurs in the field; there is some very slender evidence for reciprocal interaction between predator and prey which, as in the simple laboratory experiments, could produce a tendency to oscillation. The many additional factors involved in any natural system should not, however, be forgotten.

4. The Significance of Parasite-Host Relationships. Here the relevant examples are the study of the knapweed gallfly by Varley, Schwerdtfeger's and Voute's work on forest insect pests, Clarke's on the St. John's wort beetles, and Taylor's on the coconut plam leaf miner. These and many examples of biological control of pest insects leave no doubt that parasites can control animals in both the ecological sense of being the density-dependent factor which determines the long-term level of the population and also in the economic sense of preventing significant damage to crops. The fact that all these examples are derived from the class of insects is presumably due to the greater economic importance of this vast group and because widespread, lethal and intricately-adapted examples of parasitism abound among insects, while in the vertebrates at least parasites appear to have been more successfully resisted.

The case of Promecotheca retchei throws much light on the production of fluctuations in numbers and Varley's views on the greater tendency to fluctuations in systems involving control by highly specific parasites will be remembered. There can certainly be no doubt that when few species are involved and the system most closely resembles an experimental one, the results are also closest to those predicted by theory and imitated in experiments. The additional complications caused by greater numbers or more catholic habits of parasites tend to reduce the likelihood of violent oscillations; a conclusion which appears to have been so far appreciated by practical men that there is today a distinct tendency to avoid the culture of vast areas of single species of plants and to re-create habitats more akin to those of the undisturbed vegetation of the earth.

5. The Relative Significance of Inter-and Intra-Specific Factors. Hitherto, this chapter, all the examples have been based on systems in which population limitation has been attributed to the operation of inter-specific factors. It will be remembered, however, that the existence of powerful intra-specific limiting factors was revealed, not only in experiments such as those involving Tribolium and the blowflies of Nicholson's work, but also in field populations of small mammals where, as in the case of Chitty's study, the species' own physiological processes appear to determine the limits of population size. In the experimental examples too, other factors than shortage of food appear to be responsible. In the voles and also probably in the snowshoe rabbits of Canada these intra-specific factors synchronize with the cycles which are such a conspicuous feature of the biology of these animals.

In many birds and also in mammalian predators behavioural limitations to population size are common, usually in the form of territorial behaviour. It is usually considered by ornithologists but disputed by mammalogists, and has been demonstrated in some cases, that the population density which results is the maximum which the food resources of the habit can support in normal years at some critical stage in the life history; either the breeding season or an unfavourable winter period. In this sense, the territorial behaviour is only a behavioural adaptation ensuring that a, presumably less fit, proportion of the population dies rather than that food shortage has a more widespread effect on the whole population. Whether these speculations are true or not, it would appear that territorial behaviour is only a mediator, as it were, and not the controlling factor, of the density-dependent food shortage.

If intra-specific factors are sometimes prominent in determining the control of common and important species such as these small mammals and possibly certain insect pests, it is nevertheless true, as has been shown above, that the interaction of predators and parasites with their prey and hosts respectively can also both bring about oscillations and determine population levels. The one type of inter-specific interaction which does not appear to be important in bringing about oscillations is competition, although it is obvious that it is often extremely important in controlling population size.

It cannot be too strongly emphasized that the conclusions reached in this chapter are based on extremely few studies, that the accounts given are very much abbreviated and simplified. It follows not only that the reader should refer to the original sources and draw his own conclusions from them, but also that this branch of population dynamics, even more than others, is especially likely to change as more studies are made and more information obtained.

Furthermore, the situations which exist in nature are almost always extremely complex, and it is quite probable, therefore, that in many of the examples given different results might be obtained in different areas or on different occasions owing to the chance effect of unpredictable factors. The Promecotheca story and that of the effect of goshawks in reducing the predation pressure of sparrow-hawks on house sparrows should not be forgotten. Nor should the effects of introducing myxomatosis into the rabbit population of Britain, the repercussions of which are still becoming manifest in changes in the vegetation, in the diet of predators and in the sizes of populations of potential competitions such as the hare and the rat.

Conclusions: The Control of Population Size in Nature

The very meaning of the word control is far from agreed among ecologists but we shall assume it to mean the maintenance of a population level which does not differ from a long-term average by more than one or two orders of magnitude. This chapter should have illustrated clearly that the mechanisms of population control vary enormously with different species and under different circumstances. Ultimately, in reasonably stable and widespread species there can be little doubt that density-dependent checks operate quite regularly, even if only perhaps during a short season of the year as in Hartley's titmice. There still remains the possibility that in unstable situations in which dispersal and rapid rate of increase are all important, refuges are few and whole populations frequently succumb to castarophe, such checks will not operate at least for long periods, especially when the populations in question are constantly recruited from others in more stable circumstances. Of the density-dependent factors thought to control more stable populations some may be inter-specific and take the form of parasitism and predation. Among birds Lack's thesis seems to apply generally: that all aspects of reproduction including territorial behaviour and clutch size are so governed as to leave a maximum number of progeny under average prevailing conditions and that they can be modified by variations in these conditions. Under these circumstances most density-dependent factors operate by changing the death rate. On the other hand there is abundant evidence that this is not an invariable rule applying to all groups of animals. As Southern has argued, it is sometimes rather an academic point whether the death of an egg in the female is to be counted as an increase in mortality or a reduction in fertility and perhaps the general term "repression" should be used when we are uncertain of the mechanism involved.

The generation of oscillations can undoubtedly arise through interaction between species as predicted by simple theory, but occasions when this happens seem to be confined to rather simple situations involving few species and there is no doubt that cycles can also be associated with intra-specific population-control mechanisms, even though the cycle may "spread" as it were from the prey to its predators. In fact the potentialities of biological rates of increase, as Hutchinson has stressed, would lead us to expect far more violent population explosions than actually occur and the remarkable stability of most natural communities provides the clearest evidence for the strength of these bonds between populations which are the framework of the community, and which human interference can so easily rupture with such devasting consequences.

We have now completed a review of the subject of population dynamics. It has been demonstrated that the size of populations is determined by a set of factors which can, despite our lack of experience in this field, be analysed and measured. It can hardly be disputed that, for an understanding of the success and local abundance of animals in nature, the methods which have been described constitute a tool which is superior to anything hitherto devised. The kind of work which we have discussed in the last chapter reveals clearly the powerful forces which are at work when a community of populations are brought into spatial and temporal contact with one another.

14

Food-Chain Relationships Between Populations

This chapter is concerned with "vertical" or feeding relationships between populations. These are considered under three headings, the numerical population aspects, the general metabolic aspects of food chains and the special metabolic characteristics of the principal trophic levels. Field evidence on effects of predators on their prey and parasites on their hosts is mainly discussed earlier, but some important ideas derived from field and experimental studies are discussed here.

Theoretical Studies

Ross's work was directed to an analysis of the practical problems associated with malaria control. He derived equations describing the progess of a malaria epidemic in a situation where mosquitoes and man were present, and was able to predict that the very practical question of whether or not malaria would persist in an area is mainly determined by the ratio of mosquitoes to men: below a certain level of infestation (about 40 mosquitoes per man in the circumstances considered) the disease will die out. This prediction was in opposition to those who regarded malaria control by the elimination of mosquitoes as impracticable and it gave a great impetus to campaigns to reduce the infestation of mosquitoes.

Thompson treated the interaction of insect parasites with their hosts in a more general way with special reference to biological-control methods. He was able to demonstrate that the slow establishment of an intentionally introduced parasite does not mean that biological control will fail since the spread of the parasite may increase rapidly in time; also the effects of different reproductive rate of the two species and of variations in egg-laying behaviour of the parasite. His earlier work was restricted in its usefulness because the calculations were based on generation frequency and not on an absolute time scale, making it more difficult to treat cases when host and parasite generations are not of the same length.

In constrast, Lotka, in a more general treatment, employed an absolute time scale. He made assumptions such as that one host killed is equivalent to one parasite born and derived the equations

$$dN_1/dt = r_1N_1 - kN_1N_2$$

$$dN_2/dt = kN_1N_2 - d_2N_2$$

Where

N_1=number of host population

N_2=number of parasite population

d_2=death rate per head of parasites

r_1=birth rate minus death rate from causes other than parasitism per head of hosts

Thus,

$K_1N_1N_2$=rate of consumption of hosts by parasites

$K_2N_1N_2$=number of hosts available

d_2N_2=death rate of parasite population

Lotka then derived a graph representing the course of a parasitic invasion based on various simplifying assumptions.

In graphs of this sort, which are used also by Gause and others to express relationships between two species, the ordinate and abscissa represent the numbers of the two species and the points are plotted as in a correlation or regression curve. Time is not represented along the axes but could be indicated by measuring the angle of a line drawn at right angles to the tangent of the curve through the "singular point" in the centre. Alternatively, the same data can be plotted with time as the abscissa and the numbers of both species as ordinates. For the same data and it will be seen that the result is a distorted "sine curve" for each species, the numbers of the predator lagging behind those of the prey; this presentation is rather more familiar to and more easily understood by some biologists.

From the first, simplified, treatment it will be seen that interaction of the two species results in oscillations of constant amplitude; the time taken for the system to "go round" one of the circles is determined by the host reproductive rate and parasite death rate as $2\pi/\sqrt{r_1 d_2}$. The different circles, representing different amplitudes of oscillation, result from the use of different values of an integration constant M^1 which depends on the relationship between the rates of increase of the two species. It seems most improbable, on common-sense grounds, that a system of two species would, in fact, continue to oscillate with a constant amplitude. The more general treatment what probably happens in nature; namely that, while the period of oscillation remains constant, the ratio of reproductive rates does not, and thus the amplitude of oscillation tends to change progressively so that the system either "unwinds," the oscillations becoming greater and greater until one species reaches zero and the system collapses or, alternatively, the oscillations tend to die down and the system comes to rest at the singular point in the centre. In the first case, the collapse of the system, if the host is the species to die out the parasite (in the limited universe which is postulated) will rapidly follow suit. If the parasite is the first to reach zero, the host population will presumably increase until controlled at a new level by a new density-dependent factor such as parasitism or food shortage.

The implications of these results for experimental populations will be considered below, but it is worth pointing out at this stage that when using them to interpret experiments it is best to regard the curves of the above figures as curves joining points of greatest probability. What reasonably large population systems are involved they may represent quite accurately what happens, but with small populations the effects of chance factors, chance variations of reproductive and mortality rates and chance variations of the intensity of density-independent factors will cause appreciable departures from predicted values. The result of this is that, as soon as the population of one species reaches zero the whole system collapses, and the poor success of many experimental systems intended to demonstrate this theory seems to be attributable, in fact, to the small populations involved and the greater liability of the systems to become extinct as a result.

Lotka's treatment only slightly antedated the independent work of Volterra, a mathematician who became interested in the results of fisheries statistics in the Adriatic which were obtained by his compatriot D'Ancona, following the First World War. Volterra defined a "coefficient of increase" which is the same as Lotka's "intrinsic rate of natural increase" and he worked out the case of predation rather than that of parasitism, treating the results of competition between two species in more detail than Lotka did.

For a predator-prey system Volterra derived three "laws":

1. The law of the periodic cycle. Prediction that oscillations will occur, depending only on the coefficients of increase of predator and prey and on the initial relative numbers.

2. The law of conservation of averages. The mean numbers within each species remain constant, whatever the initial conditions, provided the intrinsic rates of increase and the efficiency of predation are constant.
3. The law of disturbance of averages. If two species are uniformly destroyed (*e.g.*, by human fisheries operations) in proportion to their numbers, the mean numbers of prey increase and of predators decrease.

Further generalizations applicable to small fluctuations only are:

1. The period, T, of the cycle is not affected by the initial numbers nor by the efficiency of predation.
2. The period T is proportional to the mean number of times by which the prey population alone would double itself and the predator population halve itself. $T=9.06\sqrt{t_1t_2}$.
3. Uniform destruction of predators accelerates (and of prey retards) the rate of fluctuation. If destroyed uniformly the ratio of amplitude of prey to predator numbers increases.

The close similarity of many of these results to Lotka's is at once evident. There is, of course, no fundamental difference between the predator-prey and the parasite-host situation, except that the order of magnitude of the constants involved is very different because one predator in its lifetime usually consumes a large number of prey, whereas a single parasite is usually but one of many individuals to draw its nourishment from a single host. Some hymenopterous prasites of insects, in particular, which have but a single host for each parasite individual are to this extent intermediate between parasite and predator. One result of the different proportions between host and predator or parasite is that the predator is potentially a more useful biological-control organism since, other things being equal, it should become established faster. However, in practice the more specialized feeding habits of parasites and their greater reproductive rates usually outweigh this advantage.

Nicholson developed a different line of approach to the problem of prodation and parasitism when he put forward his theory of the "balance of nature" based especially on the concept of the "area of discovery." According to Nicholson the searching by a population for food, unless organized on a basis of co-operation by the whole population, will inevitably be random. (This premise is disputed by Thompson, who assumes systematic searching, and is evidently untrue in certain cases, as when an individual's searching activities reduce the probability of other members of the species searching the same ground. An extreme case of this occurs when a herbivore removes food which would otherwise be searched for and eaten by its fellows.)

Assuming random searching, which must occur very generally in nature, Nicholson illustrates the result on the consumption of prey by a "competition curve" of area against time which indicates the progressive difficulty of finding food as the population increases. This, is an exponential curve in which the rate of increase in inversely proportional to the extent of departure from the upper asymptote, which represents 100 per cent searching success. It is rather unfortunate that Nicholson used the word "competition" to include parasitism. Williamson has used it to cover predation as well; but most subsequent authors have followed the practice observed in this book of restricting the word to include species exploiting the same food or shelter, etc., but excluding the interaction of predators and parasites.

Nicholson then correlated the "area of discovery", or area effectively traversed by the predator in its lifetime, with the power of increase of the species. (This "power of increase" is not r, the intrinsic rate of natural increase, but R, the number of times the population can multiply itself in a generation under the specific set of circumstances.) The power of increase and area of discovery were then related to the "steady state" at which the first is balanced by the mortality of the population.

On the basis of this equation Nicholson produced an elaborate series of conclusions about the interaction of parasite and host, some of which are the same as Volterra's, while others concern more specifically the "balance of nature."

Nicholson's first two conclusions, which link "competition" with the "balance of nature", foreshadow two ideas which have come to play a predominant role in theories about population. The conclusions are:

1. "Intra-specific competition automatically regulates the severity of its action to the requirements of each population provided the inherent resistance of the environment is sufficiently low to permit the species to exist." This is fundamentally the concept of "density dependence" which was developed further by H.S. Smith.
2. "For the steady state to exist each species must possess some advantages over all the other species with respect to some one, or group, of control factors to which it is subject." This is a direct corollary of the idea that two species with the same ecological requirements cannot exist together which was first stated by Darwin, but has frequently been attributed to Gause and has been a potent influence in certain fields of ecology.

Nicholson criticized Lotka's treatment to some extent and developed it further in the predator-prey and parasite-host field, and came to a number of conclusions about the results of displacement from a "balanced" situation, and of introduction of further parasites or predators. He then considered the conditions for oscillation in a very simple host-parasite system and concluded that oscillations are likely to increase indefinitely but will, however, be reduced as more alternative hosts or parasites are introduced. Territorial behaviour and catholic feeding behaviour on the part of predator or parasite are recognized as further damping factors.

Nicholson considered that physical factors such as climate are non-density-dependent and contrasted their operation to that of density-dependent biotic factors, although these terms were not used. Certain, rather trival, criticism of Nicholson's work is based on the fact that it has since been shown that climatic factors are not always density-independent and nor are biotic factors always density-dependent, in their action in any given instance.

The most serious defect of the earlier equations of predator-prey and parasite-host interaction is that they all make one or more of the following assumptions, namely, that changes in the population size of the one will be instanteneously reflected in a response by the other, that the appetite of the predator is is infinite, that searching by the predator is truly random and that reproductive rate of the predator is proportional to the whole prey population size. These assumptions have been critised by a number of authors some of whom have suggested improvements for example Beverton and Holt, Thompson; Chitty and Wangersky *et al.* The latter for instance explore the possibility of introducing time lags in the interaction of the prey and conclude that it is only possible to predict the outcome of particular instances with the aid of a computer. A more radical revision of the whole matter is due to Watt who constructed a model on the basis of practical experience and tried it out by substituting data from a wide variety of field populations. His basic equation is:

$$N_A = PK(I - e^{-AN_0 P^{1-b}}) \text{ where}$$

N_A=number of prey attacked

N_0=initial number of prey liable to attack

P=number of predators searching

A=coefficient of attack, the instantaneous rate N_A/P

K=maximum number of attacks that can be made per P during the period that N_0 are vulnerable.

Means of obtaining these measurements are discussed.

The inherent feedback in the system which may or may not lead to stability is demonstrated and it is shown how different biological properties of predators will increase their effectiveness at different densities.

A closer approach to natural conditions is also apparent in the attempts to develop methods of estimating morality in populations which are changing rapidly due to continual or intermittent recruitment of new individuals and where the incidence of attack is not independent of age.

Newer Theoretical Ideas Derived from Laboratory and Field Experiments

More complex examples of predation phenomena in the field are discussed in the next chapter, but the recent tendency to base theoretical models on field experience and to carry out field experiments on specific aspects of population dynamics makes it more convenient to consider here a few examples more directly relevant to theoretical problems, especially as they throw light on the validity of assumptions made in earlier models. Varley and Edwards tested the assumptions of Nicholson's approach in the case of experiments on the house-fly and its Chalcid parasite Mormoniella by De Bach and Smith and found (in contrast to the latter's conclusion that Nicholson's theory was well fitted by the data) that in fact, the "area of discovery" falls to low values when the host population is high and the number of hosts attacked then becomes independent of host density.

Varley and Gradwell made use of an idea due to Haldane to analyse the successive effects of mortality in the field. Haldane expressed the differences between the logarithms of the population numbers before and after each of a series of mortality factors had operated as k_1, k_2, k_3 . . . Then the total mortality over a period K is equal to $k_1+k_2+k_3$ . . . Applying this to field populations of the Winter Moth it was shown that k_1 (=winter disappearance in the soil) over a number of years is largely related to K and accounts for most mortality. The mortality due to pupal predators, although less than that due to winter disappearance, operates in a density-dependent manner, being higher when the over-wintering population was larger.

Rather different approaches to the analysis of mortality are those of Watt, who was able to incorporate simultaneously-acting factors in a single model by measuring their effects independently, and Lloyd who related various aspects of breeding success over a series of years to ranked estimates of a number of m rtality factors. Thus we have three quite distinct techniques suitable for the analysis of mortality factors acting in succession through one season, all together and to different extents in different seasons, respectively.

Another study derived from Varley's work is that of Dixon on the searching behaviour of the Coccinellid bettle Adalia decempunctata and the reactions of the aphids on which it feeds. Contrary to earlier suppositions aphids have a variety of very effective defence mechanisms and, unless the beetle is at least as large as the aphid, are not usually eaten. Once a young Adalla has received its first meal however its chances of survival are greatly increased. The behaviour of Adalia is such that simple suppositions about the relations between random searching, rate of supply of aphids and rate of increase of predators are invalid. The aphids occur in clusters and adult Adalia search for these and lay their eggs there, thus effectively increasing the area of discovery of their young.

An example of predation which varies in intensity with the age of the prey is provided by Connell's study of predation by Thais lapillus on Balanus balanoides. The predator (more widely known to ecologists as Nucella or Purpura, according to their generation) requires more time to remove the barnacles than to eat them and so, by concentrating on the larger barnacles, increases both its own production and that of the barnacles (which are largely limited by competition for living space). One result of this agedependent mortality is that earlier estimates of growth rate were seriously at fault.

This very rapacious predator is responsible for a very high proportion of barnacle mortality in summertime and is also very destructive to Mytilus. These authors refer to an old record reporting that population cycles can result from Thais-Mytilus interaction but are unable to throw light on the quertion.

Experimental Studies

Gause was the earliest worker to do extensive experiments on predator-prey systems with the intention of testing whether Lotka's and Volterra's predictions can be applied to laboratory populations. He was particularly interested in the maintenance of oscillatory systems, and these he was able to achieve on a number of occasions. In none of the experiments quoted in his book was this possible without arranging for small daily additions at a regular rate of one or other of the species. This was regarded is immigration by Gause. It is generally agreed that the numbers of animals present were too small to avoid accidental extinction of one or other species, as discussed above.

However, in the experiments on Paramecium bursaria feeding on Saccharomyces pombe and P. aurelia on S. exiguus, several complete cycles of oscillation were maintained before extinction occurred. On the basis of the figures he obtained, Gause calculated the values of the parameters in Volterra's equations:

$$dN_1/dt = b_1N_1 - k_1N_1N_2$$

$$dN_2/dt = k_2N_1N_2 - d_2N_2$$

In the first case the birth rate, b_1 of Saccharomyces pombe was artificially maintained by doubling the numbers of unconsumed yeast cells daily, while in the second the numbers of Paramecium aurelia were reduced to 0.45 of their level at the end of each day, thus achieving an artificial mortality d_2 of 0.45. To this extent even these experiments were not free from objection, but presumably they show that, had animals with these particular vital characteristics been available, oscillations would have occurred.

As far as is known to the author, no laboratory predator-prey populations have been maintained in oscillation for long periods without manipulation of some sort and it is evident that such "simple" systems are not easy to achieve. A usual fate of such systems is shown by the experiments with Didinium feeding on Paramecium in which the prey population was usually completely destroyed and the predator then starved. When, however, "cover" in the form of sand grains was provided, a few prey individuals usually survived even the most intense predation and the predator then starved—to be followed by a new outburst of the numbers of the prey.

Certainly in nature small amounts of migration between relatively isolated populations are frequently found and under these circumstances, at least, the perpetuation of erratic oscillations could be predicted. However the role of emigration and immigration in modifying the effects of inter-species interactions has been studied very little experimentally.

An interesting experiment on two mite species on strawberry plants is reported by Huffaker *et al.* and summarized by Andrewartha and Birch. The predator is Typhlodromus cucumeris and the prey, which is an economic pest is Stenotarsonemus pallidus. Experimental plants kept free of the predator by means of chemicals bore heavy infestations of the pest while those in which the predator was allowed to flourish remained very lightly attacked. These results were repeated under field conditions and in no cases did either predator or prey disappear (except because of the insecticide). This is thought to have been due to the availability of refuges for the prey which prevented complete extermination. Further experiments with different but related mites and a "universe" of oranges were designed to investigate the effect on the outcome of an experiment of this kind of rates of dispersal and of refuges. Huffaker was able by using many oranges and with the aid of petroleum jelly barriers to maintain the predator-prey system in operation for six months (or about twelve prey generations).

An example of a more complex situation is provided by the report of Flanders *et al.* of an experiment in which the moth Ephestia and its parasite Idechthis were cultured together. The parasite normally kept the number of its prey down to a low level but when a mite predator Melichares which fed on the eggs of Ephestia was introduced, the number of the host rose. This can be interpreted theoretically as an examples of Volterra's law of the disturbance of averages or of Nicholson's compensatory reactions.

METABOLIC ASPECTS OF THE "VERTICAL" (FEEDING) RELATIONSHIPS BETWEEN POPULATIONS

The first attempts to describe the functional relationships between populations were based on the food chain. This idea goes back to the French "Physiologists" of the last century, and, according to Brewer to Semper in America. But it was neglected until Elton extended its use. He pointed out that the number of links of successive predators and parasites is quite small, not usually more than three, giving a total of about five stages when the plant and herbivore links are included. He also drew attention to the great disparity between the numbers of species in the successive stages or steps of the food chain. This he expressed by saying that in a single community there is a "pyramid of numbers," whose basal layer is the abundant mass of plants, while the herbivores and successive carnivores constitute successive layers of rapidly decreasing size until there are quite small numbers of the larger predators. This conception has come to be called the "Eltonian Pyramid" by American workers in particular. The construction of food chains or "food webs" is an essential preliminary to an understanding of the biological structure of a community. The expression "food web" is American in origin as is "trophic level," meaning a "horizontal" stage in the pyramid of numbers, consisting of organisms consuming the same kind of food; both have been generally adopted in this country. The concept of the "pyramid of numbers" has sometimes been misunderstood in America, where papers have been published describing pyramids whose successive steps represent simply different species or size groups arranged in order of abundance. This seems a rather elaborate way of comparing the abundance of different species and has nothing to do with the original "Eltonian" conception whose essence was a functional analysis of the community. In work of this sort the food habits of the animals are frequently not studied at all which must make things much easier for the experimenter, but correspondingly less informative for the reader.

Examples of valuable ecological work based on the food chain and the food web are many and references to them will be found in the standard textbooks. Three recent examples will be mentioned here, however, because the authors have been able to combine studies of feeding habits with observations on the effect of time on the changes in the structure of the food chain. Thus, Glynne-Williams and Hobart worked on the crevice fauna of rocky seashores. They found that there are two complete faunas, both involving scavenning organisms, herbivores and predators, and both showing zonation in two respects, namely a vertical zonation determined by the time of submergence by the sea and a horizontal one depending on the distance from the mouth of the crevice. These two faunas, the "terrestrial" one with Collembola as the main detritus feeders and mites and fly larvae as predators, and the "marine" one with Polychaet worms as the main predators, replace one another when the presence of water or air permits the activities of the swimming and walking animals in turn. In addition there are a number of links between the two sets of animals; the

Collembola, for instance, include forms which feed on dead Mollusca.

Robinson's work on the animals of a brown flux on elm trees introduces a community dominated by a very different time scale. In place of a twelve-and-a-half-hour cycle of change, the animals have their "exits and their entrances," as the melancholy Jaques noted of mankind, the time unit being the annual cycle of the seasons. The different forms replace each other, taking up their ecological functions where their predecessor left off and with their active and resting and migratory stages so interwoven that both the interrelationships of the community are maintained and also the life cycles of the members are fulfilled. Taken from Robinson's paper, illustrates the changing seasonal aspects of the food chain.

A third example, typical of recent German work, is illustrated from Brauns. This shows in a striking way the sort of complexity which is to be expected in communities such as forest soils. However, it should be noticed that this particular "food web" is confined to fly larvae only and that these are listed only in family groups. If any attempt were made to cover all animal phyla and to differentiate the many hundreds of species such a diagram would become impossibly complex.

Making the Food-chain Quantitative: Numbers and Biomass

The food chain is not, as it stands, a device for measuring the magnitudes of the effects of the feeding of species on each other or in competition with each other. In practice, however, the more we learn about an assemblage of animals the more numerous do the biological ties prove to be and the more complicated does the food chain become. There is a great need, therefore, for a means of distinguishing the important from the irrelevant "links" in the chain and for distinguishing those species whose absence would profoundly alter the whole structure from those whose importance is trivial. The necessity of being able to select the important species is most apparent in the complicated terrestrial communities such as those of grassland and woodland which often contain nearly a thousand species of animals.

Since the organisms exploiting a given food resource may range in size from bacteria to elephants, the census must be converted to estimates of total protoplasm or biomass. However there are two fundamental limitations to the use of biomass in food-chain studies. The first is that the measurements we want are not represented by the standing crop of the species at all. As stressed standing crop is, for many reasons, a very poor index of production because metabolism is not directly related to it. Secondly, while it is possible to measure the food-chain links in terms of the quantities of organic matter which pass along each in unit time, it is difficult in practice to label and recognize particular parcels of food. Phillipson's ingenious technique of presenting different coloured foods to a predator is one possible approach. Another is used by Cushing who worked with a rather simple food chain in the sea in which two types of Copepod fed on virtually a single species of Alga. Here measurements of algal division rates, Copepod grazing capacity and reproduction and mortality figures sufficed to calculate productivities and to show how the spring outburst of Algae, the nutrient cycle in the sea and the capacity of the Copepods to "overgraze" the Algae are related.

Another method which has been used in the plankton community is to follow the levels of particular chemical elements in the water.

However, there is a fundamental difficulty which was originally realized by Thienemann, in using chemical elements such as phosphorus or nitrogen in estimating production. If one considers the fate of a certain atom of the element taken up by a certain planktonic plant, say a diatom, it may be excreted into the surrounding water and taken up again by a second plant almost immediately; alternatively, it may be eaten with the plant by a water flea, there to remain for an indeterminable period, short or long, before excretion; or again, it may remain inside the diatom until the latter dies and sinks to the bottom of the water, there to remain perhaps until the annual "turnover" of the lake in autumn, or to become incorporated in the bottom mud in such a way that it is not made available to living organisms again until the lake dries up in thousands of years' time. These are but a few of the many possible fates of such an atom and, since the extent to which any given element may circulate in the community is both unpredictable and also irrelevant to the productivity of that community, it follows that the concentration or chemical substances cannot be used for estimating productivity of whole communities.

Furthermore, in most communities of any complexity, such as a lake, the generation times of the members of the different trophic levels are very different. Thus the liberation of substances at death occurs with different frequencies according to the species and it is not possible to sum the productivities of the different species when these have been estimated with biomass as a basis. Bacteria and Algae have several generations in a day, Nematoda and Parasitid mites live for a few weeks, while springtails have three or four generations a year and many larger organisms live for a year or more. Again, organisms such as fungi, which do not produce discrete individuals are difficult to treat quantitatively for obvious reasons.

The Advantages of Using Metabolic Activity in Studying Foodchain Relationships

There is, however, one measurable entity which does not recirculate in the community, which is independent of the biological idiosyncrasies of the organism and which appears to be a suitable basis for estimating productivity of whole communities and to permit comparisons within them; this is energy. Energy enters the community as a proportion of the radiations emanating from the sun and intercepted by the earth. Of that which penetrates the atmosphere and enters plants a high proportion, probably about seven-eighths, is liberated as heat even in the processes of photosynthesis. However, the remaining one-eighth is transferred by the plant to organic substances of high energy content—carbohydrates, fats and proteins.

The plants themselves, in addition to their photosynthetic activity, are continually respiring, that is to say, are liberating energy, breaking down organic substances and reducing oxygen to carbon dioxide. In this way a proportion of their energy intake is dissipated. However, a larger proportion is made available to other organisms in the form of fresh green stuff which is grazed, of storage organs which are plundered or of decaying remains of plant matter which are decomposed. These sources of energy are the supply for the herbivores, predators are decomposing animals of all the remaining "trophic levels."

Similarly, the members of the successive trophic levels partly break down their ingested organic matter and liberate energy either in the direct production of heat or in the form of mechanical work used in locomotion and similar activities which ultimately produce heat in their turn. But, and this is the important point, the energy does not recirculate in the community as chemical substances do. An individual stores part of its energy intake in its tissues and liberates part as heat or mechanical work. This happens throughout the food chain. It is thus possible to draw up a balance sheet for a community in which the debtor side is the sum total of all the "packets" of energy liberated by the individuals, plant and animal, in the community, while the credit side is almost entirely made up of the single vast contribution which the plants make by their photosynthetic activities.

The measure of the energy flow can be taken at all the different points in the community at which it passes from one population to another when the productivity of components is being compared; it can be taken at its entrance to the plants and the sum total of its exits from all the members when that of the whole community is to be measured. It is directly related to the common-sense idea of the "productivity" or "activity" of a community. An "active" community is one in which the plants are capturing a high proportion of the incident solar energy; a balance is achieved between the activity of the plants and the rate of supply of nutrient substances to these (through the liberation of mineral matter from the animals and decomposer organisms). The photosynthetic activity of the plants, which should equate with the energy liberation of the "heterotrophic" element, is thus a measure of the activity of the whole community, for, if the non-photosynthetic organisms were to disappear, the plant life would rapidly die out for lack of nutrients.

From the point of view of the constituent organisms in a community, the relative importance of different species can be compared more simply on the basis of their contribution to the energy flow of the community than in terms of biomass. The population which exploits the greatest quantity of stored energy is contributing most to the rapid liberation of nutrient substances. These will ultimately find their way back to the plants and thus contribute to raise the energy intake of the community as a whole.

Efficiency and the Food Chain

The concept of efficiency was first associated with the food chain by Lindemann who used it to compare the production (*i.e.*, the total calorific content of the food exploited by all members) of the successive trophic levels in a hypothetical food chain. Efficiency in his sense is therefore the ratio of food made available to a "higher" trophic level divided by that available to the level under consideration. This has been termed the Food-chain efficiency by Slobodkin in contrast to the ecological efficiency in which the denominator is the energy content of the food actually consumed by the lower level. Lindemann suggested that efficiency is greatest at the lower levels of the pyramid and least at the higher but most later authors doubt the validity of this generalization.

A great number of other energy ratios can of course be recognized and some of these have been tabulated by Pattern who also introduces the concept of entropy in a stimulating paper. At the present stage of ecological theory and practice, however, there appears to be little to gain from the substitution of the concept of increase of entropy for that of reduction of energy as he has done.

Some Examples of Quantitative Food-Chain Studies

During the past few years a number of pioneer quantitative studies of food chains have been carried out in the United States: these have mainly derived from the precept and example of E.P. and H.T. Odum and have now been summarized, together with much valuable information on the energy dynamics of natural and man-dominated communities. This work has provided us with practical figures of total production, trophic level production and efficiencies in the case of two freshwater communities, the Spartina marsh already mentioned and an oceanic coral reef. Since then H.T. Odum has published some figures for tropical rain forest and Golley for a grassland community. Some of these communities are reasonably self-contained and show normal types of numerical, biomass and energy pyramids, but others such as the spring investigated by Teal are based on the import of practically all plant-produced energy in the form of dead leaves and show a very restricted photosynthetic layer and a great preponderance of decomposer organisms. The coral reef is interesting especially because the corals containing zoochlorellae combine in one individual two, or possibly three, trophic levels. The relative importance of the Algae from the production angle was not measured and has since been disputed on the grounds that it is insignificant—although the Algae are important as a kind of excretory organ for the corals and to this extent contribute to increased turnover. The section to follow serves to illustrate the approximate outcome of a study of this kind when applied to one particular community—in this instance to grassland.

The Structure of the Grassland Ecosystem

No terrestrial ecosystem has yet been studied throughly enough to provide material for a complete synthesis. However, it is possible, by drawing data from a number of separate studies to compile a general energy flow picture for a permanent grass field grazed by cattle and to arrive at a number of, admittedly tentative, conclusions about its main features. In the standing crops of the main groups of organisms are represented by "boxes" whose area is proportional, on a logarithmic scale to their calorific content. These are joined by energy flow "channels" whose width is also proportional, on a logarithmic scale, to their mean annual energy flow in calories per unit time. Obviously since the time unit is arbitrarily chosen the size relationship between the boxes and the channels is also arbitrary and of no special significance. If we first compare the ratio of the boxes to the channels we see that there is a progressive increase in ratio of flow/standing crop, *i.e.*, of rate of turnover of organic matter as one passes from grasses to man through the above-ground diagram. No such generalizations are possible for the subterranean diagram. That is to say a stock containing about 1,000 calories of plant, 200 calories of beef and 160 calories of man respires a calorie per day. The micro-organisms require 400 calories content to respire one calorie, the large soil invertebrates about 1,000 but the smaller ones about 200. This means once more that biomass (proportional to calorific content) is not a good indicator of activity. The main features to notice about the widths of the flow channels are that about one-sixth of the energy of photosynthesis is respired by the plants, about two-thirds goes in the form of dead plant matter to the decomposer organisms, mainly in the soil and less than one-quarter is consumed by all herbivores. Even of this proportion about half is returned to the soil as non-assimilated faces. So in a field which the farmer believes himself to be fully using in this way less than one-half per cent of total biological production is harvested by man and about three-quarters goes to the soil. Obviously, the soil is in a privileged position. Of the energy flow entering the soil all is ultimately respired provided no appreciable accumulation as peat, etc., is taking place. About five-sixths of the respiration is microbial and one-twelfth (a very uncertain figure this) is by Protozoa and the remainder is shred between the larger fauna

TABLE 14.1

Figures from a Hypothetical Grassland Community Showing Order of Magnitude of Numbers, Biomass and Metabolism for the Main Invertebrate Groups. Numbers and Biomass are Typical for Areas in which the Respective Groups are Abundant, i.e., Most Groups would be Uncommon in Some Grasslands and none of them would be as Common as this in One Place Together. Also the Balance of Groups would be Quite Different in, say, Moorland or Woodland. Figures in the Same Horizontal Row do not Necessarily Correspond where Independent Estimates are Available.

	Results of Metabolism Experiments				*Approximate Estimates Grassland, per sq m*		
Group of organisms	*Weight of organism, mg*	*O_2 uptake Cu mm per hour per individual*	*Temperature °C*	*Calories per day per g at 16°C*[b]	*Numbers*	*Mass g*	*Calories per year*[d]
Bacteria[a]	10^{-9}	7×10^{-9}	20	575	10^{15}	1,000	—
Fungi[a]			20	161		400	—
Protozoa[a]	0.05	10^{-3}	20.5	14	5×10^{8}	38	113
Nematoda	0.001	$1{,}2\times10^{-3}$	16	144	10^{7}	12	355
Lumbricidae	5,000	300	16	7	10^{3}	120	180

Enchytracidae	0.14	0.117	16	100	10^5	12	160
Mollusca[c]	1,500	1,000	28	29	50	10	62
Myriapoda	25	10	19	36	500	12.5	96
Isopoda	22.7	9	19	36	500	5	38
Opiliones	18	10.5	19	53	40	0.4	5
Acari—							
Parasitids	0.2	0.5	16	280	5×10^3	1.0	64
Oribatei	0.25	0.12	14	72	2×10^5	2.0	30
Araneae	10	3.0	19	27	600	6.0	34
Coleoptera	250	60	13	39	100	1.0	8
Diptera	610	105	13	29	200	1.0	6
Collembola	0.46	0.55	16	144	5×10^4	5.0	153

[a]Figures for micro-organisms are only available per g of soil; these have been multiplied by 10^7 to give an extremely approximate estimate on a square metre basis.

[b]Metabolism figures are corrected to 16°C by means of Krogh's graph. Values are in small (g) calories.

[c]Molluscan figures based on snails, weight of soft parts only.

[d]Daily metabolism figures for 16°C are multiplied by 212 to correct for annual temperature cycle in N. Europe; (based on Neilsen). Values in large Calories. Owing to effects of microbial antagonism rates for natural conditions are far lower than in the laboratory, but no valid estimates are possible. Note that the total from field experiments on undisturbed soil is around 5,000 Cal/year.

such as earthworms and Myriapods on the one hand and the "mesofauna" (mites and Collembola) on the other.

SOME SPECIAL PROPERTIES OF SPECIFIC TROPHIC LEVELS

Plants and Photosynthesis

Green plants are almost entirely responsible for the primary production of the whole community and for this reason can be set apart from the animal and other consumer organisms. (Chemosynthesis by Bacteria is generally agreed to be a relatively unimportant phenomenon, in comparison.) Because the energy intake of plants in a closed system is equal to the integrated total of respiration throughout the community photosynthesis is the best and most practicable measure for determining the productivity of the whole community. The proportion of photosynthetically bound energy which is liberated again in the plants' own respiration seems to vary over a wide range amounting to well over half in the tropical rain forest studied by Odum *et al.*, and in tropical plankton communities according to Ryther. In Golley's study of temperate grassland on the other hand it amounted to about one-sixth. Since the photosynthetic efficiency and total primary production are higher in the tropics this may have the effect of reducing the non-respired energy available to the rest of the food chain to a more even level.

The most direct method of determining primary productivity is to measure the difference between (a) oxygen production, carbon uptake or mineral uptake associated with apparent photosynthesis in sunlight, and (b) the quantities of these elements exchanged in the reverse direction due to the simultaneous respiration of the plants, as measured in the dark. This is the basis of the dark and light bottle method for plankton and some gas train methods on land. A less direct method of determining gross productivity is the chlorophyll method in which the quantity of chlorophyll present is multiplied by an independently determined factor which related it to photosynthesis.

The net primary productivity, *i.e.*, that remaining after plant respiration and animals and other heterotrophic organisms have taken their toll, can be measured by one of several harvesting methods. These are mainly applicable to seasonal plants. When feeding by other herbivores is reduced to a minimum in man-controlled systems such as vegetables and corn crops the human harvest may represent quite a high proportion of the difference between gross primary productivity and plant respiration, *e.g.*, man may harvest up to two-thirds of the gross primary productivity in an alfalfa crop during the six-month growing season.

Primary Productivity Methods Based on Chemical Exchange

The classical method used to determine gross productivity of plankton is the light and dark bottle method already mentioned. The rates of increase of oxygen in a clear bottle and of decrease in a dark bottle are added. They are determined by the Winkler oxygen-concentration method.

Alternatively the rates of carbon dioxide absorption in light and production in dark can be determined, either chemically by titration or with a pH meter. The validity of results obtained in this way has been disputed on the grounds that the inner surface of the dark bottles are sites of increased bacterial respiration, and that the lower turbulence inside the bottle reduces photosythetic rates. Both these effects would appear to indicate that published figures of productivity are too low. Early estimates in which nanoplankton were neglected are also far too low. Steemann-Nielsen introduced an alternative method of measuring carbon uptake. He added solutions of sodium bicarbonate containing C to the samples in bottles and after exposure to sunlight measured the quantity of radioactive carbon present in plants filtered and centrifuged from the water. Short exposures were used to reduce the passage of carbon to heterotrophic organisms. Neilsen's estimates were lower than those generally accepted. According to Ryther, Nielsen was really measuring net and not gross productivity because photosynthesis and respiration are virtually contemporaneous and only storage in excess of immediate requirements is measured. The practical details and comparative advantages of these methods have been usefully reviewed by Goldman, who demonstrates that measurement of primary productivity is not an easy matter. Rohde has helped to explain discrepancies on the grounds of the rapid variation in productivity during the day.

Similar methods have been used for the measurement of gross primary productivity of terrestrial communities; carbon dioxide, absorbed in alkali from a stream of gas, is estimated by titration. A simple field technique for low herbage, and some results have been published by Wallis and Wilde, and the use of transparent plastic bags in conjunction with a similar apparatus to determine production of tree leaves is mentioned by Odum *et al.*

Alternative direct methods employ the measurement of mineral uptake, particularly phosphorus by chemical analysis. As already explained these are really methods of measuring net primary production and are only applicable in rather simple systems such as plankton communities during the period of seasonal increase.

The Use of Chlorophyll as an Index of Primary Productivity. The quantity of chlorophyll present in a sample of water can be determined by a number of elegant physical and chemical methods and on the whole is surprisingly constant at particular latitudes. As a practical measure of productivity in the sea it has been criticized by Gillbricht because up to five-sixths of the chlorophyll may in fact be in the faeces of Copepods and not living in algal cells. The quantity of chlorophyll on an area basis is much more difficult to determine for land communities but it appears to lie within a narrow range of values. Bray surveyed a wide range of native and managed plant communities and obtained values between 0.3 and 3.0 g dry weight per sq m. The quantity of chlorophyll was higher in forest than in grassland, in lowland meadows than upland grassland and at the later stages of succession. Odum *et al.*, give figures around

3.0 g per sq m for tropical rain forest and Gessner, 1.0 for grassland and 1.2 for forest. When suitably related to measurements of productivity chlorophyll appears to be a useful index of potential productivity, but, of course, actual short-term productivity is always subject to the influence of environmental conditions and particularly light and temperature.

Harvesting Methods of Productivity Measurement

As already explained harvesting methods measure net primary productivity and are best adapted to seasonal crops. Harvesting methods have been applied to timber production by Muller and Moller by Ovington and by Pearsall. These all involve fairly straightforward, if technically difficult, measurements of weight or volume of organic matter produced.

The Range of Primary Productivity Values Found in Practice

Total annual gross productivity values expressed in g carbon per sq m per year range from less than 45 g for deserts to 450 g for favourable agricultural areas and up to 2,250 g in exceptionally fertile situations. A very approximate calorific content (in gramme calories) can be obtained if these figures are multiplied by 10,000. Many figures have been published for net production of managed and natural communities although it is not always clear whether roots are included with aerial parts of plants. When reduced to the above basis of g carbon per sq m per year (1 g carbon $\simeq$2.5 g organic matter) normal values for temperate conditions range from 150 g to 370 g but tropical crops range up to about 2,500 for Cassava. English coniferous and deciduous woodland each produce up to 540 g and 270 g, a reflection of the seasonal interruption of production in the latter. Golley obtained grassland production figures of 580 g gross and 490 g net. The most recent estimates of world annual photosynthesis put the activities of marine plants at 2.5×10^{10} tons and of land plants 1.7×10^{10} tons of carbon equivalent to a total world turnover of some 0.63×10^{24} calories.

The Food-Chain Efficiency of Primary Production

Primary production efficiency is low when based on comparison of calorific content of plant products with the energy in solar radiation. As already mentioned some 30 per cent of gross production is dissipated as plant respiration, but this is a small loss in comparison with the 50 per cent loss of radiation between arrival at the plant and absorption by chlorophyll and an efficiency of only about 10 per cent in the actual chemical changes in photosynthesis itself. Estimates of overall efficiency from reception of solar energy over large areas of the earth's surface to production of organic matter usually vary from about 1 per cent to 5 per cent; Golley estimates 1.25 per cent for grassland and Ovington and Heitkamp quote about 2.7 per cent for pine wood or 3.5 per cent when roots are included as a maximum for British woodland. Kalle quotes 0.04 per cent for the North Sea where photosynthetic efficiency is reduced by turbidity of the water but assisted by mechanical mixing of nutrients and water with the Algae; however for the reason given below this should probably be nearer 0.14 per cent.

H.T. Odum has pointed out that, since conditions for maximum photosynthesis are usually below the optimum and since production rate rather than efficiency is likely to be of importance to the survival of a community, selection of those systems with higher production but lower efficiency is likely to occur. This point is discussed below.

A variety of figures for incident solar radiation can be found in the literature referred to above, in papers by Fogg and Strickland and also in climatological and physical publications, for example that by Hilsum. Ovington and Heitkamp give a formula for calculating radiation at the earth's surface in terms of climatic factors. Typical figures of radiation available to plants are 2.5×10^8 calories per sq m per annum in Britain, 47×10^8 in Michigan. Kalle gives 8.4×10^8 for the North Sea, but this is based on a midsummer value and is not corrected for seasonal changes in radiation.

The Herbivore Trophic Level

This is the secondary level which is most widely exploited by man and as a result has perhaps received an undue prominence. Of the net production of a typical meadow grazed to capacity by cows, for instance, less than one-third is consumed by herbivores, and the general impression from unexploited terrestrial communities is that a much smaller proportion of plant energy goes to the herbivores. In the marine plankton community the herbivores appear to ingest a higher proportion of plant production but to assimilate less of it.

The food-chain efficiency of domestic herbivores suitably managed can reach high levels but hunted prey are considerably less efficient. Productivity maxima for U.S. sports fisheries are about 5 g per sq m per annum compared with a primary production of about 300 g. It will be remembered that Allen obtained a figure of about 17 per cent as the overall efficiency for trout of which never more than 10 per cent was harvested by man, the rest going to natural predators. In both this and Watt's studies the high proportion of production which is lost as young fish to other predators points to the need to eliminate predators or to accept a smaller fish in the catch if productivity is to be increased.

The Carnivore Levels

These have been very little studied. Lindemann was of the opinion that carnivores are less efficient than herbivores because they use more energy in searching out and catching their prey. Certainly among mammals the basal metabolism of carnivores appears to be no higher than that of herbivores in relation to size. The index might not be the ideal measurement if it were shown that carnivores are capable of periodically raised metabolic rates and herbivores not. Among terrestrial invertebrates there is some evidence of greater total metabolic activity among carnivores than herbivores. At present Slobodkin's opinion that food-chain efficiencies vary little from one trophic level to another seems to be the safest generalization, and values of the order of 10 per cent are to be expected.

Decomposer Levels

These have been much neglected in the past but in terrestrial systems, at least, probably process a higher proportion of plant production than do the herbivore and carnivore levels. As Hairston has pointed out, wherever a community is relatively static the impression of abundant food available to herbivores is deceptive; ultimately all the food is consumed and much of this by the "reducer or decomposer" organisms. The same opinion has been expressed by other biologists also. Aquatic communities vary in this respect: most of the animals in Teal's spring were decomposers and in Odum's Silver Springs study most metabolism was through decomposers although their biomass was less than that of herbivores. In typical grassland communities both standing crop and production of decomposers appear to exceed that of herbivores as illustrated above.

It appears that terrestrial plants as a whole have evolved methods for keeping their energy stores away from herbivorous organisms during their lifetimes; many grasses are armed with silica spines and store their food reserves below ground in winter. Trees are only exploited by defoliating insects and some specialized browsing mammals. Thus Ovington and Heitkamp point out that in managed forests the energy content of the litter is small compared with that of the timber. However, this is no index of the energy turnover in the litter and it is not, of course, any guide to the state of affairs in natural woodland where the whole of man's harvest is added to the intake of the decomposer organisms as each tree dies and releases its hoarded wealth of organic matter. The estimates of tree litter contributed to the soil vary considerably; Pearsall quotes 1.5 to 1.6 kg per sq m per annum or about 6×10^6 calories for oak wood; Ovington and Heitkamp's figures for oak indicate about

1.5-2.5 × 10^6 as a standing crop (no annual figures are given but in such woods the annual fall disappears during the year).

Liberation of 5 × 10^6 calories per sq m per annum spread over the year represents the amount of heat required to boil about 150 cc of water per day—or enough to make a cup of tea.

15

Competitive Relationships Between Populations in the Same Trophic Level

This chapter is concerned with inter-specific relationships, that is to say with the biological properties of interacting populations. These properties have been studied in the past through simplified theoretical models, through laboratory experiments and through field observations on population interactions and their effects. More recently these approaches have been unified to some extent by attempts so base more realistic mathematical models on observation of nature rather than on more or less plausible assumptions.

It is convenient, if not strictly logical, to divide this vast field of work into (a) relationships between animals which potentially share common resources at one trophic level. These we shall call competitive or horizontal relationships. On the other hand (b) we shall consider relationships involving feeding, predation and parasitism which we shall term vertical. It can be justifiably argued that the effects of parasities on their hosts for instance are similar to those of "competitors", in that they are competing for the same food supply. However the nature of the population reactions in two cases is usually different because the growth of the parasite population tends to lag behindthat of the host. Both horizontal and vertical relationships can be studied at the theoretical, laboratory and field levels, but since the latter usually reveal situations of great complexity containing both intra-specific as well as inter-specific relationships it is more practicable to deal with field examples separately.

Competition means here inferred or proven interaction between species which share a common population-limiting environmental resource such as food or shelter. It is not extended, as Nicholson extended it, to cover parasitism and predation, despite the fact that the bodies of prey are, in a sense, a common resource for the two component populations of such a system. Elton and Miller have distinguished from the general term competition those cases where "it is known that one species affects the population of another by . . . reducing the reproductive efficiency or increasing the mortality of its competitors." This more limited use is termed by them "interference". They justify the introduction of this term because "competition" has already acquired many shaaes of meaning for geneticists and others concerned with evolution and the isolation of species and because many examples of implied competition "just suggest that allied species tend to evolve into different ecological niches, which is another way of stating that the facts exist and that one believes in natural selection." Harper seems to use these two terms in opposite senses.

(*a*) *Theoretical Studies.* Volterra considered the case of two species competing for a common food resource and showed that if ε_1 and ε_2 are the coefficients of increase of the two species with unlimited food ($N_t = N_0 e^{\varepsilon t}$) and if h_1, h_2 are the constants of food consumption and if γ_1, γ_2 are the constants of susceptibility to food shortage and N_1, N_2 the numbers of the two competing species, then:

$$\left.\begin{aligned} dN_1/dt &= \varepsilon_1 N_1 \\ dN_2/dt &= \varepsilon_2 N_2 \end{aligned}\right\}$$

The total removal of food by both species is then $h_1N_1+h_2N_2$, and this removal reduces ε_1 and ε_2 to a new lower level such that ε_1 and ε_2 change respectively to:

$$\left.\begin{array}{l}\varepsilon_1-\gamma_1(h_1N_1+h_2N_2)\\ \varepsilon_2-\gamma_2(h_1N_1+h_2N_2)\end{array}\right\}$$

Therefore $$\left\{\begin{array}{l}dN_1/dt=N_1[\varepsilon_1-\gamma_1(h_1N_1+h_2N_2)]\\ dN_2/dt=N_2[\varepsilon_2-\gamma_2(h_1N_1+h_2N_2)]\end{array}\right\}$$

Then if $\varepsilon_1/\gamma_1>\varepsilon_2/\gamma_2$, the second species population approaches zero, the first approaches $\varepsilon_1/\gamma_1h_1$, *i.e.*, the species which shows the greater susceptibility to food shortage will be eliminated in competition with that which shows the lesser susceptibility.

Volterra then traced the possible fates of the two species under different combinations of initial conditions and vital parameters, the consequences of competition between more than two species, of combinations of competition with predation and of the effects of adding new species to associations already in a steady state.

Gause produced similar competition equations in the form:

$$dN_1/dt=N_1b_1\frac{(K_1-N_1-\alpha N_2)}{K_1}$$

$$dN_2/dt=N_2b_2\frac{(K_2-N_2-\beta N_1)}{K_2}$$

where α and β are the coefficients of competition and are proportional to the population-depressing effect of the one species on the other.

Gause recognized three cases:

(i) When $\alpha>K_1/K_2$ and $\beta>K_2/K_1$.

The population-depressing effect of each species is greater on the other species than on its own population. In this case only one species will survive, depending on the initial populations.

(ii) When $\alpha > K_1/K_2$ and $\beta < K_2/K_1$ or when $\alpha < K_1/K_2$ and $\beta > K_2/K_1$.

Competition results in the elimination of that species which is most harmed by the other; the case considered by Volterra above.

(iii) When $\alpha < K_1/K_2$ and $\beta < K_2/K_1$.

Both species survive in equilibrium indefinitely. This situation arises when both species have a niche in which they are more favoured than the other species. This condition was realized by Gause in experimental conditions when two species of Paramecium (P. caudatum and P. aurelia) were able to live together in culture.

(b) *Gause's Theorem.* Although Gause himself does not appear to have enunciated any "law" or "theorem" to summarize these results, "Gause's theorem" is the term very widely used to signify the theme which was, perhaps, first enunciated by Darwin.

"As the species of the same genus usually have, though by no means invariably, much similarity in habits and constitution and always in structure, the struggle will generally be more severe between them, if they come into competition with each other, than between the species of distinct genera."

A similar idea, without reference to taxonomic affinity, is contained in Nicholson's second corollary. To most ecologists the attribute of taxonomic affinity is not a sine qua non of "Gause's theorem," but only a corollary of it, and to this extent Hartley would appear to be wrong in equating Darwin's and Gause's statements. The theorem is usually rather loosely stated in some such form as "two species with the same ecology cannot co-exist." A recent comparison and a criticism of different formulations does not include Hutchinson's and Deevey's version that "two species with the same niche requirements cannot form steady-state populations in the same region."

Quite obviously the theorem as usually stated is not intended to be a rigid law, and its main implication that competition is more severe between animals whose ecological requirements are most similar can very easily become a circular argument. It could be made "water-tight" if three principles were followed: first, any formulation must be adequately supported by definitions of words such as "niche," "the same," "co-existence," etc. Secondly, it is patently obvious that taxonomically-related species often do occur in close physical proximity. It is therefore desirable to separate from the theorem itself, which is concerned with ecological affinity, the further corollary that ecological affinity implies taxonomic affinity, for this may not always be the case. For the same reason, and because spatial proximity may be no real index of ecological overlap, this part of the theorem needs much more careful consideration, if only because Gause's formulae apply only to species in competition under conditions where the total population is limited. Both Nicholson and later Crombie emphasized the latter point, that the theoretical results of competition will only ensue if a density-dependent, population-limiting factor is actually shared by the two species. Crombie emphasized the importance of Gause's third case when two species each have their own part of the environment in which each can increase more successfully than its competitor. He concluded that the elimination of one species by the other will not follow when both populations are held at a low level by some alternative morality factor so that competition too is at a low level.

Hutchinson also pointed out that continual fluctuations in environmental factors night effect the coefficients of competition in such a way as continually to reverse the theoretical outcome and, thus, the establishment of one species at the expense of the other would be prevented. He believed that such a situation may arise with the succession of phytoplankton algae in lakes. It is also widely believed that the marked climate changes which occur in temperate latitudes through the seasons may operate in a similar manner and allow the continuance of species in seasonal succession which might otherwise compete more severely if weather conditions remained constant.

Although a satisfactory formulation of "Gause's hypothesis" presents difficulties and does not seem to have been achieved, there can be little doubt that the fundamental idea behind it has been useful as a stimulus to research and especially to field observation. For instance. valuable contributions to the natural history of titmice and of coarse fish by Hartley would appear to have been stimulated and oriented by this theme and, especially among ornithologists, "Gause's hypothesis" has been most useful in providing a unifying principle, a feature all too rare in ecology.

The Lotka-Volterra-Gause theory of inter-specific competition has been attacked on the grounds that it contains a fundamental logical contradiction. The objection seems to be based on the definition of K, the saturation density, as defined by the "logistic equation"—

$$N_t = N_0 e^r \frac{K-N}{K} \quad \text{or} \quad \frac{dN}{dt} = rN \frac{N-K}{K}$$

In the logistic equation, which applies, of course, to a single species population, this saturation density is not exceeded by N, the number of animals present at any one time. On the other hand, K, as used in Gause's equations, is exceeded by the number of individuals in one population plus the equivalent number of units of the competing population. This, claim Andrewartha and Birch, is inconsistent because it implies that the maximum density can be exceeded by two species but not by one species alone. The inconsistency is due to the critics who compare the two sets of equations and not to the separate originators of the two sets. There is nothing magical about the saturation value, K, which is simply the upper asymptote of the logistic curve to which the single species population trends. It is well known, both in theory and from experimental results such as those of Gause and of Park that, owing to statistical variations this "upper asymptote" is, in fact, frequently exceeded both in single species populations and in experiments involving more than one species. Arbitrarily to give different definitions of K in the two cases obviously results in an inconsistency, and is not a valid method of argument. A more mathematical refutation of Andrewartha's and Birch's argument has also been published.

The theoretical implications of statistical study of dispersal rates of populations do not yet seem to have been applied to competition problems. Skellam demonstrates that competitors with different dispersal rates may continue to coexist if that with a faster dispersal rate has a lower reproductive rate. This indicates the need for a further modification of simple models, a need which has been met to some extent by Williamson's theory of competition, which incorporates stochastic influences. It is evident, however, that many of the predictions of simple theory are borne out by experimental studies and that there is room for more theoretical work which will help to bridge the gap from the very oversimplified models which have been used so far.

(*c*) *Experimental Results.* Gause himself was the author of many experiments on competition and he succeeded in demonstrating a number of the theoretical predictions in experiments on micro-organisms. The Yeasts Saccharomyces cerevisiae and S. kephir were grown together with a limited daily ration of sugar, Paramecium aurelia and P. caudatum and Glaucoma scintillans among Protozoa were used as competitors. One of the most interesting cases was that of the two Paramecium species competing for the same yeast food; P. aurelia can survive at a greater depth in the culture tubes employed because it can tolerate higher concentrations of waste metabolites. P. caudatum has a superior competitive ability in less-crowded conditions and thus survives better near the surface of the tubes. This demonstration that even a small glass tube can encompass two "niches" is particularly significant in relation to Hairston's analysis of five species of Paramecium in a 40-ft stretch or river.

A very extensive series of experiments was performed by Park using cultures of Tribolium confusum, the flour beetle, to which was later added T. castaneum for experiments on competition (Plate III). The outstanding feature of Park's later experiments is that these were continued for very long periods (four years) and were replicated many times. In this way Park was able to demonstrate three important principles—

1. Even apparently simple competition situations can, in reality, be complicated. In earlier work, for instance, the presence of a Sporozoan parasite, Adelina, was neglected and under these conditions the outcome of experiments in which small quantities of flour were started with both species of Tribolium in equal quantities was always that T. confusum replaced T. castaneum. However, when a technique of rearing the beetle free of Adelina infection was developed it was found that the situation was completely reversed. This was apparently due to T. castaneum being much more susceptible to the parasite, as was shown also by single species experiments in which its total population size was much less in the presence of Adelina, while that of T. confusum was reduced to a smaller extent.

The complexity is further shown by the persistence, even in single species populations, but more so in the competition cultures, of fluctuations in total numbers and of the numbers in the different life stages—egg, larva, pupa and adult. These fluctuations, which presist through the longest cultures of 70 generations and more, have not so far been analysed; they presumably involve such density-dependent factors as cannibalism and interference with breeding, feeding and copulation of each species as well as intra-specific factors. The implication of the complexity of the Tribolium experiments for the interpretation of field results is, of course, that it would be naive to expect to be able to demonstrate the operation of the simple mathematical situations by simple inspection except in the most exceptional circumstances. An impression of the manifold possibilities from the point of view of invertebrate herbivores, in the case of soil, for instance, can be gained from a review of the enormous number of fungi which occur there.

2. Park's repetition of his experiments many times, over long periods, enabled him to demonstrate that the outcome of a particular competition situation, beginning with certain definite proportions of the two species in a fixed amount of flour medium, is not rigidly fixed, but, like any other population system is subject to statistical probability. He was forced to express his results in the form that a given outcome has a given probability of occurring. Again the implication for field-data interpretation is obvious and important.

3. The great length of Park's experiments in relation to a life-cycle enabled him to show that the outcome of competition is frequently not decided for many generations; that again affects the probability of a given outcome occurring and might well permit, under more "natural" conditions, the survival together of two species although theoretically one of them was doomed to be eliminated by the other.

Park's later work has been mainly concerned with the analysis of the mechanism of population dynamics in the cultures, and in the relations between the outcome of single species population experiments and those in which the same species are in competition. Thus, the equilibrium population level for two species of Tribolium living separately was determined in six different combinations of temperature and humidity. The two species were then cultured together in the same sets of environmental conditions, but in competition. Under these circumstances, Park found that in some case the species which alone maintained the higher density in a particular environment was also the "winner" in competition, while in other cases the reverse was the case. Experiments of the effects of a wide range of initial ratios of the two species, but under equally favourable conditions for both, gave the surprising result that, in 92 out of 97 cases T. castaneum "defeated" T. confusum, which event was heralded by the disappearance of all females in the latter species. Many further questions are posed by these experiments which serve to illustrate the great complexity of superficially simple systems.

Among many experiments which have been done with grain-insect populations, those of Crombie covered a considerable range of competition phenomena. Crombie worked with the beetles Oryzaephilus surinamensis and Rhizopertha dominica as well as Tribolium confusum and with the moth Sitotroga cerealella. O. surinamensis can normally survive in the presence of the other species owing to its rather different ecology. In a Rhizopertha dominica—S. cerealella system the only part of the life cycle at which competition appears to operate is in the larval stages, when space for development is short. Lack of food or oxygen or "conditioning" of the medium by excreta did not act as density-dependent factors of any importance.

One of the most interesting of Cromble's results was obtained from a T. confusum-O. surinamensis system. In single species populations of both these beetles the rate of increase of the population is determined by the balance between oviposition, rate, and rate of cannibalistic eating of the eggs and pupae by the active larval and adult stages. This, in a dense population. accounts for over 99 per cent of the mortality and also determines the maximum population density reached. In wheat both species live together but in flour T. confusum survives on account of its more voracious cannibalism. When, however, glass tubes of a suitable bore are mixed with the flour, O. surinamensis is able to survive in addition to T. confusum because the larve are able to find refuge in the tubes. Crombie claimed that this small change in the medium is, in fact, not only able to provide "niches" for two species instead of one. but that Rhizopertha dominica can also survive there as well.

Crombie's work, which suffers in most cases from being based on experiments with very few replicates, preceded Park's later long-term experiments and was of much shorter duration; it is possible that he might have had fewer examples of "Gause's third case" to report had his experiments lasted longer.

Crombie fitted his experimental results to Lotka's and Volterra's equations and calculated the constants involved. He stated that the basic assumptions of these equations, namely

(a) that the potential rate of increase remains statistically constant, and (b) that factors inhibiting increase are linearly related to density are, in fact, justified in his experiments. He was careful to stress, however, that this is frequently not the case and that one should not accept these assumptions without first testing them in any given case. In addition to his experimental work, Crombie also contributed a useful survey of competition experiments up to that date.

In experiments on intra- and inter-specific population systems of water fleas, Simocephalus vetulus and Dephnia pulicaria, Frank was unable to determine the population-limiting factors operting in single species populations, but the outcome of mixed cultures was always the same, namely that S. vetulus was always eliminated and D. pulicaria survived. One of the most interesting aspects of Frank's work was his analysis of the effects of density on birth rate and death rate, the changes in which accounted, at least in part, for the outcome of competition. Another contributory factor, which reveals a new extension of the Gause's theorem principle, is the greater inhibitory effect of high densities of D. pulicaria on its own populations than on those of S. vetulus; on the other hand, S. vetulus populations are almost equally harmful to populations of both species. Frank, once again, was able to detect some differentiation of even the small culture bottles into two "niches," S. vetulus tending to concentrate at the sides and bottom of the bottle and D. pulicaria more in the centre. These habits correspond with their distribution in their natural habitats.

Although a great deal of work has been done on the above level of description, the subject has hardly begen to develop methods of causal analysis. Park's attempts to relate fecundity and mortality factors to success in competition which have already been mentioned, and Birch's demonstration that r was, in some cases at least, related to success in competition are important pioneer developments.

In the present state of ecological science it would be premature to draw any but very tentative conclusions from the results of experimental studies on population systems. The application of demographic analysis and of a more detailed treatment of such systems has only recently begun, and it is almost certain that great advances will be made in this field in the near future. The theoretical work of Lotka, Volterra and others has undoubtedly been an important stimulus to research, even when the basic assumptions have been too simple and arbitrary, but the need is now appearing for further methods of analysis and there is no doubt that valuable work could be done by mathematicians.

Metabolic Aspects of Competition

A consideration of the relative metabolic activities of competing populations is a useful and valid way of assessing their relative importance. The very approximate analysis of grassland communities clearly indicates the value of this approach to field problems, but this aspect of competition appears to be completely unexplored in the experimental field. It might be suggested that the great complexity of the approach just discussed, using data from population dynamics and demography, would be further complicated if combined with studies of metabolic activity. That the reverse may be true is indicated by the success of Frank's and of Slobodkin's work, from which rather simple generalizations emerged in the case of single species population.

Again, in so far as metabolic rate is likely to be a major factor in a high intrinsic rate of increase, the rather simple relation between this and success in competition which was discovered by Birch poses questions which are an obvious challenge to the experimenter. The complete analysis of such experiments may demand the compilation of "metabolic life tables," but the existence of useful generalizations at the inter-population level could be revealed by quite simple experiments.

16

Marine and Terrestrial Environments

Primary production refers to the production of organic matter by photosynthesis. As the principal producers of all of the Earth's carbohydrates, terrestrial plants have been the traditional focus of biogeograph. In this chapter the focus will be on biogeographic processes by describing both marine and terrestrial plants.

The differences in physical composition of the surface of the Earth provides a compelling reason to subdivide the planet into two systems: the terrestrial environment and the marine. This chapter broadly outlines and compares these two major environments. It reviews potential primary productivity and its distribution. Finally, it considers some of the biogeographic implications of the interplay between the marine and terrestrial environments of the Earth.

The Earth's two Major Environments

The marine and terrestrial environments differ in several respects. The most obvious difference relates to the supporting medium and its relative stability. The medium of the marine environment is liquid and in constant motion. Water transports both physical properties and biota from one place to another. As a result, short-lived plants dominate in the marine environment. These plants have evolved to cope with constantly changing physical conditions of the ocean. In contrast, the terrestrial environment has a soil which has provided plants with a far more stable medium. The stability is due to the fact that physical properties and geographic positions persist over long periods of time. The terrestrial environment has promoted the evolution of long-lived plants with specialized multicellular appendages, such as root systems and above-ground structures. The also have adaptive features for the particular physical conditions of the place where they reside.

The temporal and structural nature of the organisms of the two environments is directly related to the type of organic compound which prevails in the sea and on the land. Proteins are the dominant compound of life in the seas and carbohydrates are the major compound of land life. The type of organic compound dictates the basic operating differences of the two environments. The protein-based life of the seas is capable of growing rapidly but it does not have an ability to store energy for long periods of time. On the other hand, the carbohydrate-based life of the land is relatively slow growing but able to store energy for extended periods of time.

There are several ways in which the physical properties of the marine and terrestrial environments influence the development, structure, and styles of existence of organisms. For example, the influence of gravity plays a significantly different role in the lives of organisms of the sea and land. In the sea, pelagic organisms exist within an amniotic-like fluid which supports and sustains life. The fluid buoys the mass of the organism and is the source of the required minerals. In contrast, terrestrial organisms must also have a structural framework that supports both the weight of the fluid and of the structure itself. This structure must be able to withstand the rigorous strain of the pull of gravity. Furthermore, the structure requires an added expenditure of energy that is borne by terrestrial organisms for development and maintenance.

As a result of the different environments, the energy requirements for locomotion are distinctly different for the organisms of the marine and terrestrial environments. In the seas, many organisms move with the fluid and require ltttle or no energy. Even in active vertical or horizontal movement, much less energy is required by sea organisms than by land organisms. On land, each step that an animal takes requires the lift of both the mass of the structure and the mass of the self-contained fluid against the force of gravity. The majority of none-marine animals have adopted flight as a means of locomotion and have thus confronted the difficulties of tetrestiral travel.

Besides influencing energy expenditures by organisms of marine and terrestrial environments, gravity also plays another role; it influences the nutrient status of each environment. Nutrients do not accumulate in the surface waters of the ocean. Instead, due to gravity, they tend to become depleted from the surface waters and sink to the ocean bottom. It has been estimated that without mixing and oceanic circulation, primary productivity of the marine environment would cease in less than a year due to the depletion of the nutrient supply at the ocean surface. In contrast, the soils of many terrestrial sites tend to accumulate nutrients. During the life cycle of terrestrial plants, some of the production undergoes decay and becomes part of the soil medium. Over time, an accumulation of nutrients may occur which in turn enhances the stability of the terrestrial environment. In essence the marine and terrestrial organisms and their supporting environments differ markedly from each other. The next two sections provide additional details on the physical basis of the biogeographic patterns of organisms of the two major environments of the Earth.

Physical Properties of the Marine Environment

Physical properties of water, such as high heat capacity, latent heat of fusion, thermal conductivity, and dissolving power, are characteristics which are important to organisms. Water is densest at 39°F (4°C). Becoming lighter as it freezes, ice forms at the surface of water and insulates the underlying water from continued rapid cooling that results from evaporation. The Earth would not be a moderately warm environment if water behaved like other substances.

The temperature extremes of the ocean are relatively narrow compared to the extremes within terrestrial environments. Temperatures range from about 28°F (−2°C) at the ocean bottom near. Antarctica and the Norwegian Sea, to about 86°F (30°C) in shallow waters at the Equator (Smith 1976). Also due to its salinity, ocean water does not freeze as readily as freshwater lakes. The freezing point and temperature of maximum density of sea water both decrease with increasing salinity. At the normal salinity of 32-38 parts per thousand, vertical convection continues until the whole body of water has been cooled to the lower freezing point. This extra cooling requirement causes a lag in the formation of marine ice compared with freshwater ice. This phenomenon contributes to the climatic dynamics operating between marine and terrestrial environments.

The relative proportion of elements that make up the major constituents of 99.9 per cent of the sea water is nearly constant. Because of the constancy of elements it is possible to measure the salinity of sea water by measuring any single constituent of the sea. The total quantity of dissolved salts, or salinity, is variable from one place to another. Salinity, used to be calculated by measuring the quantity of chlorine, but now electrical conductivity is the most commonly used measurement of salinity.

In addition to the major constituents, sea water contains virtually every other natural element, but they are in small concentrations and highly variable from one place to another. The surface waters, where photosynthesis is most active, tend to be depleted of dissolved nutrient elements, while deep waters are enriched in these elements through settling of detrital matter (Smith 1976). In view of the limited nutrients at the surface, mixing and circulation of ocean water become critical to the understanding of the biogeochemical system which supports organisms of the sea.

TABLE 16.1

The Principal Constituent Elements of Sea Water (After Smith 1976)

Ion	*g/kg of water of salinity 35%*
Chloride	19.353
Sodium	10.762
Sulphate	2.709
Magnesium	1.293
Calcium	0.411
Potassium	0.399
Bicarbonate	0.142
Bromide	0.0673
Strontium	0.0079
Boron	0.00445
Fluoride	0.00128

There are many types of perturbations of sea water, including internal movements which resemble the massive hurricanes or tornadoes of the atmosphere. One source of these movements is the Gulf Stream. As it meanders along the east coast of North America, large loops of water break off and form either warm-or cold-core "rings", or mossive eddies of water which move across the open ocean. There are about ten cold-core rings and three warm-core rings at any one time in the North Atlantic Ocean. These eddies of water extend from the sea surface to a great depth in the ocean. Cold-core rings last two to three years while warm-core rings last about a year. Rings have been recorded as coalescing and breaking up. Eddies and rings of water have been described in many places and apparently can transfer large pools of water over long distances. The rings and eddies of water play an important dynamic role in mixing and creating variety in what might otherwise be stagnant marine waters such as the Sargasso Sea.

Marine waters of different temperature and salinity intermingle and thereby help to mix the marine waters. The intermingling has been likened to the meeting of fingers in a hand clasp and referred to as salt fingers. Atmospheric storms apparently churn up the near-bottom cloudy layer (or nepheloid layer) in the deep ocean. The ocean has many processes which operate to stir and mix water and sediments. These types of findings counter the notion gained from casual observations at the surface that the ocean is mostly a large and not very dynamic pool of water.

The distribution pattern of marine organisms is closely intertwined with the physical properties and dynamics of water. Several properties of water pose paradoxes for marine organisms. For example, heat and dissolved oxygen and carbon-dioxide are important in growth processes. However, the amount of gas absorbed by water is inversely related to its temperature. For that reason, warm tropical waters tend to have a low level of absorbed gases, while the cold waters of high latitudes tend to have high quantities of absorbed gases. Tropical waters are suitably warm for optimum growth but their low levels of absorbed gases, such as oxygen, limit the number of organism that can be supported. For example, the level of oxygen drops from a supersaturated 5 ml/l to a nearly anoxic 0.2 ml/1 within the top 150 m of tropical waters. At high latitudes, plants are not as limited by dissolved gases as they are by low temperatures.

A second property which poses paradoxes for marine organisms is that photosynthesis can only take place within a narrow range of light intensity. Light intensity in water is limited by adsorption and turbidity. Water àbsorb light energy and stores it as heat. The major factors which influence turbidity are terrestrial runoff, bottom mixing on the continental shelf, and the quantity of living and dead plankton, or seton. Table 16.2 compares the light penetration of sea water at near-shore and offshore sites. Note that the penetration of light into water rapidly decreases in the top 20 m. The nearshore areas have the greatest potential for plant growth because of mineral availability, however, the increased seton contributes to the reduction of the penetration of light within these same waters. Also, day and night periods differ with latitude. In the Tropics there is a diurnal pattern in the light, while there is a seasonal gap in light within Arctic and Antarctic latitudes.

TABLE 16.2

Comparison of Light Penetration Depth (m) of Sea Water at Nearshore and Offshore Sites

Per cent Light absorption	*Nearshore*	*Offshore (16 km)*
50	2	4.5
75	4.5	9.0
90	8	17.0

Finally, as described above, nutrients at the surface of the water are limited because of gravity. Organisms absorbs nutrients, die, and sink. The loss of required nutrients, especially available phosphorus and nitrogen, has been found to severely limit the growth of marine phytoplankton. Successive layers of organisms live on the organic detritus as it settles toward the sea floor. Eventually, the materials arrive at the sea floor where some elements may be decomposed by microorganisms and become available for reentry into the marine system.

Only a few regions of the marine environment support a large biomass. Meteorological conditions, the dynamics of the atmosphere—ocean system, and the geographic arrangement of mountain ranges and coasts as described earlier dictate most of those regions. Four regions are particularly productive. Within the middle latitude area of the Earth, coastal estuaries are exceptionally productive. As part of the interplay of polar and tropical air masses, the seasonal flooding cycle operates to transport, via stream systems, substantial quantities of sediments and organic matter from the floodplains to the estuary. Mid-latitude estuaries undergo wide swings in productivity because of seasonal temperature fluctuations and variations of the amount of runoff.

A second type of condition that dictates high productivity is turbulent regions of the ocean where warm and cold currents meet. One such place is the Grand Banks. This is the site of turbulence created by the collision of the Gulf Stream and the Labrador Cold Current. The turbulence brings nutrient-rich waters from the sea floor to the surface and supports a locally large biomass.

A third phenomenon that creates a productive region is where upwelling of cold nutrient-rich waters occurs along the west coast of continents as a result of the Ekman Spiral. These areas are subject to large swings in productivity as a result of El Nino—Southern Oscillation phenomena. At irregular intervals that vary from two to ten years, the Trade Winds are unusually weak over the tropical Pacific Ocean. As a result, upwelling becomes disrupted. When upwelling ceases, nutrient sources are no longer available and the growth of marine plants virtually ceases, causing devastation to the biological systems of the crystal zones of Ecuador and Peru. In the 1982-83 El Nino, chaos occurred within extensive regions where organisms depended upon upwelling. But, in addition to a collapse of the fisheries of Ecuador and Peru whole bird populations were reported as abandoning en masse Christmas Island in the equatorial Pacific.

A fourth, extremely productive region in the marine environment, is the polar seas. During polar summers, glacial melt water enters into the high-latitude oceans. Melt water carries large quantities of nitrates and silicates in suspension which are a major limiting nutrient for the algae inhabiting these regions. In the polar environment the biomass undergoes a large seasonal fluctuation.

Only a small percentage of the marine environment has the suitable combination of physical conditions to support a large biomass. The distribution pattern of production of biomass found in the marine environment. Productivity is based on the growth of single-celled algae that cannot store large reserves of carbohydrates. Instead they rely on rapid reproduction and growth to replenish and maintain their mass.

Physical Properties of the Terrestrial Environment

The terrestrial environment is much more amenable to plant structure ramification and productivity than is the marine environment. In contrast to the marine environment, it consists of a relatively evenly matched combination of solids, liquids, and gases. Terrestrial plants are in contact with all three phases of matter: their roots are enveloped by minerals, liquids, and air spaces of soils, and more or less free from the paradoxical physical constraints which characterize the marine environment. Terrestrial plants have the possibility of obtaining a near optimal mixture of gases, water, minerals, and light for growth in large geographical regions of the land areas of the Earth.

The supply of plant requirements in terrestrial environments is reasonably direct. The penetration of light, for example, is not limited by the turbidity or absorption of the air immediately surrounding the plant as much as by the competition resulting from the growth of other plants. For example, in the tropical rainforest, the tallest trees are often 100 m high. There are several layers, or stories, of canopy. Each story absorbs light and thereby reduces the penetration of light. At the surface of the ground, light may be less than 1 per cent of the light at the top of the canopy. Light is being absorbed directly by the plants, and not by the medium as in the case of the marine environment. Thus, the plant structure, which has been produced by the gradual accumulation and storage of carbohydrates, plays a key role in the productivity of the terrestrial environment.

The major physical limitations for terrestrial plant growth is meteorological, or more specifically, the delivery and supply of energy (light and heat) and water. Global productivity of terrestrial plants coincides directly with the combined delivery of heat and water. The extreme range of terrestrial temperatures between the Tropics and the Poles is around 295°F (146°C). Libya has recorded a high temperature of 136°F (58°C) and Antarctica has recorded a minimum temperature of −126°F (−88°C). Likewise, precipitation varies from near zero in some rain shadow and west coast deserts to above 300 cm in some mountain ranges. In general, the Earth's atmospheric system tends to distribute heat and precipitation in a manner which promotes plant growth over large land areas. With few exceptions, this map closely resembles Koppen's map. This isreasonable because, as has already been discussed, Koppen's map was based upon vegetation response to heat and precipitation.

Exceptionally productive areas of the Earth are floodplains and deltas. They occupy only 3 per cent of the Earth's land area, but they are responsible for over 12 per cent of the total terrestrial productivity (Rodin *et al.*, 1975). The productivity is the result of an adequate hydrological regime and the continual input of geochemically fresh minerals from terrestrial sources. These areas are so important that they merit special attention.

Biomass studies provide evidence for the idea that forested lands are the major site for photosynthesis on the Earth. Instead of being dominated by short-lived, single-celled plants, the terrestrial environment is the realm of long-lived, multi-cellular plants. The terrestrial plants are enormously varied and have become adapted to almost all of the diverse range of terrestrial habitats. There are over 14,000 times more reserve carbohydrates in the terrestrial environment than there are in the marine environment. While the terrestrial environment is less than 30 per cent of the size of the marine environment, the structural elaboration of land plants more than compensates for the reduced area of the land they inhabit. The structures of trees support a leaf surface area that is many times greater than the land area they inhabit. This fact, plus a far more favourable combination of physical conditions for plant growth explains the enormous productivity of the Earth's terrestrial plants. The total land area is estimated to outproduce the total sea area by 25-50 per cent.

Soil as a Major Evolutionary Development

Soils are an integral part of the terrestrial system and that reason deserve special emphasis. One way to humus soils is to view them in relationship to other component of the Earth's biogeochemical system. Definition of soils was exposed by V.V. Dokuchaev at the 1893 Chicago World's Fair. Using Dokuchaev as a starting point, Hans Jenny refined the ideas over a period of 40 years (Jenny 1980). His definition is as follows:

$$soil=(cl, o, r, p, t, \ldots)$$

where:

cl=climate

o=flora and fauna (organisms)

r=topography (relief)

p=parent material

t=time

. . .=other factors such as dust influx, coastal salt spray, or fires not contained in cl and o.

This equation introduces the major soil-forming factors but does not assist with understanding of their relationship with one another. In order to provide some insights into the dynamic nature of soils, two examples will be developed, one involving the interrelationship of organism (o) and time (t), and the other involving climate (cl) and organism (o). The parent material (p) and topography (r) componcnts will be described in more detail in the next chapter. For purposes of explanation, soil-forming factors are dealt with here in combinations of two. In reality, these factors cannot be separated from each other.

Soil Evolution: Organisms and Time

One means of understanding soils is to consider the impact that organisms have on them. Soils are in part a product of the metabolism of the organisms which inhabit them. One could expect that during various geologic eras, soils would be changed as a result of the kinds of organisms, and probably more importantly, the metabolic qualities which plants and animals imparted to them during various geologic eras.

The earliest soils were inhabited by microorganisms such as cyanobacteria (bluegreen algae), fungi, and bacteria that are similar to marine microorganisms. The products of their metabolism become constituents of the soil. There are a variety of ways in which microorganisms exist which characterize their relationship with one another and with their environment. Some microorganisms are producers, others are consumers. The consumers may be herbivores, carnivores, or they may eat the remains of other organisms as scavengers. The relationships are as complicated and varied as any found on the Earth. Most of the biochemical pathways exhibited by microorganisms probably evolved very early in the marine environment. The terrestrial environment provides an enormous range of conditions, and microbes have demonstrated anazing tenacity and capacity to adapt to virtually all conditions on the Earth.

Microorganisms are capable of surviving the harshest terrestrial conditions. For example, small colonies of fungi, as well as bacteria, have been found existing on or beneath the surfaces of rocks from many of the most arid deserts of the Earth, including Antarctica. With the evolution of the multicellular plants and animals of the terrestrial environment, many new habitats were opened up for microorganisms. For example, roots provided a habitat which microorganisms fully exploited. Some bacteria live only on the growing up of the root and apparently assist the root in moving through the soil. Other bacteria exist within nodules of the root and assist with the supply of basic plant macronutrients such as nitrogen. Still another organism, fungus, lives both within the root cell and in the soil and acts as an agent for the transfer of nutrients into the plant. These root-soil inhabiting fungi are called mycorrhiza. The nutrients brought into the plant by mycorrhiza are conducted into other structures of the plant such as leaves where they are converted into various organic substances which enter into the plant structure. As leaves die they fall back to the soils where they become food and substance for soil microorganisms. The nutrients are thus available for plant uptake and recycling. But that is only a small part of the relationship between multicellular organisms and soil microorganisms.

From the perspective of plants and soils, the intestinal tract of animals is an extension of the soil environment. Larger animals provide a mechanism for the rapid movement of large quantities of organic matter and nutrients from the standing mass of plants to the soil. The speed of the movement depends upon the type of metabolism of the organisms that are involved. During the Mesozic Era, the amphibian and reptilian organisms dominated the land. As ectotherms, these animals consumed food required for growth, sustenance, and reproduction, but little else. They consumed food in amounts which were more or less dictated by the size of the organism and the temperature of the environment. They ate infrequently and thus land relatively limited impact on plants. Likewise, at all stages of the nutrient clain, the same slow metabolism was in evidence. Ectotherms are generally efficient at the conversion of plant matter into protein. However, the biological system as a whole undergoes a relatively sluggish movement of materials from soils through the plants (via nutrient uptake and conversion into plant matter), to animals and back (via the intestinal tract) to the soils.

During the late Mesozoic and throughout the Cenozoic another metabolic system evolved, first among some of the dinosaurs but especially among birds and mammals. Endothermic organisms produce internal body heat. Heat production requires a substantial increase in the amount to food intake. When the warm-blooded animals began to consume large quantities of the productivity of plants (leaves, fruits, seeds, and so on), every component of the system also underwent change. The overall effect was an increased pace of the Earth's biosystem that was revolutionary to the Earth's plants, animals, and soils. The warm gut is a nutrient-rich environment where microorganisms act much more efficiently to break down plant materials. As a result of high rates of consumption and rapid decomposition during passage through the intestinal tract, greater quantities of exereta are produced. The soil benefits from the increased output of organic matter and nutrients. The benefits are passed on to soil microorganisms which feast on the large increase of organic material. Those plants which adapted to the richer soil environment—and the increased herbivorous activity resulting from endothermy—became dominant in the Earth's pattern of life. Although the efficiency of conversion of plant matter into protein by endothermic animals is significantly lower than that by ectothemic animals, the overall biosystem benefited from the increased metabolic rate of the warm-blooded animals. In this way, the constituents of soils have evolved as a result of the deposit of residues in the soil from countless generations of organisms (Nikiforoff 1938).

It is often stated that the conversion of plant matter to protein is less efficient in the terrestrial environment than in the marine environment. This is probably due to the energy requirements of terrestrial existence and the result of the large number of mammals that inhabit the terrestrial environment compared to the marine environment. However, it seems likely that conversion rates of plants to protein, which is generally based only on conversion by large herbivores, are underestimated in the terrestrial environment. The underestimation would be due to the fact that soil micro—and macroorganisms are excluded from the tally. These small soil organisms are the largest part of the animal mass of the terrestrial environment and should be recognized as a constituent part of the terrestrial organisms which convert carbohydrates into protein.

Pedogenesis: Climate and Organisms

Pedogenesis is an approach to soil classification which takes into account the interaction of climate and organisms in the development of soils. This approach has been criticized for being too theoretical to define the actual field characteristics of a given soil. However, the pedogenic approach provides a framework for developing an overview of the geographic pattern of the Earth's soils. The following relationships are key concepts the pedogenic approach to classification.

(a) The rate of decay of organic matter is directly related to temperature. At low temperature there is a preservation of organic matter, while at elevated temperatures, organic matter is rapidly consumed by microorganisms.

(b) The products of metabolism vary enormously, depending on the type of soil microorganism in the soil, which in turn is determined by the soil atmosphere. Water is especially important in determining the soil atmosphere. Under saturated conditions, anaerobic microorganisms dominate the soils, while aerobic microorganisms dominate the soil in unsaturated conditions.

(c) The movement of minerals within the soil is a function of the balance between precipitation and evaporation. In a humid climate, soil cations such as calcium, magnesium, and potassium tend to be flushed from the soils. On the other hand, in dry climates, the same soil ions tend to move upward through the soil profile and accumulate on or near the surface.

(d) Soil pH is inversely related to the number of cations present in the soil. Leached soils tend to have a low pH while unleached soils tend to have a high pH.

(e) The types of minerals present in a soil are related to climate in that temperature and precipitation determine pH and chemical reactions that will take place. Thus, dry, cation-enriched soils tend to have minerals such as montmorillonite, while humid, leached soils tend to have minerals such as bauxite.

Soils which support large quantities of vegetation are located in humid environment. Within the tropics, under warm conditions, laterization (or ferratillization) is the pedogenic process which forms the soil profile. At the surface of a lateritic profile, almost no accumulation of organic matter occurs. Leaching is intense and various iron and aluminium sesquioxides remain in the surface, while silica and most cations are removed from the soil profile.

Under humid conditions within mid-latitude climates, where there is a seasonal fluctuation of temperatures, soil undergo podzolization. During the warm season, growth is rapid, but during the cold season, decay processes are retarded and an accumulation of organic matter takes place on the soil surface. The overall result is the development of a soil profile which has an organic-rich horizon on the surface. The organic matter tends to undergo slow decay and form some acids which percolate through the upper horizons of the soil and leach them of cations. The cations tend to accumulate in deeper horizons and move out of the soil profile into the water table.

In high latitudes, the cold season is so extended that little or no decay occurs, leading to gleivation. In this condition organic matter accumulates at the soil surface. The thick horizons of organic matter are called peat. Peat absorbs and retains large quantities of water but excludes air. The cations are strongly leached from the soil and the minerals left behind include iron in its reduced or oxygen-free form, Under anaerobic conditions, the iron-containing clary minerals appear blush-gray. Thus, the term gley is applied when these soils are first exposed. Under marshy conditions, nitrogen is depleted. Carnivorous plants such as the pitcher plant (Sarracenia), sundew (Drosera) and Venus's-flytrap (Dionaea muscipula) are plant adaptations tothe nitrogen deficiency.

In semi-arid climates at low-to mid-latitude locations, calcification is a predominant pedogenic process. In dry climates, the growth of vegetation is extremely slow. An accumulation of calcium, magnesium, and other cations takes place at the soil surface when potential evaporation becomes equal to, or exceeds, precipitation, During droughty weather, ground water rises into the soil via capillary action. The water evaporates and the dissolved cations remain behind in the upper horizons of the soil. These soils also tend to have an accumulation of organic matter in the upper horizons due to a lack of water for the support of microorganisms of decay.

Salinization occurs under conditions of extreme aridity. Like calcification, cations are brought to the soil surface through capillary action. However, in the case of salinization, a pooling of cations takes place. An arid landscape with basins is a physical setting which enhances the process of salinization. Basins exist where there is too little precipitation to accumulate enough water to carve a drainage channel to the sea. Thus, in a basin, dissolved minerals from the surrounding landscape accumulate and, under evaporation, create a highly saline (*i.e.*, chloride-enriched) soil. Plant growth is generally restricted to a few extremely salt-tolerant species on these soils.

The response of soils to climate, and to the way in which precipitation and temperature influence organisms, and the movement of minerals in the soil profile, are all characteristics of soil pedogenic regimes. The pedogenic approach to soil classification provides a global overview of soils which matches the scale of the Koppen climatic scheme as well as the physiognomy scheme developed by botanists and phytogeographers. In the case of soils, the map of pedogenic regimes is a general overview of the biogeochemical regions of the Earth. It groups the Earth's pattern of productivity according to major zones of dominant nutrients of the biogeochemical cycle.

Plants, climate, and soils all have similar map units. That is, any one of these three major components shows a pattern major the other component parts of the relationship. This is reasonable because all of these systems are integrated with each other. A map of a part of the relationship produces a pattern similar to the whole relationship.

Productivity and Structures of the Marine Environment

Although large areas of the marine environment are characterized by low productivity and few species, there are some important exceptions to that characterization. The exceptions provide insights into the factors which contribute to increased productivity and species richness. Structural complexity is one factor which serves to increase productivity and acts to support large number of species per unit area. In the marine environment, there are such large difficulties in the establishment of structural complexity that within most areas of the ocean, organisms have been unable to overcome them. In shallow areas where the bottom is rocky and relatively stable, however, structural complexity has been possible. Both corals and kelp succeed in developing structural complexity that surpasses all other marine communities.

Reefs

Coral reefs are the most massive structure ever produced by organic activity on the Earth. Reef-forming organisms include a number of species that deposit calcium carbonate. Stony corals are the predominant reef-building organisms in shallow, tropical seas. In addition, calcium carbonate is also deposited by protozoan foraminiferans, coralline algae, serveral types of molluses, echinoderms, and tube-secreting annelid worms. In general, a community formed by one or more of these calcium carbonate depositing organisms is called a coral reef. Coral reefs are the oldest known biological community on the Earth. Archaeocyathidae formed barrier reefs during the Cambrian, over 500 million years ago, in western Antarctica and eastern Australia (Hurley 1960). Coral reef communities have spanned the Phanerozoic age.

The coral is a small coelenterate animal with tentacles adapted for feeding on zooplankton in the water. Each coral secretes a calcareous skeleton. Lodged within the tissues of the coral animal and the calcareous skeleton are numerous algae (Zooanthellae) and small plants. The function of the algae and plants within the coral community has been a topic that has aroused much curiosity, C. M. Yonge conducted early studies on coral colonies that were raised in large darkened tanks and fed ample diets of zooplankton. Under these conditions the coral colonies thrived. Later studies showed that the supply of zooplankton in the infertile tropical seas was inadequate to account for the rapid growth and large population of the coral colony. Gradually, the complexity of the symbiotic relationships between coral and algae became better understood: algae embedded in the coral community provide the food and oxygen supplement necessary to account for the energetics of the coral colony. The coral in turn provides the nutrients for algal growth. This conclusion has been reinforced by many studies which have demonstrated that food manufactured by algae is absorbed by coral, and further, that algae are important in the formation of the skeletal structure of the coral community.

Reef-building corals generally grow in waters 100 m deep or less and require water temperatures between 73°F (23°C) and 84°F (29°C) and thus thrive mainly in tropical and subtropical areas. There are some reefs found in northern latitudes off the west coast of Norway. Compared with most corals, these occur in unusually deep water and are part of the plethora of organisms which accompany the Gulf Stream. They are not symbiotic corals and grow more slowly than their tropical counterparts.

Coral reefs are massive structures which range in thickness from tens to hundreds of meters thick. They have developed as a result of the growth of coral colonies over hundreds to thousands of years.

The two most important regions of corals are the Indo-Pacific area from the east coast of Africa through the islands of the western Pacific to Hawaii, and the western Atlantic region from Bermuda to Brazil. The Indo-Pacific corals have been studied the most intensively and that region has long been regarded as the superior site for the most vigorous coral development. Recent studies on the long ignored western Atlantic reefs have revealed that during the Holocene these coral structures have developed at twice the rate reported for the Pacific coral structures.

Three types of reefs have been identified; fringing reefs, barrier reefs, and atolls. A reef growing in shallow water along the shore of the coast is called a fringing reef. It has a site of active coral growth on the seaward side called the reef front. Behind the reef front is a reef flat which consists of dead coral, coral sand, mud, and debris. The reef flat is exposed during low tide.

The barrier reef is very similar in structure to the fringing reef. It differs from the fringing reef in that it is much farther from land. Hence, the barrier reef can be distinguished from a fringing reef because of the presence of a lagoon between the reef flat and the shore. The most famous reef is the Great Barrier Reef off the northeast coast of Australia. It extends about 2000 km along the coast and occurs up to 160 km from shore in some places. Over 200 species of corals make up the Barrier Reef. There are numerous gaps and inlets along the entire length of the reef.

A circular or horseshoe-shaped reef which surrounds a lagoon is called an atoll. In an atoll there does not appear to be a connection between the reef and land. Coral alone makes up the materials of the above-water segments of the land. In all cases, coral reefs create a living breakwater. In the case of fringing and barrier reefs, the breakwater protects the coasts from wave-generated erosion. On an atoll, the living part of the reef provides materials for reef flat accumulation as well as serving as a breakwater for the atoll.

In tropical waters the most common corals are colonial cnidarians called stony corals. These organisms grow at rates varying between 0.5 and 2.8 cm per year. Reef-building corals have their greatest growth rates in waters no more than 45 m deep—they cannot grow in waters that are more than 300 m deep. Since many reefs exceed the maximum depth at which the coral can function, some explanation is required. Several theories have been advanced to explain the development of massive reefs. Currently, the subsidence theory proposed by Charles Darwin (1842) has been accepted as a plausible explanation for coral reef development. The theory accounts for most of the massive coral reefs.

Darwin postulated that various types of reefs develop as the result of the subsidence of the lands they are growing on. The corals begin as a fringing reef, but as the land sinks below sea level the fri nging reef becomes a barrier reef. As the land sinks below sea level and the reef continues to grow, it becomes perched above sea level as an atoll. Support for Darwin's hypothesis became especially strong as geologists learned more about the nature of the movement and subsidence of volcanic islands as they more away from an oceanic ridge or hotspot during sea-floor spreading. Darwin's hypothesis is clearly appropriate for the Indo-Pacific reefs where subduction has been occurring for a long geologic timespan. But massive reef accretion also occurs in the western Atlantic on continental shelves where subduction has not been a long-term geologic process.

New information also supportssome of the ideas of Louis Agassiz and Reginald Daly who suggested in the mid-1800's and 1910 respectively that glaciers controlled reef growth. This theory postulates that reefs were established on wave-cut platforms during the period of maximum glaciation and began to grow rapidly as glaciers melted. Reef development kept pace with the rising sea level. This theory is especially applicable to reefs in the western Atlantic which developed on continental shelves and were not subject to long-term subsidence.

Recent research stresses the concept that several processes, such as subsidence, glacially controlled sea-level changes, and wave action must all be considered with respect to the influence they have on the growth of coral reefs. The Indo-Pacific reefs have more massive development and more species than the Caribean reefs, but the western Atlantic reefs are presently growing more rapidly than the Indo-Pacific reefs. The difference of the two regions is related to different mechanisms of sea-level rise. The glacial mechanisms of rising and sinking sea level have periodically created conditions which either enhance or impede coral growth, while crustal subsidence mechanisms of the Indo-Pacific promoted a longer geologic history of coral growth. A multidimensional model is required to understand coral reef growth.

Sponges are the second largest biomass component of the coral reef fauna. They live in deeper waters than corals, and are composed of the protein spongin, a form of keratin that is related to horn. Sponges resemble corals in that they frequently (nine out of ten of the most common species on Davies Reef of the Great Barrier Reef) contain symbiotic cyanobacteria. Over half of the most common species produced three times more oxygen than they respired. Sponges are believed to be crucial to the coral reef community in that they make possible substantial augmentation of photosynthesis within the reef. The oxygen and nutrition produced via the sponge and cyanobacteria symbiotes assists to promote coral reef growth and maintenance.

The cold, deep upwelling waters along the west coast of Africa and America do not support the growth of coral reef structures. These areas are the chief regions of another type of organic structure, kelp.

Kelp

Several groups of multicellular marine algae produce a complex structure. Instead of being composed of calcium carbonate, or protein structures, as in the case of coral and sponge, seaweeds are composed of carbohydrates. Agar, for example, is a polymer made up of subunits of the sugar galactose. The most elaborately developed forms of seaweeа have three parts including a basal disk called a holdfast, a stem called a stipe, and fronds of varying shapes and lengths. The order Laminariales is a group of brown algae that includes many large and advanced forms of seaweed called kelp. Kelp thrive in much cooler waters than is optimal for corals. Kelp communities make up "forests" of the sea. They share with corals the characteristic of high productivity and of harboring large numbers of organisms. Several genera of kelp are especially important.

Laminaria is a kelp that attains a length of about 1-3 m. It is abundant along the Pacific coast and around the British Isles. This kelp consists of a holdfast, a stipe, and a blade. The stipe lives several years, although the blade dies each year. Meristem tissue which supports active cell division occurs between the stipe and the blade and produces new growth during the annual warm season. Nerocystis, "sea otter's cabbage," grows in deep water and rapid tideways and attains lengths up to 40 m. It anchors to rocks via a holdfast and has a stipe and blade and bladders that provide buoyancy for the blades. The largest known kelp is Macrocystis; it grows up to 65 m long and its structure is similar to Nerocystis. Macrocystis is limited in distribution by the fact that it can reproduce only at temperatures below 64-68°F (18-20°C). It is kelp inhabit only a narrow strip of the marine environment where temperatures are suitably cold, and the bottom is shallow and stable enough for the attachment of the holdfast.

An intertidal kelp community on Tatoosh Island, Washington has been estimated to yield up to 14.6 kg organic matter/m^2/ year. This is seven times more productive than a tropical rainforest which measures production of about 2 kg of organic matter/m^2/year. The most productive terrestrial plant community ever recorded is 8.5 kg/m^2/year for a species of grass in Hawaii. The enormous productivity of kelp is due to its ability to capture both sunlight and benefits from wave energy. A rainforest has enormous standing crop but most of it is nonproductive woody tissue. Kelp fronds are composed mainly of photosynthetic surfaces which are supported by the water. Thus, rainforests bear about 8 square metres of leaves per square metre of growing surface, while kelp bears about 20 square metres of frond per square metre of growing surface.

Kelp growth is enhanced in other ways. Waves have been estimated as delivering at least twofold and up to 15 times more energy than solar radiation delivers to the intertidal kelp community. Wave energy enhances kelp growth in several ways. The wave energy protects kelp from starfish and mussels which are its principal predators, and these animals cannot survive in rough seas. The kelp fronds are kept in constant motion by waves, and the rushing water enhances nutrient exchange between sea water and kelp. During wave action, fronds are alternately exposed to light and dark conditions as they move above and below each other with rising and falling wave crests. The waves dislodge competing organisms from spaces that kelp readily occupy. All of these dynamic factors contribute to the productivity of kelp in the intertidal community. While the local productivity is extremely high, the distribution of kelp is confined to a relatively small area of cool intertidal marine coastline. Thus, geographic constraints impose serious limitations on kelp distribution and its ability to strongly influence the overall productivity of the marine environment.

Productivity and Structural Characteristics of Terristrial Biosystems

In the recent past, it was commonly held that the Earth's oceans were the major site of photosynthesis. The area of the ocean is so large that it seemed logical to attribute immense powers of productivity to it. Research presented here on productivity suggests that forested lands are the major site of photosynthesis. However, this issue is not resolved and, as described for kelp, there are major surprises. The Sargasso Sea may be more productive than generally recognized because of the symbiotic relationship between nitrogen-fixing bacteria and diatoms that form a floating diatom mat. Several underlying factors which contribute to the high productivity of the terrestrial environment have already been described in the section on the physical characteristics of the two environments. The most productive areas of the marine and terrestrial environments are associated with structures which enhance photosynthesis, increase productivity, and have the added advantage of creating a more complex habitat. While the marine environment is dominated by short-lived, single-celled plants, land areas are the realm of longer living multicellular plants. The most distinctive land plant is the tree. A tree is an elaborate network of tissues composed of carbohydrates. The structural elaboration of trees helps explain why there are over 1400 times more carbohydrate reserve within the terristrial environment than in the marine environment.

Roots

The study of roots provides further insights into the relationship between plants and the terrestrial environment. Roots connect the plant with the soil environment. They carry on a number of important funtions including supporting the aboveground portions of the plant and supplying water and minerals to the plant for growth and maintenance. The abundance of elements found in the Earth's crust does not correspond with the amount required for plant growth. Table 16.3 shows the relative abundance of elements in the Earth's crust compared with the abundance typically found within plants. Often, roots work against amazing odds. For example, in serpentine soils, there are many toxicity problems because nickel levels are very high and pH levels are low at the soil surface. Nutrients required for plant growth are in extremely short supply, and magnesium levels are so high as to be extremely toxic. One wonders how plants are able to survive at all in these types of soil conditions. But, they do.

Roots work against odds in several ways. During the season of growth they undergo progressive branching which leads to a huge surface area of direct contact with the water and minerals of the soil. An understanding of root dimensions war arrived at by laborious investigation. For example, one experiment involved one rye plant grown in a box about 30 cm square and 56 cm deep. After the plant reached maturity roots were separated from the soil and their dimensions measured and area calculated. The main roots excluding root hairs were found to have a length of 623 km and an area of at least 237 m^2. When the root hairs were included, the total length came to nearly 11,265 km and the area increased to about 650 m^2. This example gives a measure of how much contact is maintained between the plant and the soil through its roots. Daily growth of the roots provides the mechanism for maintenance of fresh mineral and water contact between the plant and the soil.

The plant undergoes active transport of ions. This pumping action of ions operates with the aid of evapotranspiration. Active transport is extremely complicated and involves the secretion of chlorides by the plant roots as well as the selectivity of specialized membranes such as the Casparian strip. Active transport is capble of increasing ion concentrations by between 200 and 10,000 times above the ion concentration within the soil solution. In this way, plants are accumulators of particular elements. As shown in Table 16.3, these elements end up becoming more abundant in the surface of the soil than would be expected from a simple consideration of the abundances of elements in the Earth's crust. If phytomass production is 171.54 thousand million metric tons then about 5 per cent of the total or 8.58 thousand million metric tons of high quality, elements necessary for the maintenance of the Earth's organisms, are being brought to the soil surface through the action of selective pumping of ions by roots.

TABLE 16.3

Abundance of Elements of the Earth's Crust Compared with the Abundance of the Elements in Plants

	Soil abundance (μg/g)	*Crustal abundance* (μg/g)	*Land plants dry weight* (μg/g)	*Enrichment ratio plant/soil*
1	*2*	*3*	*4*	*5*
Calcium	24,900	43,000	18,000	0.7
Carbon	20,000	170	464,000	23.2
Chlorine	100	100	2,000	20.0
Hydrogen		1,000	55,000	
Magnesium	9,200	24,000	3,200	0.3
Nitrogen	1,000	20	30,000	30.0
Oxygen	490,000	466,000	410,000	0.8
Phosphours	420	1,000	2,300	5.5
Potassium	23,000	18,00	14,0000	0.6
Sodium	12,000	23,000	1,200	0.1
Sulpher	700	300	3,400	4.9

Arsenic	6	1.9	0.2	0.03
Chromium	53	120	0.23	0.004
Cobalt	10	34	0.5	0.05
Copper	25	65	14	0.6
Fluorine	200	600	0.5—40	0.003—0.2
Iodine	5	0.5	0.42	0.08
Iron	25,000	57,000	140	0.006
Manganese	560	900	630	1.1
Molybdenum	2	1.3	0.9	0.5
Nickel	20	95	3	0.2
Selenium	0.2	0.1	0.2	0.1
Silicon	330,000	277,000	200—5,000	0.0006—0.02
Tin	10	1.9	<0.3	<0.03
Vanadium	76	190	1.6	0.02
Zinc	54	87	100	1.9
Aluminum	66,000	81,000	500	0.008

(Contd.)

TABLE 16.3 (*Contd.*)

1	*2*	*3*	*4*	*5*
Antimony	3	0.2	0.06	0.02
Barium	554	450	14	0.03
Beryllium	1	1.5	<0.1	<0.1
Boron	34	7	50	1.5
Bromine	5	2.0	15	3.0
Cadmium	0.06	0.19	0.6	10.0
Lead	20	9	2.7	0.1
Lithium	30	20	0.1	0.003
Mercury	0.03	0.046	0.015	0.5
Radioactive (U)	6	7.3	0 038	0.006
Rare earth (La)	41	25	0.085	0.02
Rubidium	30	90	20	0.7
Silver	0.1	0.09	0.06	0.6
Strontium	240	380	26	0.1
Titanium	3,000	6,000	1	0.0003
Zirconium	240	130	0.64	0.003

Tree Morphology

Along with the study of roots, the study of tree morphology provides additional insights in the reason for the productivity of the terrestrial environment. Tree morphology has some common ties to early works in physiognomy, and life forms. Recent studies have advanced several new approaches that provide insight into the development of trees. Research is being carried out on the ways that tree shape is established according to principles of construction, and in combination with deterministic and opportunistic processes. Also, theoretical aspects of crown shape have provided insights into developmental processes. And, finally, experimental research on hydraulic processes within a tree has demonstrated mechanisms which favour water conduction along pathways that assure survival of some limbs over others. These themes are being integrated into the "architecture model" of tree growth. The arhitecture model of tree growth considers the total relationship of the components of the tree to each other and to the environment.

The study of tree architecture began with the examination of a few easily observed morphologic features in a large number of tropical tree species. All trees can be fitted into 23 different models which are named in honour of the botanists who contributed information on the development of the species that closely resembles the model. The work has led to several useful generalizations. For example, the site of the largest diversity of tree forms is the Tropics, where representatives of all 23 forms. exist. At sites in middle latitudes many forms are not present. Broad systematic relationships do not directly correlate to architecture. Conifers are architecturally poor, displaying only 4 recognized models; but so are the Leguminosae. Gymnosperms are generally undiversified architecturally, while the angiosperms contain representatives of every architectural form. One of the most important insights that the study of tree architecture has produced concerns the way which tree species survive drought stress and repair a damaged crown. In drought, it is universal for trees to cut the supply of water to minor limbs and to preserve water for the growth of the tallest parts of the tree. In nature few trees are undamaged; understory trees may be damaged by falling branches, wind storms, ice, or even animal predation. Most trees undergo repair through replication of the original architectural model from one or more active buds. This type of repair is termed reiteration. As a group, the tree form is one of the most successful structural features found among the Earth's organisms. By linking insights of the components of this form, it becomes possible to derive some useful understanding into why the terrestrial environment is such a productive habitat of the Earth.

Interaction of Marine and Terrestrial Systems

None of the Earth's systems stands alone. All systems interact with each other to make a larger, more elaborate system. The marine and terrestrial environments illustrate this point. The hydrologic cycle and global circulation patterns of the Earth provide a rational basis for an overview of the interaction of the two systems.

From the point of view of the hydrologic cycle, and the movement of organic matter and minerals, the terrestrial environment is essential to the productivity of the marine environment. The marine environment supplies water and the evaporative surface necessary to maintain the hydrologic cycle. The land is a major recipient and benefactor of the water but ultimately the water returns to the ocean. As it returns through channels of branching streams, water carries a load of minerals and organic materials which provide the nutrient base for many organisms of the marine environment. In this way we can understand that productivity of the marine environment has its roots in the terrestrial environment. An exchange is also found which influences the organisms of the terrestrial environment.

The global circulation pattern involves the movement of large volumes of surface water toward the Tropics. The returning water becomes organized into an extremely persistent current. These currents strongly influence the climates of the southeastern coasts of continents of the Northern Hemisphere and the northeastern coasts of continents of the Southern Hemisphere. They also influence the distribution of organisms of the Earth. The Gulf Stream, also known as the Florida Current, illustrates the influence which these phenomena have on the organisms of the Earth.

The first good chart of the Gulf Stream was made by Benjamin Franklin and Timothy Folger and printed in 1769 or 1770. The chart was created when Franklin was in London as Deputy Postmaster General for the American colonies. One question which arose concerned why mail packets took a fortnight longer to sail to America under British captains than did merchants ships under sail by American captains. Franklin sought the advice of his cousin Folger who was a Nantucket ship captain then visiting London. British captains were unaware of the Gulf Stream and frequently stemmed the current,

TABLE 16.4

The Major Currents

Location	*Currents East of continents warm current water*	*Currents West of continents cold upwelling water*
North Atlantic	Gulf Stream (including Florida Current)	Canary
South Atlantic	Brazil	Benguela
North Pacific	Kuroshio	California
South Pacific	East Australian	Peru
North Indian	Somali (during S.W. monsoon)	—
South Indian	Agulhas (including Mozambique)	West Australian

Folger sketched the Gulf Stream on a chart and Franklin had it printed. The chart continues to be an accurate summary of the course, breadth, and strength of the Gulf Stream.

The Gulf Stream carries more water than all of the Earth's terrestrial rivers combined. At Bimini in the Bahamas the flow averages about 20 million cubic metres per second or about 1,728 cubic kilometers per day. The greatest strength of flow of the Gulf Stream occurs off the coast of Florida. It flows in a northeasterly direction off the east coast of North America. At Cape Hatteras (35°N), the Gulf Stream swings eastward and gradually broadens as it flows toward the coast of Europe. As the Gulf Stream flows toward Europe a splits into a northern segment which flows past Norway and Splitsbergen, and a southern segment which flows along the coast of Britain and France.

The Gulf Stream has an enormous impact on the marine and terrestrial environments. It acts, to transport large pools of tropical marine waters, intact, complete with flora and fauna, to mid-latitude locations. In the course of its flow, the Gulf Stream undergoes meandering and large portions of the stream break off as warm- or cold-water eddies and massive "rings" (Backus *et al.* 1981, Cresswell 1982, Wiebe 1982). Schematically illustrates the distribution and movement of rings in the Atlantic. Most of the organisms in the eddy gradually succumb to predation and the change of temperature as the mass of water dissipates. The contact zone between the eddies and rings and surrounding water is dynamic and a highly productive area of the marine environment which is rich in fauna and flora.

For land areas the impact of the Gulf Stream is equally profound. The Gulf Stream transports enough heat to the Scottish coast to support-palm trees at Oban (56°N) and Ullapool (58°N). The flora and fauna of the coasts of Scotland and Ireland have links with the Caribbean and tropical American fauna and flora. For example, sea beans are regularly found washed up on the shore. Numerous insects and a sponge from the tropics of America inhabit these coasts. Also fonnd are a number of coastal plants, including lady's tresses orchid, pipe-wort, an iris, and, Canadian blue-eyed grass.

Another organism being transported by the Atlantic marine currents is the European eel (Anguilla anguilla). Eels breed in the sea, starting their life cycle in the deep Sargasso Sea south-west of Bermuda and gradually moving with the Gulf Stream current toward the European freshwater rivers where they move inland to nature. They grow rapidly in the freshwater environment and upon maturity return to the sea to breed in the warm tropical water of the Atlantic. Eels and many other organisms are involved in long-distance movements that are an important part of the dynamic interplay between organisms of widely separated sites.

